D1301686

Organic Trace Analysis by Liquid Chromatography

Organic Trace Analysis by Liquid Chromatography

JAMES F. LAWRENCE
Food Research Division
Health Protection Branch
Tunney's Pasture
Ottawa, Ontario, Canada

1981

ACADEMIC PRESS
A Subsidiary of Harcourt Brace Jovanovich, Publishers
NEW YORK LONDON TORONTO SYDNEY SAN FRANCISCO

ACADEMIC PRESS, INC.
111 Fifth Avenue, New York, New York 10003

United Kingdom Edition published by
ACADEMIC PRESS, INC. (LONDON) LTD.
24/28 Oval Road, London NW1 7DX

Library of Congress Cataloging in Publication Data

Lawrence, James F.
Organic trace analysis by liquid chromatography.

Includes bibliographical references and index.
1. Liquid chromatography. 2. Trace elements--Analysis.
3. Chemistry, Organic. I. Title. [DNLM: 1. Chromatography, Liquid--Methods. QD 272.C4 L421o]
QD272.C447L38 547.3'0894 81-3464
ISBN 0-12-439150-8 AACR2

PRINTED IN THE UNITED STATES OF AMERICA

81 82 83 84 9 8 7 6 5 4 3 2 1

To Jessie and Jimmie

Contents

Preface xi

1 General Considerations in Developing a Trace Analytical Technique Employing Liquid Chromatography

I. Liquid Chromatography versus Gas Chromatography 1
II. Residue Analysis versus Formulations Analysis 6
III. Pump and Injector Requirements 7
IV. Detectors 7
V. Chromatography 9
VI. Extraction and Cleanup 11
VII. Minimum Detectable Levels 12
VIII. Matrix Effects 14
IX. Use of Derivatization 14
X. Microprocessors and Automation 16
References 17

2 Pumping Systems

I. Introduction 18
II. Constant Volume Pumps 19
III. Constant Pressure Pumps 25
IV. Gradient Elution 27
V. Comparison of Pumping Systems 31

3 Sampling Technique and Injection Ports

I. Effect of Sample Solution on Peak Shape 34
II. Requirements for Trace Analysis 37

III. Types of Injection Ports 38
References 46

4 Chromatography Columns and Packing Materials

I. Requirements for Trace Analysis 47
II. Pellicular versus Microparticulate Porous Packings 47
III. Commercial Prepacked Columns 49
IV. Column Packing Techniques 50
V. Stationary-Phase Materials 52
References 54

5 Detectors

I. Introduction 55
II. General Detector Requirements 56
III. Photometric Detectors 59
IV. Electrochemical Detectors 84
V. Solute Transport Detectors 93
VI. Refractive Index Detectors 98
VII. Mass Spectrometer 102
VIII. Other Detectors 102
References 107

6 Chromatography Theory

I. Fundamental Principles 110
II. Adsorption Chromatography 117
III. Partition Chromatography 122
IV. Macroreticular Resins 129
V. Ion-Pair Chromatography 131
VI. Ion-Exchange Chromatography 135
VII. Ion Chromatography 136
VIII. Gel Permeation Chromatography 138
IX. Ligand-Exchange Chromatography 139
X. Charge-Transfer Chromatography 140
References 140

7 Chemical Derivatization

I. Introduction 143
II. UV-Absorbance Derivatization 154
III. Fluorescence Derivatization 167
IV. Derivatization for Other Detection Modes 182

V. Postcolumn Reactions 184
References 189

8 Sample Extraction and Cleanup

I. Considerations 198
II. Sample Extraction 202
III. Sample Cleanup 208
IV. Column Chromatography 215
V. Thin-Layer Chromatography 228
VI. Distillation 229
VII. Low-Temperature Precipitation 232
References 232

9 Approach to Method Development and Routine Analysis

I. Choosing the Best Chromatography System 235
II. Analysis Time and Interfering Peaks 238
III. Qualitative and Quantitative Analysis and Confirmation 240
IV. Automation 241
V. Integration of LC with Other Analytical Techniques 242
VI. LC as a Cleanup Technique 244
References 245

10 Applications

I. Introduction 246
II. Clinical (Biological Fluids and Tissues) 246
III. Environmental 262
IV. Foods 272
References 282

Index 285

Preface

During the past decade, modern high-performance liquid chromatography (LC) has expanded greatly, especially in the area of formulations analysis; this includes quality control of pharmaceutical preparations, cosmetics, food colors, pesticide formulations, and other sample materials where the compounds of interest are present in relatively clean matrices at concentrations of about 1% or greater. The application of LC to trace analysis has been slower to develop because of the difficulty in isolating and sufficiently purifying the analyte prior analysis. Of the many LC books available up to now, there is none completely devoted to organic trace analysis. It is hoped that this volume will go a long way toward filling the gap.

The introductory chapter provides an overview of the special requirements of LC for trace analysis and compares the approach to gas chromatography and formulations analysis, noting the essential differences. The chapters concerning LC equipment discuss the best conditions and types of instrumentation suitable for trace analysis. The detector chapter is particularly detailed, since the choice of the most appropriate detection system, as well as optimum operating conditions are important to the trace analytical chemist—especially at present when the selection of useful detectors is limited.

The chapter on chromatography theory is brief and includes mainly descriptive accounts of the principles of the various forms of chromatography, including recent developments (mainly through the work of Scott and Kucera) in the understanding of adsorption, reversed-phase, and ion-pair chromatograpy. Rigorous theoretical treatment of chromatography is not included in light of the abundance of texts already available on this subject. Chemical derivatization is an important tool for the trace analytical chemist, since it enables him to improve detection limits by forming a suitable derivative. The chapter concerning derivatization outlines many reactions applicable to trace analysis of a large number of substances in a variety of sample materials.

Sample preparation, including extraction and cleanup, is treated in some depth. This is a very important area for trace analytical methodologies employing LC. The chapter includes many diverse approaches to sample extraction and extract purification and should provide the analyst with some guidelines for solving his analytical problems.

The final chapter illustrates typical approaches that have been used for trace analysis and shows how the various parts of a complete LC method are integrated to create a successful determination, including initial sample preparation, chromatographic separation, and detection. The chapter is divided into three major areas of organic trace analysis: clinical, environmental, and food analyses; it is intended to serve as a guide to the reader and is not a comprehensive compilation of the literature.

The book should be of value to all analysts involved in the determination of trace organics in many different substrates. It will be of particular interest as a teaching aid for those entering the field of trace analysis with the intention of employing LC.

I would like to express my sincere thanks to the Bureau of Chemical Safety, Food Directorate, Health Protection Branch, Department of National Health and Welfare, Canada for their support in preparation of this book.

James F. Lawrence

Chapter 1

General Considerations in Developing a Trace Analytical Technique Employing Liquid Chromatography

I. LIQUID CHROMATOGRAPHY VERSUS GAS CHROMATOGRAPHY

Most analysts entering the field of liquid chromatography (LC) probably have had previous experience with gas chromatography (GC). Although the concepts of "chromatography" remain the same, there are important differences between the two types. In GC, the two major variables used to achieve a separation are the type of stationary phase and the column temperature. The chromatographic separation is based on a vapor pressure phenomenon and an affinity for the stationary phase. In LC the separations may be described as solute–solvent interactions where the solute molecules establish an equilibrium between stationary and mobile phases. For LC analyses one has the choice of different stationary phases and mobile phases. Column temperature is usually ambient and any deviations from this are used for special cases. Figure 1.1 illustrates a typical LC equipment setup.

At present, over 150 commercially available stationary phases for gas chromatography are available. These range from very nonpolar types, such as OV-1 methyl gum where separations are essentially based on differences in boiling points, to polar columns, such as the ethylene glycol esters or cyanopropyl silicones. The characteristics of many of these phases for separating several types of compounds have been tabulated by McReynolds (*1*). In LC the choice of stationary phases is far more limited. For adsorption chromatography silica gel or alumina is usually used.

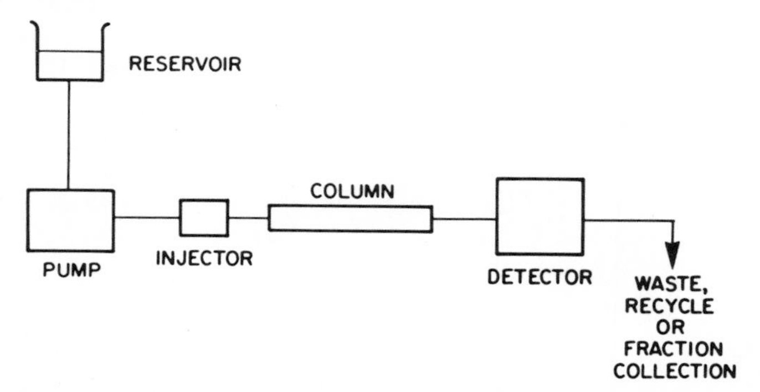

Fig. 1.1 Typical arrangement of LC equipment for organic trace analysis.

There are perhaps four or five commercially available types of polar bonded phases useful for normal-phase partition chromatography, and about seven or eight for reversed-phase partition chromatography. Research in this area is continuing at a good pace. However, at the moment, the vast majority of LC analyses are carried out on either silica gel or the C_8 or C_{18} reversed-phase materials.

From this it can be gathered that the mobile phase plays a far greater role in LC than in GC. It is this variable which gives LC its great flexibility. Such types of chromatography as adsorption, normal-phase partition, reversed-phase partition, ion-exchange, ion-pair, exclusion, ligand-exchange, and affinity chromatography make possible the separation of closely related species by a much wider variety of selective interactions. In effect, LC should be suitable for any type of compound that is soluble in a liquid suitable for a mobile phase. The choice of mobile phase is limited only by the compatibility of the solvents with the materials with which they come in contact (metal tubing, pump components, stationary phase material) and the type of detector. For example, halide salts or acids usually have to be avoided since they damage the stainless steel of most pumps and connecting tubing. If a UV detector is employed, then solvents such as benzene, acetone, or carbon disulfide cannot be used where absorbance is monitored below 280 nm, a region where most work with that detector is carried out. This illustrates the need for the analyst to understand the operation and limitations of his equipment.

In comparison to GC, LC is much less flexible in its detection systems. There are a number of very sensitive general and selective detectors for GC, including electron capture, flame ionization, and thermal conductivity as general detectors; the alkali–flame ionization detector is selective for nitrogen or phosphorus; the flame photometric selective to sulfur or phosphorus; and the electrolytic conductivity detector selective to nitrogen or halogens. Many attempts have been made to convert several of these for use with LC systems. However, since all of these operate by monitoring vapors, problems have been encountered in the phase change required. Transport arrangements have been designed where the eluant from an LC column flows into an oven or onto a moving wire, chain, or rotating disk, where the solvent is evaporated. The remaining solute molecules pass through either a detector using a flame or an electron capture cell. Because of the problems associated with the phase change, this approach to LC detection has not become popular. Thus, the most widely used detectors to date are those that carry out measurements in liquid streams. These include detectors based on absorbance, fluorescence, electrochemical oxidation or reduction, and refractive index differences, the last being the least sensitive or selective, thus generally unsuitable for trace analysis.

In GC, to analyze several compounds over a wide polarity or volatility range, temperature programming is employed. The equivalent in LC is gradient elution, where the mobile-phase composition is altered continuously during the chromatographic run. This normally requires two pumps and the necessary electronics to program them. However, ternary electronic mixing devices have been developed for solvent mixing and gradient elution with a single pump. Such an apparatus should become popular since it avoids the cost of a second pump.

The reproducibility of gradient runs and time required for reconditioning are somewhat superior in GC compared to LC. It is of course easier to make reproducible temperature-programming runs in GC since the stationary and mobile phases are not altered. However, in LC where the mobile-phase composition actually changes, the initial chromatographic conditions require a longer time to be reestablished. Also, retention values (k') are less accurately reproduced from one run to another in LC than in GC. Both systems are susceptible to detector interferences resulting from temperature (GC) or mobile-phase (LC) changes. These can be especially problematic when doing residue analysis where high detector sensitivities are required. In both types of chromatographic systems, isothermal (GC) and isocratic (LC) separations are to be preferred.

Start-up times are usually shorter for LC than for GC. Detector stability and chromatographic conditions are usually established in 30–60 min from complete shut-down. The time required for GC start-up is usually significantly longer. In some special cases in LC, such as ion-pair chromatography, start-up time also may be longer.

Maintenance requirements for both systems are of course different, but comparable in time required and frequency.

Since most analytical laboratories have more GC than LC equipment, methodology is usually directed to GC. It is the opinion of this author that LC need be used only for those compounds that are not easily analyzed by GC. There is little value in developing LC methodology for compounds such as organochlorine pesticides, for example, since the GC methodology is so well developed. The area where LC has made great gains is in the pharmaceutical industry, where many drugs because of their size and nature are not suitable to GC analysis, whereas they are very easily analyzed by LC. In the final analysis, it is up to the individual to decide which technique would be more suited to his needs. This can only be done through experimentation.

Figures 1.2 and 1.3 compare GC–EC and LC–UV results of two pesticides in foods. The chromatograms each represent results from the same sample solutions. It can be seen in the case of terbacil, a uracil type of herbicide, that the LC results are far superior to the GC ones. However, at the same time, the opposite is true for the wild oat herbicide benzoylpropethyl. Thus it cannot be said that one technique is better than the

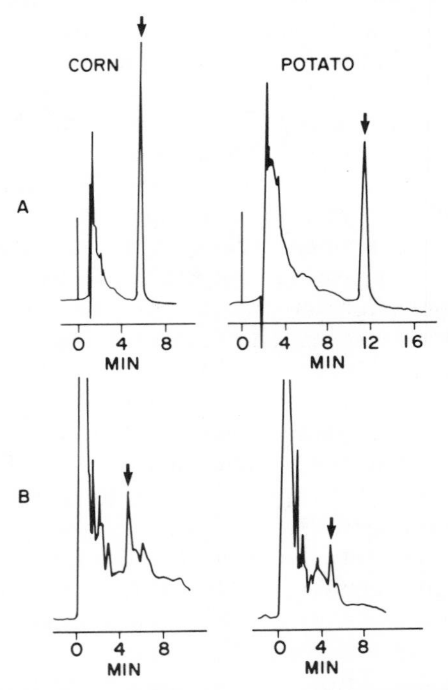

Fig. 1.2 Chromatograms of terbacil in corn (2.0 ppm) and potato (0.2 ppm) by LC–UV, 254 nm (A) and GC–EC (B). Injections were made from the same sample extracts. Arrow indicates terbacil peaks. From Lawrence (3), with permission from Preston Publications, Inc.

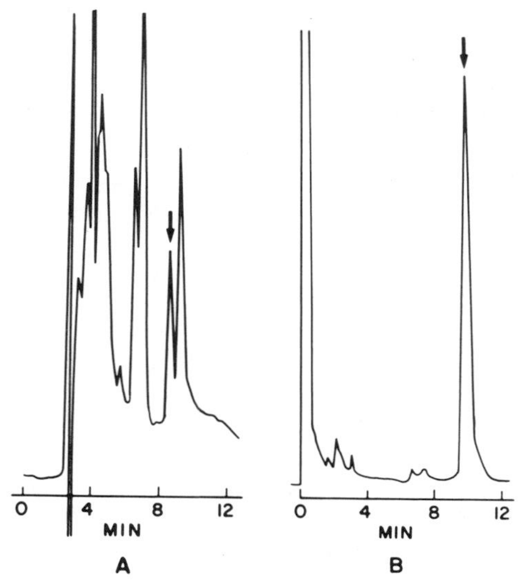

Fig. 1.3 Chromatograms of benzoylpropethyl in corn at 0.2 ppm. A, LC–UV (254 nm); B, GC–EC. Injections were made from the same sample extract (diluted in the case of the GC–EC results). Arrows indicate the peaks. From Lawrence (*3*), with permission from Preston Publications, Inc.

other. LC and GC complement one another and should be employed in that manner. Further discussion of the integration of LC methodology with that of GC is presented in Chapter 7.

Thin-layer chromatography (TLC) is not really a technique competitive with GC or LC. The necessary instrumentation cost far less than the other two but it is usually employed for qualitative or semiquantitative analysis. The major advantage of TLC is its simplicity and low expense. Most TLC plates are disposable and are used for only one chromatographic run, which for a 20 × 20 cm plate means that about 10–15 samples can be analyzed each time. Since the plates are used only once, the analyst can afford to try methods that include very little cleanup, and it is not necessary to worry about irreversible contamination of the stationary phase (the TLC layer). If a sample proves to be too dirty, the TLC plate is discarded and more cleanup of the sample extract is carried out. In LC, such a procedure could prove costly in ruined columns, especially in adsorption chromatography.

Another advantage of TLC is that selective post-chromatographic reactions can be carried out with relative ease. This simply involves spraying, dipping and/or heating the developed TLC plate to produce a colored or fluorescent spot sometimes detectable in low nanogram quantities. Such reactions in LC are far more difficult to achieve and require specialized reagents, mixing units, and extra pumps and plumbing.

The major limitations of TLC are that it cannot compete with LC or GC in quantitation, separating power (efficiency), sensitivity, and ease of automation. All of these are important criteria for the development of a trace analytical method. Because of this TLC has been reserved for use as a confirmatory test or as an initial screening procedure to indicate the presence or absence of a substance in a sample. Both of these uses, however, are important and will continue to make TLC a widely used analytical technique.

II. RESIDUE ANALYSIS VERSUS FORMULATIONS ANALYSIS

Probably the greatest use of LC at present is for formulations analyses. Commercial preparations of drugs, cosmetics, pesticides, coloring agents, etc., are often monitored for quality control by direct LC analysis with little or no cleanup. The reason is that the solutes are present at concentrated levels in relatively pure form. Even the analysis of trace impurities in such samples without cleanup is possible since these are usually in the order of 100 ppm to 0.1%.

Residue analysis, on the other hand, is far more difficult to carry out because of the nature of the samples and low concentrations of solute. Materials such as food, biological tissue and fluids, soil, and natural waters create great problems with LC analyses. The removal of a drug or pollutant from such matrices inevitably brings with it a host of compounds with many similar properties (solubility, polarity, etc.). The analyst's problem becomes one of how to remove as many of these as possible for successful LC analysis.

How this is approached first depends on the nature of the compound of interest. For example, if the compound were strongly fluorescent, one would try to employ a fluorescence detector for the analysis. The cleanup would involve removal of interfering *fluorescent* materials. It is possible to have many coextractives present in the final sample solution but if they do not fluoresce, they will not interfere in the analysis. However, it must also be kept in mind that coextracted material can interfere in the chromatography by distorting peaks or altering retention values, when too much is present. Even though chromatograms may be relatively clean, the repeated injection of "dirty" samples can lead to contamination and eventually to decreased column life. To keep the chromatographic system functioning well for as long as possible, it is good practice to inject as little sample as necessary to obtain a reliable result. This is especially important for trace analysis and of lesser importance for formulations analysis.

The major difference in trace analytical methodology and methodology for formulations as well as other commercial preparations is the concentration of the analyte in the sample and the extent of cleanup required. These differences are large ones and have been the main reason why trace organic analysis by LC has been slower to develop.

III. PUMP AND INJECTOR REQUIREMENTS

The minimum requirements for pumps are that they deliver essentially pulseless flow over the range 0.5–2.0 ml/min at high detector sensitivity. The noise of course depends upon the type of detector, but generally a satisfactory pump produces less than 1% noise at 0.01 absorbance units full scale (AUFS) on a UV detector above the detector noise itself. Most popular pumps in use today, including the dual- and single-piston reciprocating pumps, easily meet this requirement. Fluorescence detectors are less affected by flow fluctuations than are electrochemical detectors.

The most useful injection ports are those incorporating a syringe-loop configuration. These permit injection of different volumes up to the volume of the loop. They are easy to use and are preferred to stopped-flow injection or injection via a septum. Although all syringe-loop injectors are rated for operation at 3000 psi or more (which is more than adequate for most LC applications), those capable of operating at 6000 psi permit the use of higher flow rates for rapid mobile phase changes, or column conditioning or cleaning.

IV. DETECTORS

The prime requirement for LC detectors that are to be used for organic trace analysis is sensitivity. This is very important when considering the quantities of sample that must be injected to produce a reliable peak. If detector sensitivity is poor, then more sample must be injected, and if done on a regular basis this will lead to shortened column life. Sensitivity becomes increasingly important if ultralow levels (e.g., parts per billion) are to be determined. For most applications, detectors sensitive to low nanogram quantities of substance (1–50 ng, for example) are required. This of course eliminates the refractive index detector for application to trace analysis.

Selectivity is another important factor in detection. It can be considered to be the ability of a detector to discriminate between the compound of interest and the coextracted material. Usually this is done by taking advantage of some electronic or optical property of the analyte. Figure 1.4 illustrates this point by comparing the simultaneous absorption and fluorescence detection of some aflatoxins. All four compounds can be detected by absorption with similar sensitivities. However, only G_1 and G_2 are more sensitive by fluorescence; the opposite is true for aflatoxins B_1 and B_2. The merits of each detection system can be judged only after application to actual sample analyses. A detector may be capable of detecting subnanogram quantities of a compound, but that does not necessarily mean it is capable of detecting low concentrations in a sample where many coextractives are present. Usually, more sample cleanup is required as detector selectivity decreases. This applies also to the UV detector when operated at different wavelengths. One wavelength may be superior to another, not because of sensitivity but because of selectivity. The problem of choosing which detector to use is not very great since

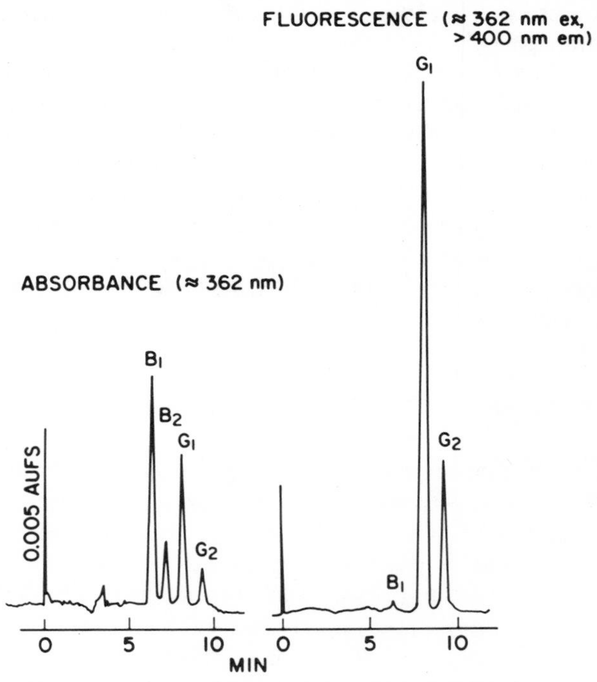

Fig. 1.4 Comparison of fluorescence and absorbance for the detection of aflatoxins B_1, B_2, G_1, and G_2. G_1 and G_2 are more sensitive by fluorescence, whereas the opposite is true for B_1 and B_2. From Pons and Franz (*4*), with permission from the Association of Official Analytical Chemists.

there are only a few available suitable for organic trace analysis. These include absorbance, fluorescence, and electrochemical (EC) detectors. All are sensitive to low nanogram or lesser quantities of many compounds. They differ as to their detection mechanism, compatibility with mobile phases, and selectivity. Detailed discussion of these three types and others appears in Chapter 5.

V. CHROMATOGRAPHY

There are no special requirements of chromatography columns that make some more suitable for trace analysis than others. Whether doing preparative work or trace analysis, column efficiency and chromatographic selectivity are both of prime importance. Efficiency refers to the ability of a column to resolve a number of peaks within a given time frame at a constant mobile phase composition and flow rate. Efficiency increases as a column resolves more peaks per unit time. This is important for both trace and formulations analysis since no matter what the concentration, the solute molecules must be separated from other components before being detected and quantitated. Figure 1.5 compares the effect of column efficiency on the resolution of two peaks.

Columns with high efficiencies permit the elution of a solute at a given retention value (k') in a minimum volume of mobile phase. Thus, as efficiency increases the volume of mobile phase which contains the solute decreases. In other words, the solute concentration increases, causing an increased detector response, measured as peak height. This is a definite advantage when doing trace analysis. Figure 1.5 illustrates the effect of column efficiency on peak height and peak area. Peak area is not affected significantly by efficiency. Details on the calculation of efficiency and the number of theoretical plates (N) may be found in Chapter 6.

One important point to consider concerning chromatographic efficiency is that high efficiency columns are not always required. For example, in Fig. 1.5B, the chromatogram may be perfectly acceptable for doing formulations since the two peaks are completely resolved. The same can be said for organic trace analyses as long as the peak height is such that it is above the minimum detectable level of the detector.

The time for a chromatographic run depends not only on how long it takes the solute molecule to elute, but also on the time required for elution for any sample coextractives as well. This of course is important since late peaks can interfere in further analyses. Such problems are well known in GC. Since the shortest time possible for a chromatographic run is preferred, the solute should elute as quickly as possible without de-

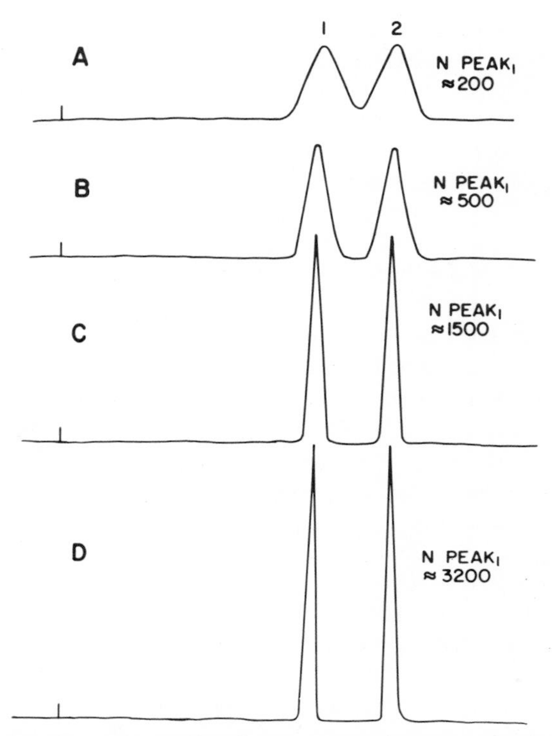

Fig. 1.5 Comparison of the resolution of two peaks at different column efficiencies. A, 200 theoretical plates (*N*); B, 500; C, 1500; D, 3200. (Calculated for the first peak.)

creasing the required minimum detectable levels in the sample. How fast this is done depends upon sample interferences. Usually, close to the solvent peak (t_o) there is an abundance of coextracted material; as one progresses further away from it, the chromatogram becomes cleaner. In practice, retention values are normally greater than $k' = 3$ (for calculation of k' see Chapter 6).

It should be noted that while mobile-phase velocity affects retention times, it does not affect k' values. Thus if one wishes to move a peak further away from the solvent peak, lowering the flow rate will not help, since the k' values of all peaks in the chromatogram remain the same. What would be more useful would be to change the mobile-phase composition so that t_0 remains the same and the selectivity of the system changes such that interfering peaks are removed from the vicinity of the analyte. Also, in order to elute the analyte as quickly as possible it is preferable to make an appropriate change in the mobile phase composition rather than increase flow rate. Two reasons are that increased flow rates unnecessarily waste solvent and they cause increased band broadening (see Chapter 6 on effect of mobile phase flow rate on chromatographic efficiency).

Chromatographic selectivity is a very important factor in LC, as it is in GC. Selectivity of a system depends upon both the stationary phase type and mobile-phase composition. The importance of selectivity (i.e., the ability of a system to separate two similar compounds) can be pointed out by the fact that it does not matter how efficient a column may be or how sensitive a detector is, if the analyte and an interfering peak cannot be separated, then an accurate analysis cannot be made. The whole process of achieving a good LC separation is done by changing the selectivity (mobile or stationary phase) until acceptable results are obtained. In LC the mobile phase has tremendous capabilities in this regard. It remains for the analyst to judge which solvent mixture would be best for a separation. Knowledge of this nature is gained not only from the literature but by practical experience as well.

VI. EXTRACTION AND CLEANUP

Sample preparation is probably the factor that most distinguishes organic trace analysis from formulations analysis by LC or GC. The degree and type of sample preparation depends upon several factors, beginning with the chemical and physical properties of the analyte itself. Then one must be aware of the type of detector system that is available and its characteristics. At this point the method for chromatographic separation must be considered. All of these factors influence the approach to the cleanup. Once a specific method has been selected or designed, the analyst must then ensure that none of the solvents or other materials used in the extraction and cleanup will interfere in the final determination. For example, if UV detection at 254 nm is employed for adsorption chromatography, then benzene or acetone should be avoided since traces of these may find themselves in the final mixture and cause unnecessary interferences. If such solvents must be employed, then it would be necessary to evaporate the solutions to dryness to remove any traces of unwanted solvent. This additional sample handling may cause losses resulting in poor quantitation.

The extraction and cleanup procedure must in general be compatible with the chromatographic system employed. For example, adsorption chromatography is very nicely suited to existing methodology for the analysis of trace levels of pesticides. Usually, cleanup involves solvent partitioning followed by chromatography on an adsorption column such as Florisil or alumina. The final residue is dissolved in a suitable nonpolar organic solvent (preferably mobile phase) for chromatographic analysis. However, if reversed-phase chromatography is used, dissolving the final sample in an aqueous solution of methanol or acetonitrile may not be possible, especially if the final residue contains some coextracted sample oils. It may be possible to dissolve the sample in pure methanol or acetonitrile, but then it is possible that chromatography will be poor, especially if the mobile phase contains about 30% or more of water. The poor chromatography is caused by the methanol (or acetonitrile) from the sample, which upsets chromatographic equilibrium as it passes through the column.

If polar solvents are used to extract trace organics from samples, much of the coextracted material will be very polar in nature. It is possible that when these are injected into an adsorption chromatographic system, they will irreversibly bind to the adsorbent. After a period of time the column performance will decrease and will not be restored upon cleaning. To avoid this, it is recommended that either a precolumn (guard column) be inserted between the analytical column and the injection port, or that a classic open-column cleanup be carried out. The latter is preferred since in complex samples some type of open-column cleanup is required anyway to remove substances that will not necessarily affect chromatography, but that will interfere in the detection. The problem of such irreversible contamination in reversed-phase or normal-phase partition chromatography is less but nevertheless possible.

It cannot be emphasized enough that sample cleanup is of the utmost importance both for maintaining the integrity of the chromatographic system and for accurate reliable results. Chapter 8 describes in depth many extraction and cleanup systems used for polar, nonpolar, and ionic organic substances. Applications of these to actual samples appears in Chapter 10.

VII. MINIMUM DETECTABLE LEVELS

There are two points to be made concerning minimum detectable levels. The first refers to detector sensitivity. As mentioned earlier, it is important in trace analysis to have sensitive detectors. In LC that means detectors capable of detecting low-to-subnanogram quantities of standards. High detector sensitivity allows injection of less substance into the LC system, thus maintaining the cleanliness; or it means that the cleanup requirement may be reduced. However, the minimum detectable level that can be detected in the presence of coextractives can be very different from that of pure standards and depends upon the nature of the sample and the amount of cleanup carried out. Figure 1.6 clearly shows this by comparing GC–EC detection with LC–UV detection of the same sample extract containing the herbicide metoxuron [ethylated (*2*)]. In this case the EC detector was 100 times more sensitive to the compound. However, as can be seen when quantities are injected such that the sample backgrounds are similar, the LC–UV system is slightly superior in terms of minimum detectable levels (mg/kg) in an actual sample. Detector selectivity and sensitivity as well as cleanup go hand in hand in determining the minimum detectable levels in actual samples.

An extraction and cleanup technique must not only remove interferences, it must also permit the recovery of the analyte in as high yields as possible. Most important is that this must be done reproducibly. The absolute recovery is seldom 100%. The minimum acceptable recovery depends upon the concentration of the analyte in the sample and minimum detectable limit of the detector. Usually in residue analysis, 80% or

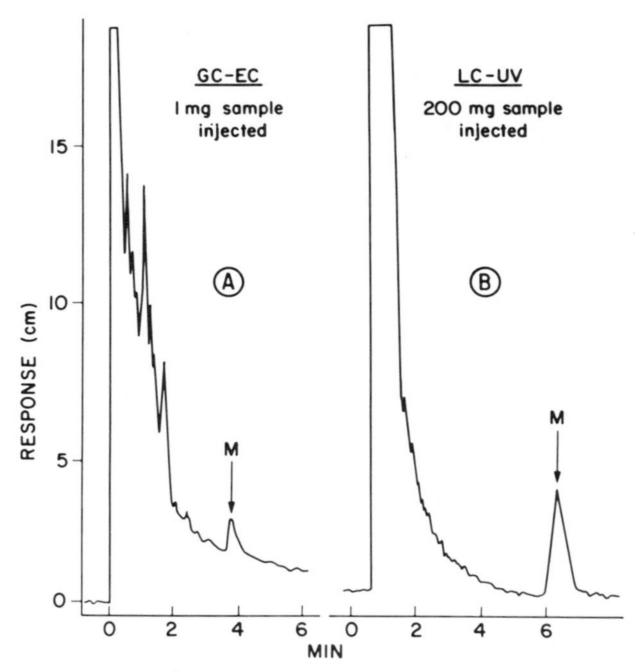

Fig. 1.6 Comparison of GC–EC and LC–UV for the analysis of metoxuron (ethylated) in soil at 1.0 ppm. A, 1 mg equivalent of sample injected; B, 200 mg. From Lawrence *et al*. (*2*), with permission from the American Chemical Society.

greater is acceptable. For many compounds, particularly polar or less stable ones, greater than 50% recovery is adequate. When recoveries are difficult to reproduce, internal standards are frequently added which can account for the variation. In this case quantitation is based on the ratio of the peak heights (or areas) of the analyte and the internal standard. It is necessary, of course, that the internal standard be similar in nature and extraction efficiency to the analyte.

VIII. MATRIX EFFECTS

The nature of the sample, i.e., both physical and chemical composition, plays an important role in determining how an analyst approaches the extraction and cleanup. A method that is suitable for the determination of some pollutants in industrial effluents may not be acceptable for the same compounds in soils or food material. The effects of adsorption onto solid materials, especially inorganic materials found in soils, should be known. Samples that contain a high percentage of oils or fats must be treated differently from samples such as many root crops or fruits which have a high water content. Not only is recovery or elimination of interferences of concern, but the technical aspects must also be considered. The method, especially the initial extraction, should be compatible with the sample. Emulsions, lack of solubility or wetting, and sample degradation when using digestion techniques are some examples of problems that may be encountered when a method for one sample material is applied to another.

IX. USE OF DERIVATIZATION

Chemical derivatization involves altering the chemical structure of the analyte to make it suitable for determination by chromatographic techniques. In GC it is frequently used to increase volatility or thermal stability so that the analyte will pass through the chromatography system as a symmetrical peak. In both GC and LC, derivatization is used to enhance detector response. In fact, for LC, increased detectability is the main reason for derivatization. Usually this involves the attachment of a highly absorbing chromophore or a fluorphore. Other techniques involve the conversion of the analyte to an electrochemically active substance suitable for electrochemical detection. There exists an abundance of reagents available for derivatization of many classes of compounds. Some have found extensive use whereas others are limited in their applications.

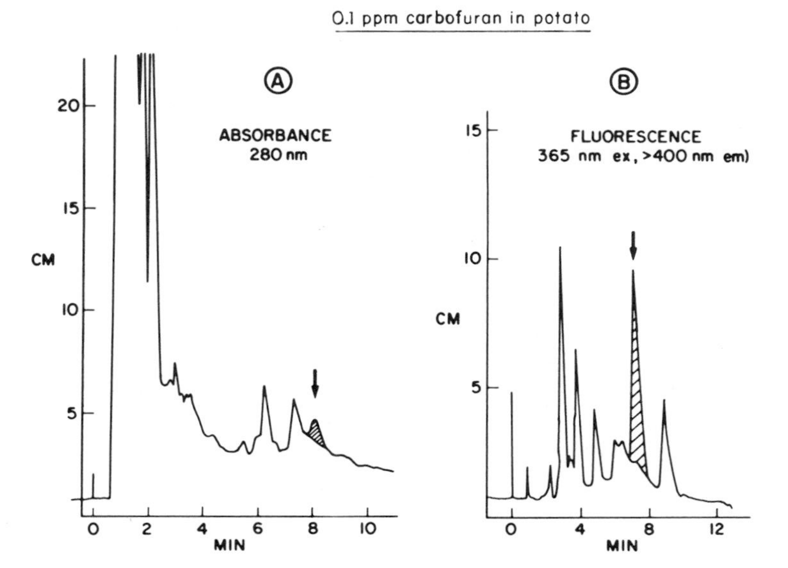

Fig. 1.7 Comparison of a potato extract analyzed by direct chromatography (A) and after dansylation (B) for carbofuran at 0.1 ppm. A, UV detection at 280 nm; B, fluorescence detection, 365 nm ex, >400 nm em. (From Lawrence (*6*), with permission from the American Chemical Society.)

Chapter 7 details many derivatization techniques that have been applied to LC.

It must be kept in mind that chemical derivatization should only be used as a last resort to solve detection problems. The additional reagents required and the increased sample handling can increase cleanup time and cause increased analytical error. The advantages of using derivatives must be weighed against the problems that may be encountered. It should be made clear, however, that in many instances derivatization simply has to be used if an analysis is to be made at all. Figure 1.7 shows the increase in sensitivity gained for the insecticide carbofuran in potato after derivatization. The same sample extract was used in each case, except that the fluorescent dansyl derivative was made for fluorescence detection. The insecticide itself is nonfluorescent. The derivatization improves the detection limit five- to tenfold in the sample.

Reactions may be carried out prior to chromatography in a classic manner, or in some cases postcolumn reactions are possible. However, for the latter, special reagent requirements are necessary and additional equipment is usually required. Details of pre- and postcolumn reactions also can be found in Chapter 7.

X. MICROPROCESSORS AND AUTOMATION

Most manufacturers of LC equipment now offer microprocessor options that can control many parameters that formerly had to be done manually. These units can be programmed for flow rate, detector sensitivity, time of run, number of injections, time between injections, gradient runs, etc. They can also interpret the data generated by each chromatographic run and change mobile-phase characteristics to achieve better separations. Quantitation and data printout also are easily achieved. All of this can be accomplished without the presence of an operator. The potential for automated analysis in LC is especially great for routine analysis where an apparatus can operate overnight. Labor costs can be significantly reduced, making the acquisition of such a system economical in the long run.

However, for such systems to function properly the basic LC components—pump, injector, column and detector—must operate well. If any one of these malfunctions, then all the results obtained via the microprocessor will be suspect. Also, it is important that LC operators know and understand their equipment so they can spot problems and solve them quickly. Microprocessors can only compensate for operational problems, they cannot correct them.

Another important factor is that when a system is automated, more equipment (usually more complicated) is required. Such equipment is also susceptible to malfunction. Thus it becomes part of the operator's work to understand and be able to troubleshoot microprocessor problems. Because of the possibility of microprocessor malfunction, the equipment should be designed such that it can be easily switched to manual operation so that analyses can still be carried out while the microprocessor unit is being repaired. In some LC systems at present this is not possible. Thus, something as simple as a microprocessor problem in establishing mobile-phase flow rate can render the complete apparatus useless.

Purchasers of LC equipment who are considering buying the accessories for automatic control should strongly weigh the advantages and disadvantages of such equipment. It is preferable that an operator have a good practical background in LC before extending into the area of microprocessors. For this reason, it is practical to buy LC equipment that offers microprocessors as accessories rather than as built-in parts of the system. In the former case microprocessors can always be purchased later.

Automation not only of the chromatographic process, but of the extraction and cleanup as well, is particularly useful in quality control laboratories where sample preparation is minimal and sample throughput is high. In residue analysis many attempts have been made at total automation. For the extraction and cleanup, a number of mechanical devices have been designed for shaking, partitioning, transferring solvents from one vessel to another, column cleanup, concentrating, etc. For the most part such systems are in very limited use.

Automation in general can be a positive step toward permitting high-frequency sample analyses. However, the analyst should never lose sight of the fact that every individual part of the whole system must function well in order to obtain reliable and accurate results. A good analyst will be prepared to spot problems in all parts of the system.

REFERENCES

1. W. D. McReynolds, *J. Chromatogr. Sci.* **8,** 685 (1970).
2. J. F. Lawrence, C. van Buuren, U. A. T. Brinkman, and R. W. Frei, *J. Agric. Food Chem.* **28,** 630 (1980).
3. J. F. Lawrence, *J. Chromatogr. Sci.* **14,** 557 (1976).
4. W. A. Pons, Jr. and A. O. Franz, Jr., *J. Assoc. Off. Anal. Chem.* **61,** 793 (1978).
5. J. F. Lawrence, *J. Chromatogr. Sci.* **17,** 147 (1979).
6. J. F. Lawrence, *Anal. Chem.* **52,** 1123A (1980).

Chapter 2
Pumping Systems

I. INTRODUCTION

Applications of liquid chromatography (LC) to organic trace analysis require the use of pumping systems that (1) are capable of delivering constant and reproducible mobile phase flow over a range of about 0.1–10.0 ml/min; (2) can operate at pressures of 4000 psi or greater; and (3) are compatible with a wide range of solvents. The first point is essential since retention of a solute in a chromatographic system is directly related to flow rate. Any inconsistencies can cause slight shifts in peaks which in trace analysis can lead to misidentification, especially when other peaks are proximate. A wide flow-rate range is also very useful. High flow rates permit rapid changeover of mobile phases and fast equilibration of the system before analysis. With the microparticulate packings commonly used at present, high pressures are often necessary to obtain suitable mobile-phase flow rates. Normally, flow rates are less than 2 ml/min for routine operations requiring pump pressures of up to 2000–3000 psi. Even higher pressures are required if increased flow rates are employed for solvent change as mentioned above.

The compatibility of the pump with a wide range of organic or aqueous solutions (the latter often containing dissolved salts) is essential if flexibility of a chromatographic unit is desired. Most LC pumps are made from 316 stainless steel with Teflon or other inert seals. Such materials are resistant to most organic and aqueous solutions with the exception of halogens, which corrode 316 stainless steel. It is important to read and keep in mind the particular restrictions on solvent usage that are supplied with most pumps when purchased.

The consistency of solvent delivery is very important for trace analysis where high detector sensitivity is required. Pulsations in reciprocating piston pumps become evident as baseline noise in the chromatograms. This can be especially critical with the UV or electrochemical detectors, whereas the fluorescence detector is much less affected. Most reciprocating LC pumps include pulse-dampeners as integral parts of the apparatus to reduce noise from pulsations as much as possible.

In selecting a pump the three principles mentioned above must be considered. How an individual's requirements relate to these will determine the type of pump needed. If, for example, pressures of only 3000–4000 psi are necessary for a method that is to be used on a routine basis, then an inexpensive pneumatic pump or single-piston reciprocating pump may suffice. As the applications become more complex and requirements more stringent, the use of better equipment becomes necessary. At present there are commercially available many pumps that cover a wide range in quality, capability, and price. Since LC has been developed on a modular basis it is relatively easy to incorporate most pumps into an existing chromatographic system. Thus, there is no need to remain always with the same manufacturer or supplier. This is a concrete advantage of LC compared to GC, because a prospective LC user can choose his components as he wishes and know that they will be generally compatible even though they come from different manufacturers. This mix-and-match ability allows one to buy exactly what suits him best.

LC pumps can be classified into two basic types—constant flow and constant pressure pumps. The former include pumps such as the dual- or single-piston reciprocating pumps, the screw-driven syringe pumps, and the diaphragm pump. Constant-pressure pumps use gas pressure either directly or through pneumatic amplification to drive the mobile phase. A discussion of these types follows.

II. CONSTANT VOLUME PUMPS

A. Small Volume Reciprocating Pumps

These types of solvent delivery systems are the most popular at present. They are available in one or two piston models and can generate flows up to 10 ml/min at pressures as high as 6000 psi. Figure 2.1 illustrates a single-piston pump. An electric motor drives a cam which withdraws a small piston from a cylinder, resulting in solvent being drawn into the cylinder chamber. On the pump stroke the piston forces the liquid through an exit valve. The check valves prevent any backflow of liquid, thus maintaining the flow in one direction. Since solvent only flows through the chromatographic system during the pumping stroke, a pulsing effect is observed.

In order to provide a more continuous flow, several designs have been

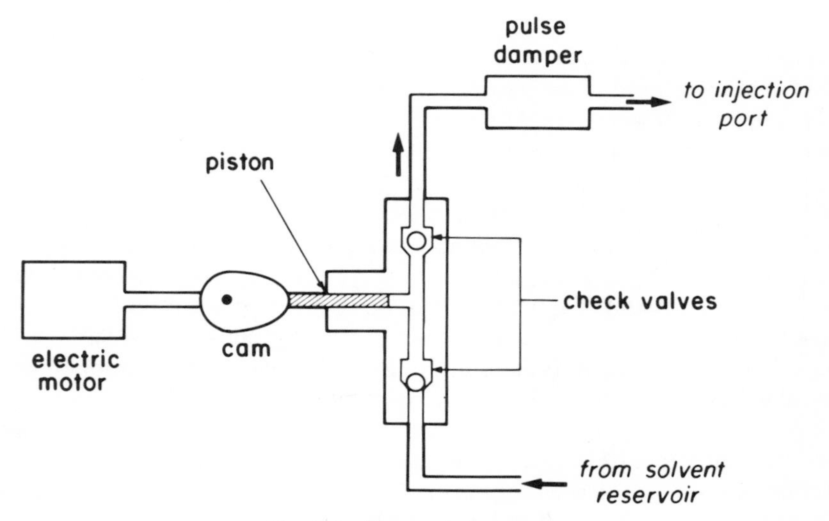

Fig. 2.1 Single-piston pump.

introduced to minimize these pulsations. Beckman (formerly Altex) introduced a spring-loaded piston on a specially designed cam which causes a rapid (about 0.2 sec) return of the piston on the return stroke, which minimizes the magnitude of the pulsations. This type of pump is capable of operating up to 10 ml/min at pressures up to about 4000–5000 psi, which makes it satisfactory for organic trace analysis. The author has found this type of pump to be satisfactory for use with UV detectors at full-scale absorbance ranges of 0.01–0.005 absorbance units with about 1% baseline noise. However, it is likely that the pulsations would result in a higher noise level in electrochemical detectors.

The Varian model 5060 LC system incorporates a single piston pumping system as schematically shown in Fig. 2.2. The pump is similar to that marketed by Beckman (Fig. 2.1) except that it has the inlet needle valve connected directly to the pump. Thus, the inlet is firmly closed during the solvent delivery stroke from the piston (90 μl volume). The return stroke is the same as that for the Beckman (0.2 sec). The advantage of the mechanically controlled inlet valve is that solvents do not need to be degassed since vapor bubbles will not inhibit the operation of the needle valve as they normally can with ball check valves.

Pulsations in reciprocating pumps can also be reduced by the addition of a second piston 180° out of phase with the first. Figure 2.3 shows a diagram of a dual-piston reciprocating pump. The second piston serves to deliver solvent to the chromatographic system while the first is drawing in more mobile phase. This action considerably reduces the pulsations ob-

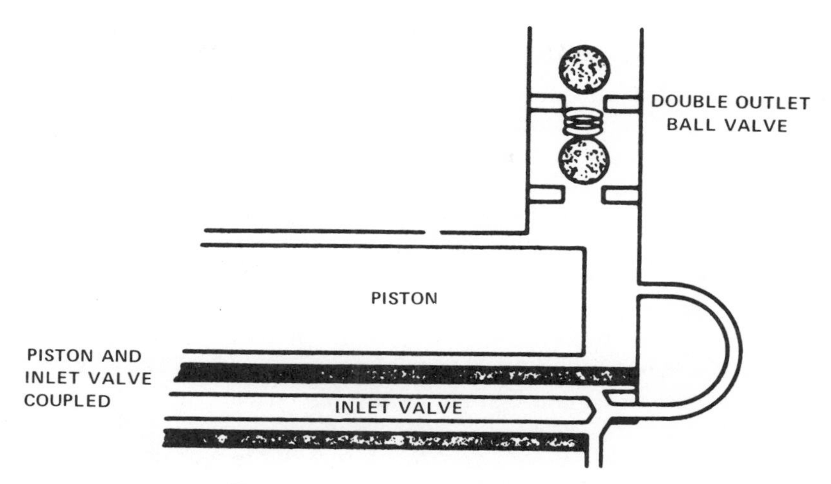

Fig. 2.2 Varian single-piston pump.

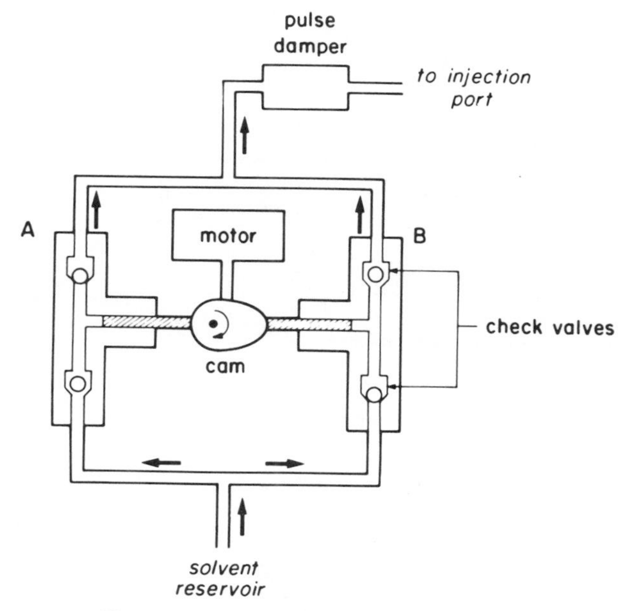

Fig. 2.3 Dual-piston reciprocating pump.

served with the single-piston pumps. However, most manufacturers of dual-piston reciprocating pumps incorporate additional pulse-dampening devices to reduce pulsations even more. Although these reciprocating pumps are fully satisfactory for working with UV or fluorescence detectors at high sensitivity, they may perform less satisfactorily with electrochemical detectors than true continuous flow pumps such as the screw-driven syringe pumps. All offer flow-rate ranges of 0.1–10 ml/min with pressures up to at least 6000 psi. The solvent delivery is more reproducible with these pumps compared to the single-piston variety, especially at pressures over 4000 psi.

Another type of reciprocating pump is the diaphragm pump produced by Orlita and Whitey. A schematic is shown in Fig. 2.4. The intention of this design is to eliminate contact of the piston with the mobile phase, therefore making the use of special seals around the piston rod unnecessary. The piston is driven by a cam that forces it into a chamber filled with hydraulic fluid. The pressure is transmitted through the fluid to the dia-

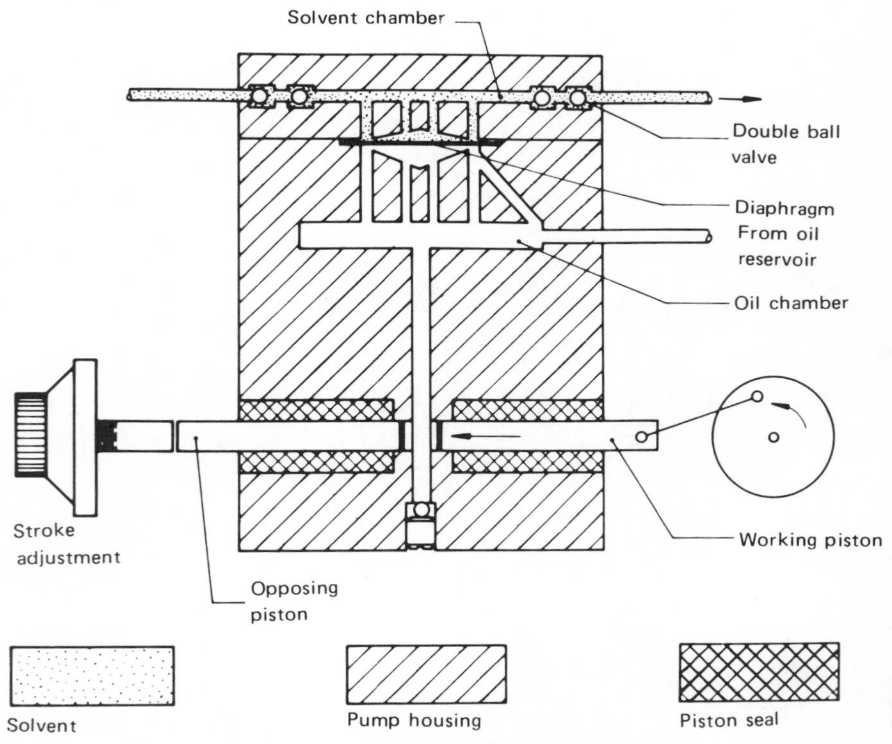

Fig. 2.4 Reciprocating diaphragm pump.

phragm (usually stainless steel) which is pushed outward, forcing solvent through the chromatographic system. On the return stroke of the piston the diaphragm returns to its normal position, drawing in solvent for the next cycle. Flow rates are changed by adjusting the stroke length of the piston, and not the frequency. These pumps can be used either singly or as dual-piston pumps where the pump strokes are 180° out of phase. The advantage of the diaphragm pump has not been evident through reports in the literature or elsewhere. These pumps in fact have not become popular. As required with the other reciprocating pumps, they also need additional damping devices.

Waters Associates has made available a dual-piston reciprocating pump that incorporates what they refer to as a "positive flow equalization" principle. The new design consists of two positions with different pumping volumes 180° out of phase. Figure 2.5 illustrates the setup. As the primary head pumps solvent (100 μl), half of that (50 μl) is taken up by the secondary head while the remainder proceeds into the chromatographic system. On the return stroke of the primary head, the secondary piston delivers the remaining 50 μl of solvent to the chromatographic system.

This pump is designed to compete with the single-piston spring-loaded pump, which is one of the least expensive reciprocating pumps suitable for use with high-sensitivity detector settings. The positive flow equalization pump is capable of delivering 0.1–9.9 ml/min at pressures up to 4500 psi. Since it incorporates two pistons 180° out of phase, it should be expected to deliver solvent with smaller pulsations. This new pumping concept appears to be less satisfactory than the normal dual-piston pumps marketed by Waters and others since Waters does not include it on their top-of-the-line Model 6000A. The positive flow equalization pumps do not

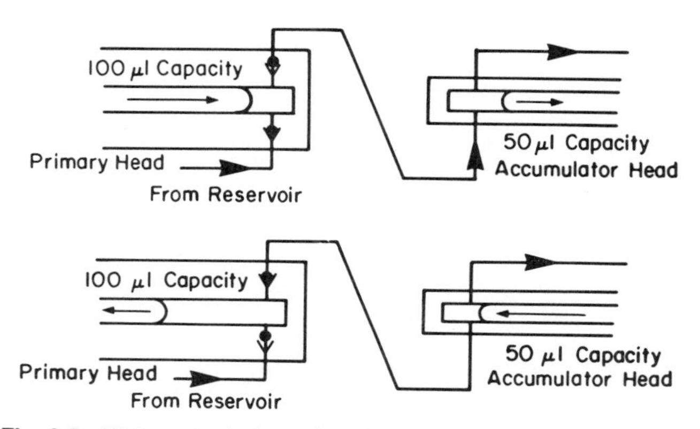

Fig. 2.5 Waters dual-piston "positive flow equalization" pump.

appear to offer any advantages over other dual-piston pumps. However, it is clearly less expensive to produce (and offers fewer features), and thus can compete with the single-piston spring-loaded pumps. Although this type of pump has not been rigorously tested for residue analysis, the author has found that it performs as well as the single-piston type.

B. Syringe Pumps

Figure 2.6 illustrates the design of a syringe pump, which, by nature delivers constant pulseless flow of mobile phase to the chromatographic system. It consists of a large cylinder that can accomodate up to 250–500 ml of mobile phase. The piston is driven by means of a screw mechanically linked to an electric motor. As the screw turns the piston forces the solvent through the system continuously until the chamber is empty. At this point, the pump must be stopped and manually or automatically refilled. Removing the cylinder for manual refilling is tedious, so newer models of this type of pump incorporate a reservoir that can hold about 1 liter of solvent, and an automatic filling capability which is activated by a switch.

Some solvent compressibility (≃3%) may cause minor flow variation at pressures of 6000 psi or higher and it should be taken into account when working at those levels. However, pressures in that region rarely are required for most routine applications.

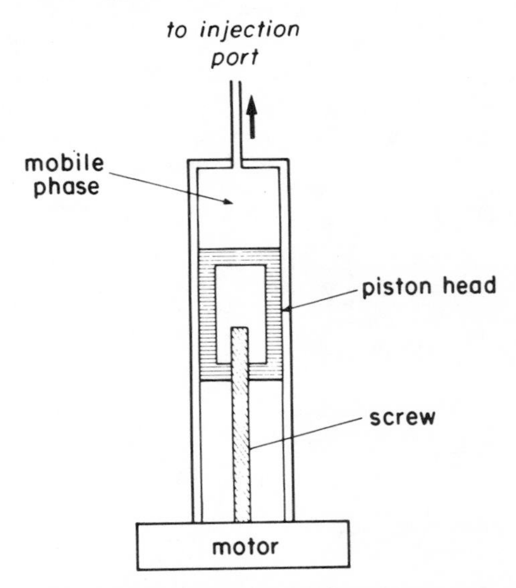

Fig. 2.6 Screw-driven syringe pump.

The syringe pump is particularly useful where flow-sensitive detectors are employed. This includes the differential refractometer and electrochemical detectors such as the coulometric, amperometric, and particularly the polarographic detector. One major drawback of this type of pump is its size and weight: it is the biggest of those currently available and is often not easily moved, compared to the reciprocating-piston pumps.

III. CONSTANT PRESSURE PUMPS

These pumps operate by means of gas pressure. The gas can act directly on the surface of the liquid mobile phase or it can drive a piston which delivers the mobile phase to the chromatographic system.

The gas-displacement pump consists of a pneumatic device in which the mobile phase is placed in a stainless steel container or coiled tube. Compressed gas (nitrogen or air) is then introduced, which forces the liquid through a valve into the chromatography system. Figure 2.7 illustrates a typical setup. Solvent from the reservoir enters a holding coil through

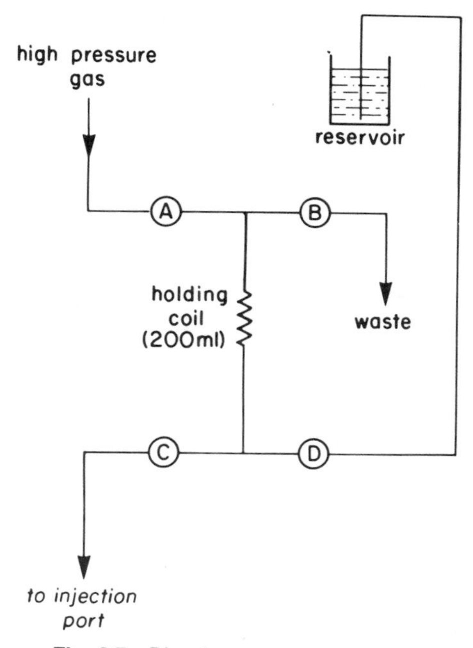

Fig. 2.7 Direct gas-pressure pump.

valve D. Gas pressure through valve A then forces the liquid through valve C to the column.

Since the gas acts directly on the surface of the liquid, dissolved gas can become a problem. However, this is somewhat alleviated by using the coil. The dissolution of the gas takes place near the gas–liquid interface while the bulk of the liquid remains gas-free. The final few milliliters of solvent, which contain significant quantities of dissolved gas, are usually discarded and are not pumped into the chromatographic system. These pumping systems are usually limited to operating pressures of about 1500 psi because of safety aspects related to highly pressurized gases. Such pressures, nevertheless, are certainly adequate for chromatographic systems that employ microparticulate packings (5–10 μm particle diameter) when flow rates of 1.0 ml or less are used. Dissolved gas in the mobile phase will pose a problem if it causes bubble accumulation or other interference in the detector.

The pneumatic amplification pump manufactured by Haskel and others is also a constant-pressure solvent-delivering system. However, unlike the direct gas-pressure-driven type, it makes use of a piston, as shown in Fig. 2.8. Gas pressure acts on the surface of the large piston, which amplifies the pressure on a smaller piston connected to it. The pressure-amplification factor is directly related to the ratio of the surface areas of the two pistons. This is normally a factor of about 50. Thus, if a 100-psi gas pressure were used it would create a pressure 50-fold greater (5000 psi) on the liquid. The Haskel pump is rated up to more than 15,000 psi and is particularly well suited for packing columns.

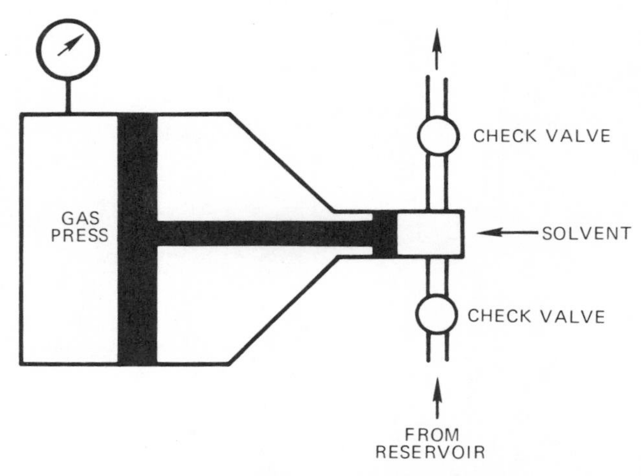

Fig. 2.8 Pneumatic amplifier pump.

IV. GRADIENT ELUTION

For routine analyses LC separations are best carried out under isocratic conditions. This approach permits continued sample analysis without altering the chromatographic system for long periods of time. It is the simplest way to analyze a large number of samples, especially if only one or two compounds are to be determined. However, gradient elution often becomes necessary when a whole series of compounds is determined over a wide range of polarities. In such circumstances an analyst has the choice of choosing two (or more) different isocratic systems (one for the less polar components and the other for the more polar compounds), or one suitable gradient program. The decision should be based on which approach will best serve the analyst's needs, especially if routine work is planned.

Time is an important factor. Gradient elution can provide a rapid and often superior analysis of a number of components relative to the same separation done isocratically. However, the gradient system requires a certain period between each sample to regenerate the original chromatography conditions before the next gradient run. This uses up time, often equivalent to as much as 30–40% of the total analysis time. If a gradient run repeatedly produces good results, then the only question for the analyst is which approach to use—two or more isocratic analyses where samples can be analyzed one after the other without altering the chromatographic system during each run, or one gradient system with its inherent recycle time.

Gradient programs must produce reproducible k' values and peak heights over a long series of cycles. Thus the gradient programmer and pumps must work effectively and precisely through repeated gradients. There are several factors that must be taken into consideration when developing a gradient elution system. First, constant-volume pumps must be employed if any versatility is desired. The pressure required to force a mixture of two or more solvents through a chromatographic system depends upon their individual viscosity and the concentration of each in the mixture. Thus, in a gradient run where the viscosity of the mobile phase is continually changing, the pressure required to maintain the same flow rate changes also. Such consistency of flow is not possible with constant-pressure pumps.

The type of detector must be taken into account. If a UV detector is used in the range below 254 nm, a shifting baseline may result because of changes in absorbance of the mobile phase as it changes in composition. Usually this is manifested by a drifting baseline. Appropriate action would be to purify or exchange the solvents.

Two types of gradient elution devices are currently in use. They are high-pressure gradient programming where the gradient is formed on the high-pressure side of the pump, and low-pressure gradient programming where the mixing takes place before the mobile phase enters the pump. Figure 2.9 shows a diagram of gradient elution using high-pressure mixing. With this type of system two pumps are normally employed, each with one component of the mobile phase. The gradient is generated by electronically controlling the pumping rate of each pump so that while the total flow remains the same, the flow rate of one of the components of the mobile phase increases while the other decreases.

Electronic gradient programmers are available from a number of suppliers and are capable of generating a number of types of gradients including linear, concave, and convex runs. The main disadvantage of this type of system is that it requires the cost of an additional pump. However, at present this system is the most popular since the second pump can always be used isocratically in another system when gradient elution is not required.

Another high-pressure gradient elution approach has been incorporated into LC instruments marketed by Micromeritics and Dupont. Although somewhat different in detail, the two systems produce a gradient program with a single pump. Figure 2.10 illustrates a generalized setup. The technique involves the addition of a high-pressure holding coil which contains the second component of the mobile phase. The coil is connected in parallel by means of two tees to the pump. A proportioning valve determines what volume of each component enters the chromatographic system at

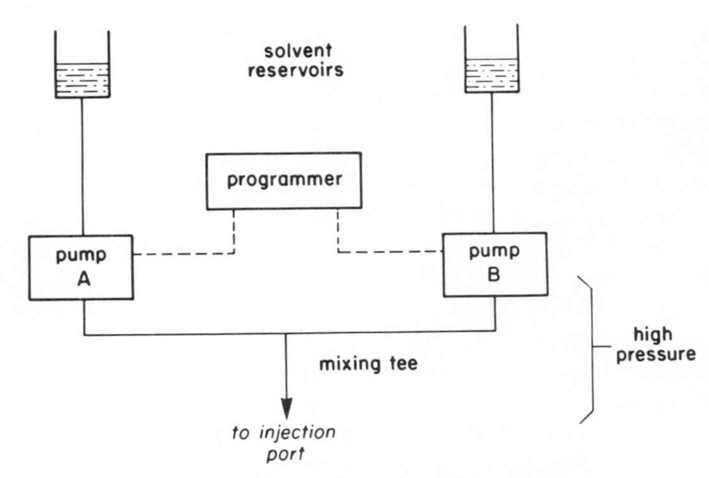

Fig. 2.9 Gradient elution using dual pumps.

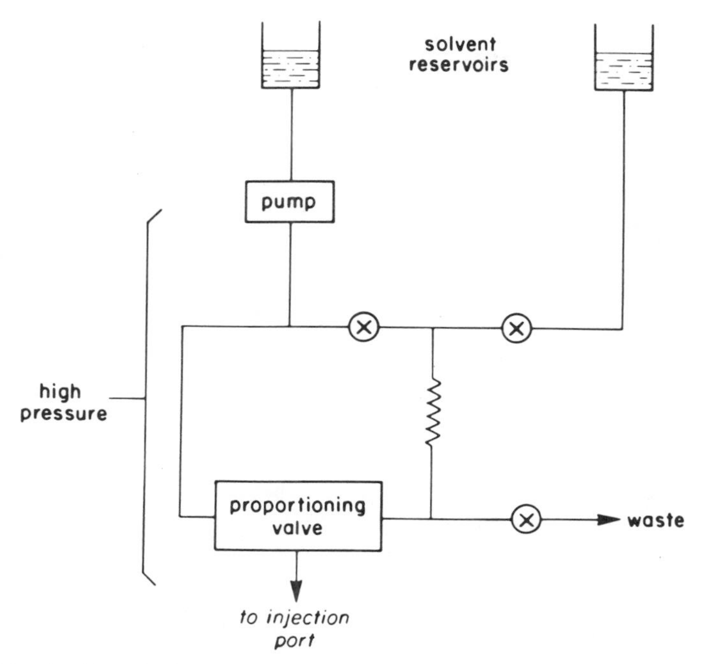

Fig. 2.10 Gradient elution incorporating a single pump with high-pressure mixing.

any given time. The holding coil is connected by means of an on–off valve to the reservoir. The proportioning valve can be programmed to create various gradients similar to those mentioned above for the dual-pump systems. The major disadvantage is that the holding coil must be refilled regularly.

Low-pressure mixing of solvents for gradient elution can be done in two ways, both utilizing a single pump. However, it should be pointed out that only small-volume reciprocating pumps are suitable for low-pressure gradient programming. This results from the fact that in the large-volume pumps such as the syringe or pneumatic pumps, there is no means to change the solvent composition once the pump chamber is filled. Reciprocating pumps that continuously draw from an external reservoir are well suited to this type of gradient formation. Step gradients can be done manually or automatically and simply involve the switching from one mobile phase composition to another in a steplike manner. A typical step-gradient system is shown in Fig. 2.11. Although this type is the least expensive of gradient-generating devices available, it can be limited, depending upon the type of detector employed. Each time a step change is made the recorder baseline may shift because of changes in mobile-phase

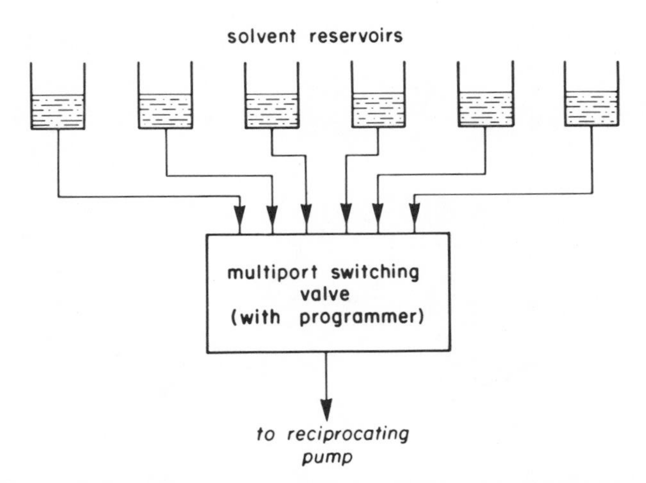

Fig. 2.11 Gradient elution incorporating a single pump and several reservoirs with different mobile-phase compositions for stepwise changes.

character. The new baseline should be attained before quantitation of peaks is attempted. Usually, mobile-phase compositions are chosen so that only one or two steps are employed for any given run.

A proportioning valve for the continuous low-pressure mixing of solvents in association with a single dual-piston reciprocating pump has been employed by Spectra-Physics in their Model SP 8000, in the Varian Model 5060, and others. The ternary mixing valve permits the simultaneous mixing of three different solvents over a wide range of compositions. Figure 2.12 schematically illustrates the principle. The mixing valve can be programmed to allow different quantities of solvent to be drawn from the reservoirs into the pump. The proportioning is done in the same manner as is carried out for the single-pump high-pressure gradient mixing. The valve consists of three small-volume solenoid switches that open and close at rates determined by the electronic control module. These regulate the volume of each solvent passing through the valve to the mixing chamber.

It should be pointed out that pulse dampeners used with reciprocating pumps delay gradient programs and can distort the gradient somewhat because of the associated dead volume. The distortion comes as a result of longitudinal mixing. Usually these are minor problems and are readily determined by running an observable gradient program. For example, with a UV detector at 254 nm and a reversed-phase column, one can create a gradient program by running 100% water to 0.1% acetone/water. The resulting recorder tracing will reflect the increase in acetone concentration

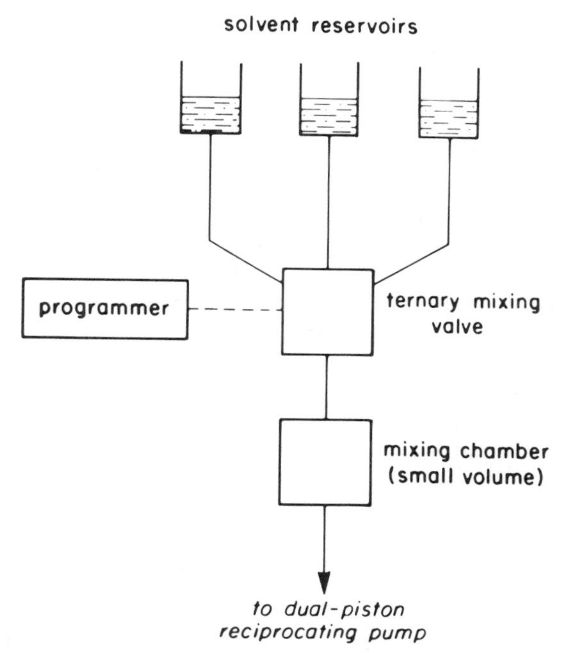

Fig. 2.12 Gradient elution incorporating a single pump with a low-pressure ternary mixing valve.

and thus the true gradient profile. This can be matched to the expected gradient profile for adjustments, if necessary.

V. COMPARISON OF PUMPING SYSTEMS

In making a comparison of the various pumping systems available, it should be pointed out that most have their advantages and disadvantages. These depend in part upon the type and frequency of use for which the pump is employed. Types of chromatography systems, analysis time, and type of detector are all important factors in determining the suitability of a given pump. However, there are some general facts that should be discussed in order that a good comparison be made, especially keeping in mind applications to organic trace analysis.

First, constant-volume pumps are to be preferred to constant-pressure (pneumatic) pumps. The latter require an additional device to measure mobile-phase flow rate. Flow-rate adjustment is done most simply by adjusting the gas pressure until the desired flow rate (measured manually with a small graduated cylinder or volumetric flask at the outlet end of the detector) is achieved. Alterations in mobile-phase composition that result in significant viscosity changes require further adjustment of pressure to maintain the same flow rate. Any increase or decrease in column permeability will also cause a change in mobile-phase flow rate because of a change in resistance. These facts should be readily understood so that troubleshooting will be made easier. As mentioned earlier, gradient elution is particularly difficult with these systems, since changing solvent composition leads to variations in viscosity and, therefore, flow rate.

The major advantages of these systems are that they are inexpensive and do deliver a pulseless flow of solvent over the pump cycle. Thus they are suited to routine analysis where isocratic conditions are employed. The direct gas-pressure pumps are even more limited than the pneumatic amplifier types since the former are only useful up to about 1500 psi. Also, because of possible dissolved gas in the mobile phase, attention must be drawn to detector problems (air bubbles) which may result.

The screw-driven syringe pumps deliver pulse-free solvent flow over a wide range of flow rates up to pressures exceeding 6000 psi. These pumps are particularly useful for flow-sensitive detectors. They do not require pulse dampening and therefore minimize dead volume between the pump and the chromatography system. This is advantageous for solvent changeover, especially for gradient elution where two pumps are employed. The low dead volume permits very little delay between the time the gradient is actually started and the time it reaches the column.

The major disadvantage of these pumps is that they are inconvenient for solvent changeover because of the necessary refilling and washing procedure. This can be particularly tedious if an analyst is in the process of optimizing a separation that requires a number of mobile-phase changes, but on a routine basis they perform very satisfactorily. In some of the older types, leakage occurred around the piston seals when pressures above 3000 psi were used over an extended period. The author has used a screw-driven syringe pump (Varian 8500) almost daily for several months with both organic and aqueous-based mobile phases at pressures of 1000–2000 psi with no leakage problems at all.

The major reasons single- and dual-piston reciprocating pumps have become popular are that they are relatively small, versatile, and can be easily interfaced with other components. The design is such that this type of pump does not need to be shut off to replenish the mobile phase. Solvent changeover is quickly and easily carried out. In some models, such as the Waters Model 6000A, a three-port inlet valve allows solvent changeover at the turn of a switch. The high pressures (6000 psi) attainable with these pumps permit rapid flushing of the system and short equilibration times for the new mobile phase. However, reciprocating pumps have more moving parts, including two sets of check valves with each piston head. Since the volume of each piston stroke is small, the piston cycle is very fast compared to the syringe or pneumatic pumps. This inevitably leads to more wear on the piston seals. In fact, it is recommended always to have spare seals in stock for any necessary changes. The check valves are an essential part of reciprocating pumps. If any one of the four (in a dual-piston pump) malfunctions, then solvent delivery will be affected.

It is important that solvents be degassed, since unlike the other types of pumps, air bubbles can accumulate in the check valves and cause a malfunction. Microparticulate matter in the solvents can have a deleterious effect on reciprocating pumps by increasing wear on seals and check valves. For this reason, it is important to filter solvents through Millipore filters or install 2- to 5-μm filters in line between the reservoir and the pump inlet to ensure that small particles are removed. Routine cleaning of the piston heads and check valves is essential to maintain these pumps in good working order. If properly maintained, such pumps will have a long lifetime. The author has used two dual-piston reciprocating pumps routinely for 6 years for applications to organic trace analysis, including adsorption, reversed-phase, and ion-pair chromatography. The only maintenance required was an occasional changing of piston seals and routine cleaning of the piston heads and check valves.

Chapter 3
Sampling Technique and Injection Ports

I. EFFECT OF SAMPLE SOLUTION ON PEAK SHAPE

Ideally, introduction of the sample into the chromatographic system should be achieved with minimum disturbance of the system. It is also best to inject as little sample volume as possible. For LC this usually means from 10 to 50 μl. The composition of the sample solution relative to the mobile phase is very important if good peak shape is required. Normally, in a reversed-phase system, the sample solution should contain a slightly larger percentage of water than the mobile phase. This will cause the solute molecules to collect at the head of the column in a very narrow band, as illustrated in Fig. 3.1. The actual chromatography takes place after the sample solution has passed into the column.

The reason the solute molecules remain at the head of the column is that when the sample solution contains more water than the mobile phase (for example, the sample solution consists of 50% methanol/water and the mobile phase is 80% methanol/water), the k' of the solute in the presence of the sample solution is very high compared to that obtained with the mobile phase. This means that there is less movement of the solute down the column during injection, which keeps band spreading because of the injection volume to an insignificant level. This same phenomenon is used for preconcentration of organics from aqueous samples (*1–4*). Aqueous samples can be pumped in large volumes through a reversed-phase column, which retains many organics because of the large k' values. The organics can then be sequentially eluted by means of a gradient comprised of increasing concentrations of methanol or acetonitrile in water.

For normal LC analyses this behavior permits the injection of volumes up to 1 ml or more of sample without loss of efficiency if the k' of the

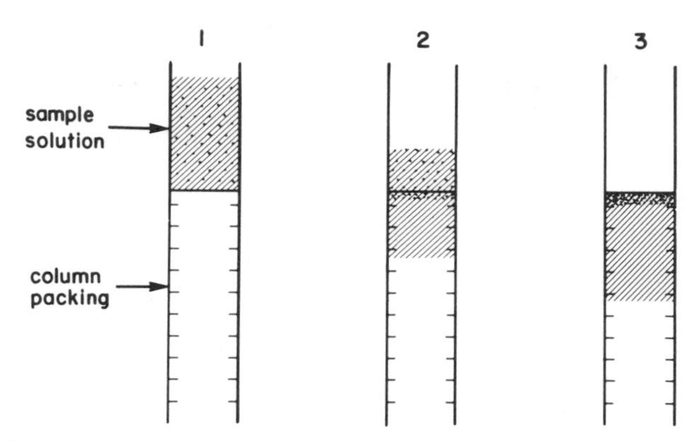

Fig. 3.1 Illustration of sample application to a reversed-phase chromatography column. Dots represent solute molecules. The diagonal lines represent sample solvent which contains more water than the mobile phase. (1) Initial application. (2) As the sample solution passes onto the column, the solute molecules remain near the top of the column. (3) Complete separation of solute molecules from the sample solution.

solute molecules in the sample solution is very high. The only restriction on the injection of large volumes is that it does not disrupt the equilibrium conditions of the chromatographic system. This is especially important if ion-pair or other pH-controlled types of chromatography are employed.

Contrary to the advantages of injecting aqueous solutions into a reversed-phase system, injection of solutions in 100% methanol, acetonitrile, or other organic solvent can pose a serious threat to chromatographic efficiency when mobile phases contain a significant percentage of water. The reason for this is that unlike the above situation, where aqueous injections left the solute molecules in a narrow band at the head of the column (large k'), the organic solvents tend prematurely to wash them down the column at a faster rate than the mobile phase. This results from the fact that in methanol or other organic solvents the k' of the solute is very small relative to the aqueous-based mobile phase. The resulting peaks appear distorted, with peak shapes and heights not necessarily reproducible. This effect is most prominent when the differences between mobile phase and sample solution are large. In a system where 80% or more methanol is used as a mobile phase, the effect may be minimal and go unnoticed.

Figure 3.2 compares chromatograms of two dioxins (1,2,3,6,7,8-hexachlorodibenzodioxin, HCDD; and octachlorodibenzodioxin, OCDD) eluted from a reversed-phase C_{18} column with a mobile phase of 90% methanol in water (*5*). The top chromatogram represents normal peaks obtained when 20 μl of sample in 100% methanol or acetonitrile were injected. The polarities of the mobile phase and the sample solution are sim-

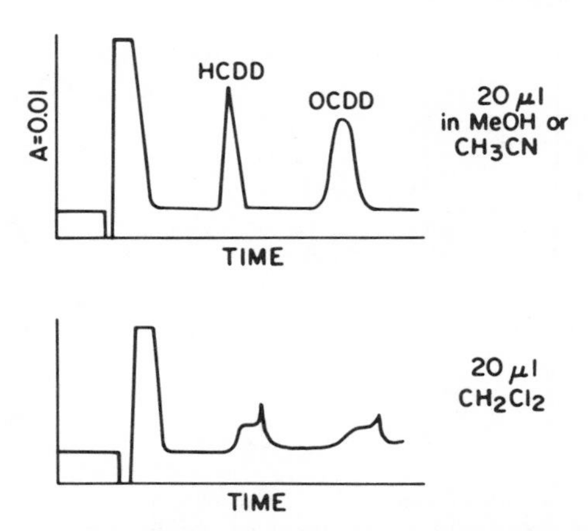

Fig. 3.2 Effect of solvent on elution of dioxins in reversed phase chromatography. Chromatograms of two dioxins injected into 20 μl of methanol or acetonitrile (a) and 20 μl of methylene chloride (b). Mobile phase was 90% methanol/water with an RP-18 column. From Ryan and Pilon (*5*), with permission from Elsevier Scientific Publishing Co., Amsterdam.

ilar, and no significant distortion of the peaks appear. However, when 20 μl of sample in methylene chloride are injected, gross peak distortion appears, since the nonpolar dioxin peaks are greatly influenced by the methylene chloride. In the same work (*5*), it was found that 50 μl of methanol or acetonitrile had no effect on peak shape of the dioxins; however, peak distortion was observed when 50 μl of sample in ethanol were injected.

The same effects can also occur in adsorption chromatographic systems. If a moderately polar compound is injected in hexane (or other very nonpolar solvent), the solute molecules will be retained in a narrow band at the top of the column, since hexane is not strong (polar) enough to elute them. In an analogous manner to injecting large volumes of water into a reversed-phase system, often large sample volumes (in hexane) can be injected into an adsorption system with little or no peak deformation. However, if the sample solution is significantly more polar than the mobile phase (for example, if the mobile phase is 5% ethanol in hexane and the sample solution is 100% ethanol), then peak distortion will likely occur just as in reversed-phase chromatography, where methanol or acetonitrile sample solutions are injected into an LC system employing a mobile phase containing, say, 30% or more of water.

In summary, for reversed-phase chromatography, the sample solution should contain at least the same percentage of water as the mobile phase, and even more if large volumes ($\simeq$100 μl) are to be injected. In adsorption chromatographic systems, the sample solution should be in a solvent that is less polar than (or equal to) the mobile phase. For large-volume injection, it is best to have the sample solution in a very nonpolar solvent. These guidelines are unfortunately limited by the solubility of the solute in the solvent. In many cases, pure water may be the best solvent to use for injection into a reversed-phase system; however, many organic compounds are not soluble in it. Thus, one is forced to use solutions containing various percentages of organic solvents. The same is true about using hexane, isooctane, or other very nonpolar organic solvents for injection into adsorption chromatographic systems.

II. REQUIREMENTS FOR TRACE ANALYSIS

There are several types of injection ports commercially available at present. They all function to apply the sample solution to the head of the analytical column in a reproducible manner with a minimum of band-spreading. The major differences between them are how this is carried out. For trace analysis, injection ports should be able to withstand normal operating pressures, which may vary up to 3000 psi. For convenience, they should also be able to withstand higher pressures that are required for high flow rates used for rapid mobile-phase change, without serious leakage.

The construction materials of the injection port should be inert to the mobile phase. This is the case for all the injection ports available with the exception of septum injectors in some instances, depending upon the compatability of the septum material with the mobile phase. Since most injection ports, like pumping systems, are made from #316 stainless steel and inert substances such as Teflon or Kel-F, they will provide acceptable service with any mobile phase that is compatible with the pumping system.

In trace analysis, low levels of analyte must be detected, often in complex biological samples. This usually means that many coextractives are present in the sample solution besides the analyte. The question then arises as to how much sample material should be injected onto the column for analysis. Kirkland (*6*) has suggested that for maximum sensitivity in trace analysis, as large a mass as possible of the sample component should be injected. However, this author suggests that the *minimum* possible amount of sample should be injected that would yield a quantitatable reproducible peak. Also, the detector should be operated at the highest useful sensitivity in order to keep sample size as small as possible. By keeping sample size to a minimum, the integrity of the chromatographic system will be maintained for a long period. Continual injection of relatively large quantities of sample extracts, in great excess of that actually required for good quantitation, only serves to speed up the deterioration of the LC column. How fast this occurs depends upon how "clean" or "dirty" the sample extract is.

III. TYPES OF INJECTION PORTS

The following describes the operating principles of several types of injectors in use at present. These include high-pressure septum injection, stopped-flow injection, loop injection and syringe-loop injection. The syringe-loop injector has become most popular because of its versatility, ease of operation, and durability.

A. High-Pressure Injection via Septa

Figure 3.3 illustrates a typical septum injector. The injection principle is the same as for GC, with some important technical differences that make it less suitable for LC. Injection via a septum permits the application of a sample to the head of the column with little or no disturbance of the chromatography system. There is no valve to turn nor any need to stop the mobile-phase flow. The sample is injected right at (or into) the top of the column, which keeps band spreading to a negligible level when small volumes ($<25\ \mu l$) are injected. This type of injection port is one of the least expensive and can be constructed in the laboratory with little difficulty.

There are, however, several problems associated with septum injection that have resulted in it becoming a less popular means of sample introduction. First, septum injection is only suitable for pressures less than about 1500 psi. Higher pressures cause septum rupture and leakage. Also, the number of injections possible with a septum is dependent upon the system pressure. Normally, no more than 20 injections can be made before a change of septum is required. This number will decrease if the syringe needle is not straight and sharp or if it has a barbed end. It should be pointed out that when changing septa one must be careful not to introduce air into the chromatographic system. This can lead to bubble formation in the detector.

The compatability of the septum with the mobile phase is very important for two reasons. If the mobile phase leaches plasticizers or other chemicals from the septum, increased detector noise may result. This effect is analogous to "septum bleed" in GC. In some cases, the septum

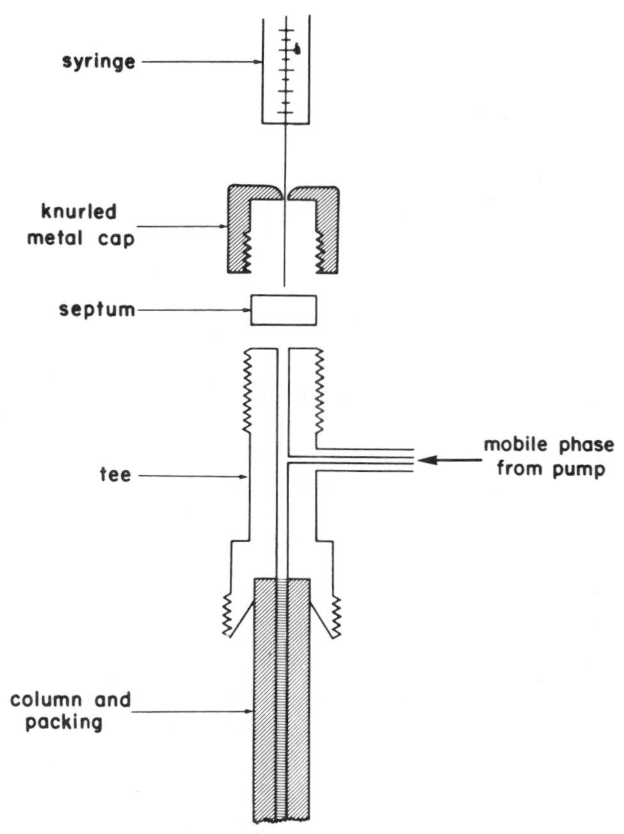

Fig. 3.3 Septum injector for LC. The syringe needle penetrates the septum and passes through the tee just to the head of the column. The sample is injected and immediately flushed onto the column by the moving mobile phase.

will absorb mobile phase appreciably and swell or become hard and brittle. Both effects weaken the septum, reducing its lifetime significantly. It has been suggested that wrapping the septum with inert material such as Teflon tape will increase septum deterioration. Table 3.1 lists some septum materials and their compatabilities with various solvents.

Since sample injection via septum is made at the pressure of the system, it is necessary to use syringes capable of withstanding pressures up to 1000–1500 psi. If not, the sample will have a tendency to be forced back between the syringe barrel and the plunger. Also, it is not uncommon to have the plunger pushed right out of the barrel if it is inadvertently released while the syringe is still in the system. Although syringe injection directly to or into the top of the column packing is the most

TABLE 3.1

Recommended Septum Material for Various Solvents

Recommended septum material	Solvents
EPR	Methylethyl ketone, tetrahydrofuran, water
BUNA-N	Hexane, dimethylformamide, alcohols, water
Viton-A	Hexane, benzene, toluene, trichlorobenzene, cresols, water
White silicone gum rubber	Most solvents

efficient, care must be taken to avoid blocking the needle with column particles. This is especially important with 5- to 10-μm particle supports. Some manufacturers have avoided this by replacing the first 5 mm or so of column with a porous plug into which the sample may be injected without fear of blocking the syringe. Glass beads (100 mesh or less) may also be used to pack the top few millimeters of the column for sample application.

B. Stopped-Flow Injection

Stopped-flow injection has been incorporated into some earlier LC systems which employed both constant-pressure and constant-volume pumps. The technique involves stopping the mobile-phase flow, allowing the column to reach ambient pressure, and then applying the sample to the top of the column. The chromatographic run commences when the solvent flow is resumed. For constant-pressure pumps all that is required to stop the flow is an on–off valve between the pump and injection port. However, with constant-volume pumps, it is necessary to stop the pump. The introduction of the sample to the column is normally done with a microsyringe via a valve, which is opened after the flow has been stopped. It is then closed before flow is resumed. Figure 3.4 illustrates two types of sampling systems using the stopped-flow technique. Both can be used with constant-volume pumps and, with the incorporation of an on–off valve, with constant-pressure pumps.

The main advantages of this type of injection over septum injectors are that injection is made at ambient pressure and there is no need of regular replacement of parts such as septa. However, since on-column injection is employed, care must be taken to avoid blocked syringe needles due to small particles of column packing material, as mentioned above for septum injectors.

Since the flow is stopped and the column is at ambient pressure during

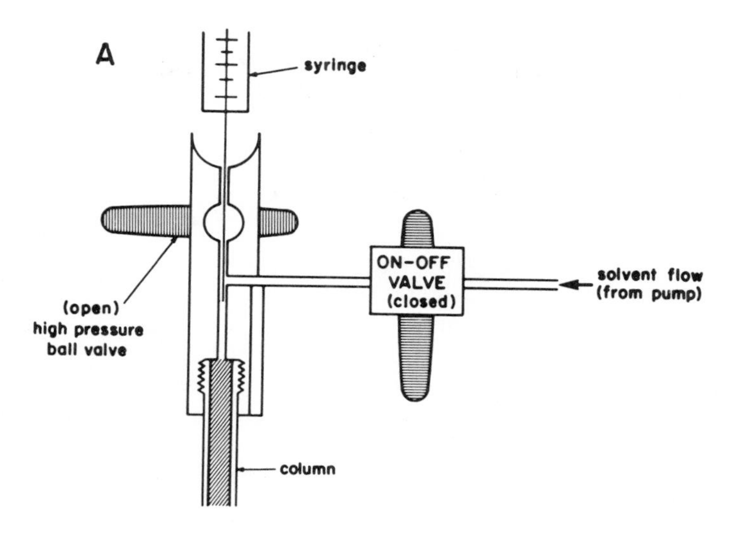

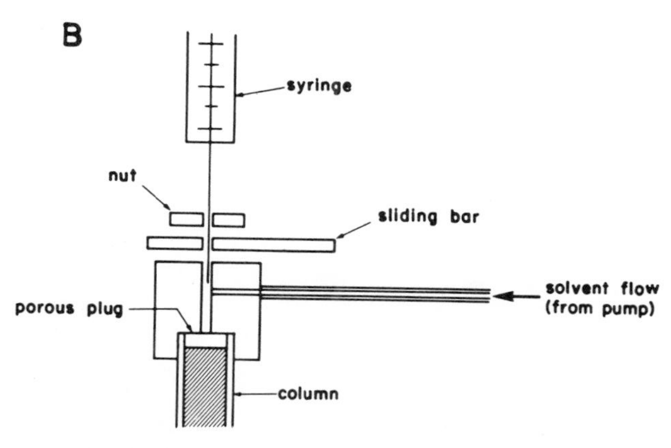

Fig. 3.4 (A) Stopped-flow injection as set up with constant-pressure pumps. The on–off valve is closed, stopping flow. The ball valve is opened and the sample applied to the top of the column. The ball valve is then closed and the mobile-phase flow resumed. (B) Sliding-bar type of injection port as set up for injection with a constant-volume pump. The pump is first turned off; then the tightening nut is loosened and the bar moved to align the hole with the injection path. The sample is applied by syringe to the porous plug. Following this, the bar is returned to the closed position, the nut is tightened, and then the pump is started.

injection, it takes some time for the chromatographic equilibrium to reestablish after mobile-phase flow is continued. This usually is accomplished within 30 sec, at which time flow rate and pressure have stabilized. The baseline can be somewhat affected by this; however, reproducibility of separations remains satisfactory.

C. Sample-Loop Injectors

Figure 3.5 illustrates a typical sample-loop injector. It consists of a switching valve which incorporates a sample loop of a given fixed volume. In the load position, the sample is drawn into the loop until it is completely filled. The valve is then turned to flush the sample onto the column. The design is such that the injection is made efficiently with little dead volume. Connections are all made with capillary tubing and low- or dead-volume fittings. This type of injection port is available for use up to 7000 psi, which is more than adequate for routine analyses. Since the injection volume is determined by the loop size, the human error associated with syringe injection is avoided, thus loop injectors are more reproducible.

One of the problems associated with loop injectors is that to fill the loop (even if it is only 25 μl in volume), a significant quantity of sample solution is lost in the process. This is usually on the order of 0.1–0.2 ml and

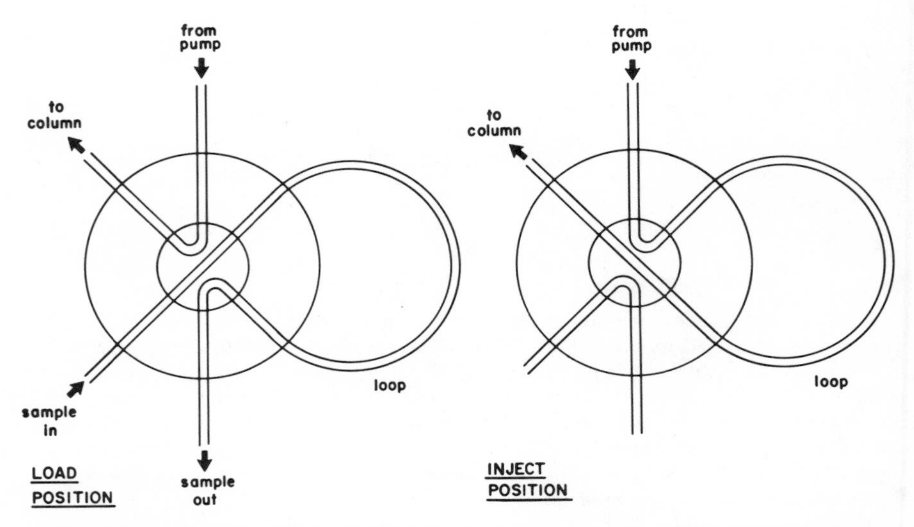

Fig. 3.5 Typical sample-loop injector. In the load position the sample is drawn through the loop until it is filled. The valve is then switched to the inject position, flushing the sample from the filled loop onto the column.

can be very critical if only that much sample is available. Of course, if 5–10 ml of sample solution are employed this does not become a problem. However, in many cases the sample must be concentrated to a small volume in order that a quantitative response be obtained by the detector.

A minor problem associated with loop injectors is that the injection volume can only be varied by changing the loop size. For routine purposes this is impractical, so the analyst is required to prepare several concentrations of his standards to make the necessary calibration curve for quantitation. This is not as convenient as syringe injection, where the calibration curves can be prepared from one solution simply by varying the injection volume.

D. Syringe-Loop Injectors

Several designs of syringe-loop injectors are now available. The major advantage of the syringe-loop injector is that it enables syringe injection at ambient pressure. This is much preferred to septum injection at high pressure. Also, the syringe-loop injector has the advantages of the loop injector except that injection error is increased because the volume is estimated by the analyst. Furthermore, sample waste is minimized when a microsyringe is used to inject the sample into the loop rather than passing a larger volume through, as in the plain loop injector. The Valco syringe-loop injector (Fig. 3.6) is essentially the same as the loop injector, but modified so that a microsyringe can be introduced to fill the loop. Although commercially available, syringe-loop injectors can be easily constructed in the laboratory from loop injectors (*7*).

Because of the design of the Valco syringe-loop injector, it is necessary to take up several microliters of mobile phase into the syringe before the sample solution to ensure that the sample is flushed completely into the loop. If not, the dead volume around the tip of the syringe needle and in the central core of the valve (see Fig. 3.6) will cause poor reproducibility. This problem is not serious; in fact, using a plug of mobile phase in the syringe is good practice in any injection system.

Although syringe injections normally are made such that the sample volume is less than the total loop volume, good reproducibility (comparable to the loop injectors) can be obtained by making sure that the loop is full. For example, if a 25 μl loop is employed, the analyst can always inject 30 μl to ensure that the loop is completely filled.

One other feature of the Valco loop injector is that the syringe need not be removed after filling the loop to switch the sample to the column. In

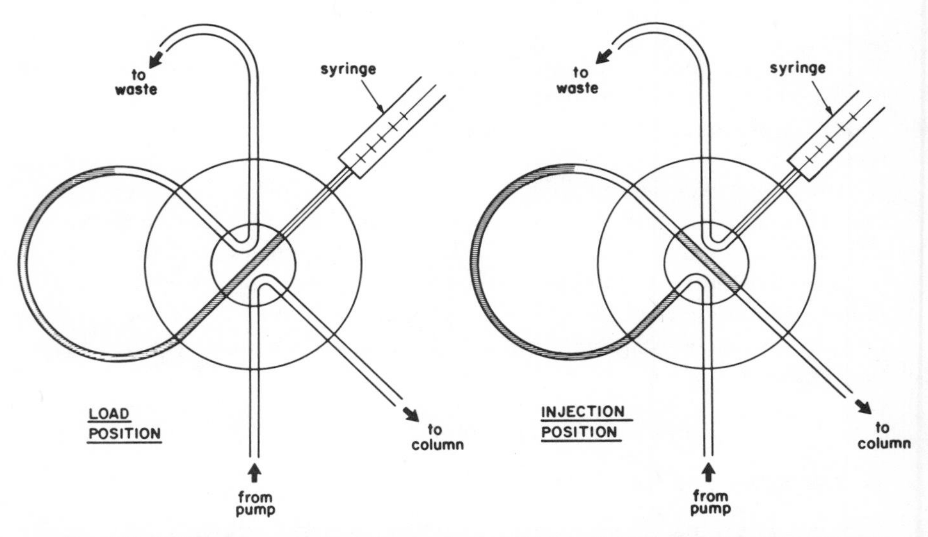

Fig. 3.6 A Valco-type syringe-loop injector. The syringe is inserted into the loop and the sample is then injected. The maximum volume is determined by the loop size. The valve is turned for application of the sample to the column, at which point the chromatogram starts.

fact, the injection port acts as a good storage vessel for the syringe, which needs to be removed only for charging with sample solution or cleaning.

The Waters model U6K and the Rheodyne model 7125 make use of sample loops that vary up to 2.0 ml in volume. Syringe injections can be made at any volume from 1 μl up to the total loop volume. These types of injectors (shown in Fig. 3.7) are designed so that little or no band-spreading or sample loss due to dead volume occurs, even if only a few microliters are injected into a loop as large as 2.0 ml. However, in the Waters injector, the syringe must be removed and a plug inserted into the injector (see Fig. 3.7) before the valve is switched to direct the sample onto the column. For the Rheodyne, the syringe remains in the injector.

The advantages of this type of injector are essentially the same as the Valco syringe-loop injector. The Waters injector, however, is several times more expensive than the Valco or Rheodyne models, although performance, in general, is similar for all three.

E. Autoinjectors

Autoinjectors designed especially for LC are included in the product lines of several manufacturers of LC equipment, including Waters,

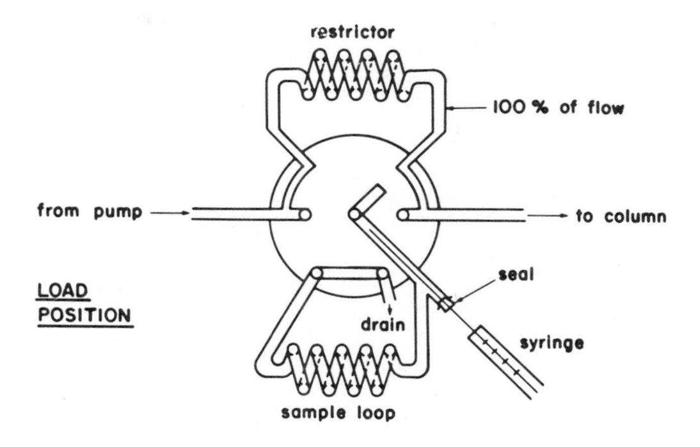

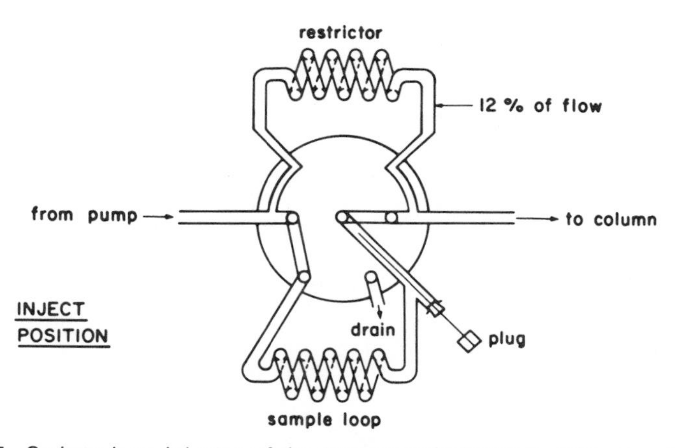

Fig. 3.7 Syringe-loop injector of the type manufactured by Waters or Rheodyne. The sample is injected into the loop, the syringe is removed, and a plug is inserted before the sample is switched onto the column.

Beckman, Spectra Physics, Hewlett-Packard, and others. These make use of compressed gas to make the sample injection mechanically. All employ sample loops and are capable of being programmed for the number of injections and the time between each run, including gradient programs. The major use of autoinjectors is to handle large numbers of samples with unattended operation. This can be of great help in trace analysis when methodology is developed to the point where routine analyses can be made either for limited survey work or for ongoing monitoring programs.

REFERENCES

1. W. A. Saner, J. R. Jadamec, R. W. Sager, and T. J. Killeen, *Anal. Chem.* **51,** 2180 (1979).
2. P. Schauwecker, R. W. Frei, and F. Erni, *J. Chromatogr.* **136,** 63 (1977).
3. J. N. Little and G. J. Fallick, *J. Chromatogr.* **112,** 389 (1975).
4. R. Kummert, E. Molnar-Kubica, and W. Giger, *Anal. Chem.* **50,** 1637 (1978).
5. J. J. Ryan and J. C. Pilon, *J. Chromatogr.* **197,** 171 (1980).
6. J. J. Kirkland, *Analyst* (*London*) **99,** 859 (1974).
7. R. M. Cassidy, *J. Chromatogr.* **117,** 71 (1976).

Chapter 4
Chromatography Columns and Packing Materials

I. REQUIREMENTS FOR TRACE ANALYSIS

The main requirement of chromatography packings used for trace analysis is that they be capable of separating the analyte peak efficiently from other sample components. This of course also depends upon other factors such as the type of mobile phase and the sample itself. The major criterion for evaluating a column is efficiency (see Chapter 6, Chromatography). A highly efficient column will produce a sharper peak at a given retention time than will an inefficient column. This has two important implications for trace analysis.

First, narrow peaks permit the separation of more compounds within a given time frame. This is of prime importance when complex samples are analyzed since there are usually many coextractives present which may elute in the vicinity of the analyte peak. Second, since the response of most LC detectors is concentration dependent, the peak area per nanogram (for example) of analyte will be approximately constant. Thus, if the efficiency of a column increases, the peak width becomes narrower, resulting in a proportional increase in peak height. In chromatography, detection limits are based on peak *height* versus background noise; thus, highly efficient columns will improve the minimum detectable quantities of analyte.

II. PELLICULAR VERSUS MICROPARTICULATE POROUS PACKINGS

Figure 4.1 compares pellicular (or porous layer) packing with small-particle (microparticulate) porous material. The pellicular packing con-

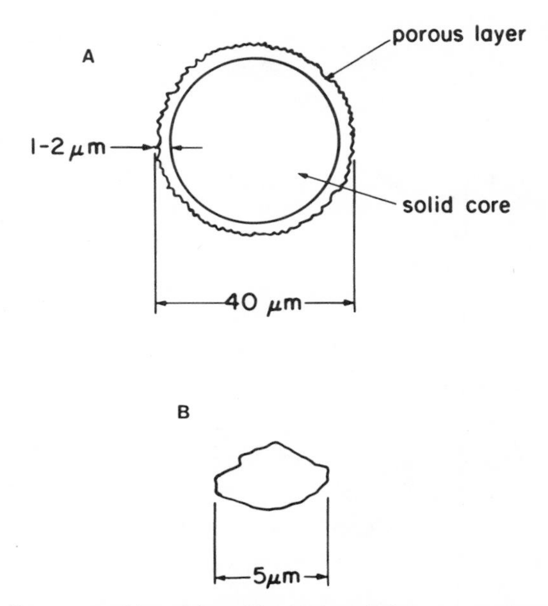

Fig. 4.1 Porous packings: A, pellicular bead; B, porous microparticle.

sists of solid spherical glass beads with mean particle diameters of 37–40 μm. On the surface of the bead is a 1–2 μm thick layer of silica gel, alumina, or other material. This thin porous layer permits rapid mass transfer, giving good efficiencies compared to porous particles of the same diameter (37–40 μm). However, the surface area of these packing materials is about 5–15 m^2/gm and has a sample capacity of only about 100 μg/gm of packing. The dense solid core and the size of the particles permits the dry packing of pellicular materials into the columns in a manner similar to packing GC columns.

The most common microparticulate packings consist of either 5- or 10-μm diameter particles, and because they are totally porous they have a large surface available, permitting applications of up to 5 mg of sample per gram of adsorbent. The small diameters of these particles prevent them from being efficiently dry-packed into columns. Several slurry techniques have been developed and are discussed in the following section. Figure 4.2 shows the effect of particle diameter on efficiency for a hypothetical compound with changing linear velocity. It can be seen that as particle diameter decreases, the height equivalent to a theoretical plate decreases, thus significantly increasing column efficiency. The 5- and 10-μm porous microparticulate packings have become most popular at present and are ideally suited to trace analytical applications. Develop-

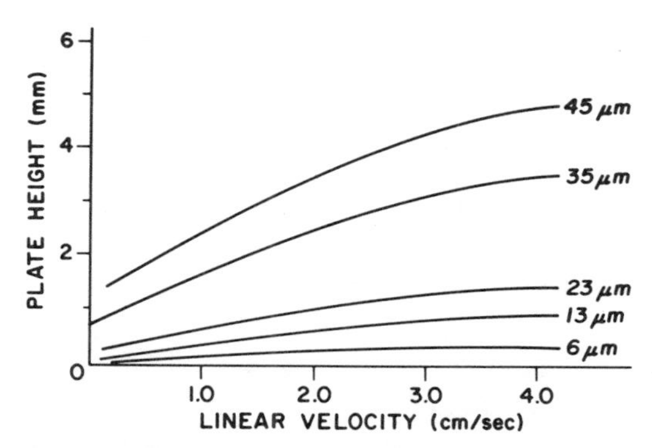

Fig. 4.2 Plate height (mm) versus linear velocity of mobile phase for porous particles of various mean diameters.

ment of new packing materials continues at good pace. Spherical porous microparticulate packings are available that increase efficiencies even further because they are more uniformly packed into a column.

III. COMMERCIAL PREPACKED COLUMNS

Most LC dealers and manufacturers supply prepacked columns in various lengths and diameters packed with many different types of stationary phases. The advantages of these are that they provide high efficiencies and are usually consistent from one column to the next with the same manufacturer. One unfortunate problem at present is the lack of consistency of column selectivity between manufacturers. Four different companies can market reversed-phase C_{18}-bonded columns which have four very different selectivities. This has caused problems with analysts who have tried to repeat work on different columns. In one instance a C_{18} column may work well; however, the method may not work when a C_{18} column provided by a different manufacturer is used. It is hoped that efforts by such groups as the ASTM, AOAC, and ACS, and analysts in general, will result in a system such as the McReynolds system (*1*) in GC, whereby column selectivity may be characterized.

The main disadvantage of prepacked columns is their expense. However, if well cared for they can last for more than a year with little loss in selectivity or efficiency. Keeping mobile phases as clean as possible and performing adequate cleanup on samples, as well as keeping the quantity of sample injected as small as possible, will serve to increase column lifetime. However, in some experimental work where exotic mobile phases are employed or where the nature of the sample to be injected is unknown, columns may be ruined very quickly. In any instances where column life may be short, it is economical to pack columns in the laboratory. The following section outlines some accepted packing techniques used for both adsorption and reversed-phase chromatography on pellicular or microparticulate support materials.

IV. COLUMN PACKING TECHNIQUES

In order to obtain efficient columns acceptable for trace analysis, it is necessary to take great care in preparation of the packing material and the column. The filling of the column is also critical if reproducible efficiencies are to be expected. It is important to minimize particle fractionation. Because of this, many analysts prefer to buy efficient prepacked columns rather than use mediocre laboratory-packed ones. Nevertheless, an analyst can, with little difficulty, pack columns with good efficiencies that function adequately for trace analysis.

A. Column Preparation

Normally, precision-bore seamless stainless steel is used for the column. The outer diameter is $\frac{1}{4}$ in, whereas the inner diameter can vary from 2.1 to 4.6 mm. The smaller diameter columns have been shown to suffer from wall-effects (*2,3*), which significantly reduce column efficiency. The trend in recent years has been to 4.6-mm (id) columns 15–30 cm in length.

The empty column must be thoroughly cleaned and dried before filling. It has been recommended (*4*) that each column be washed, in succession, with chloroform, acetone, water, 50% phosphoric acid, 10% nitric acid, water (until the washings are neutral), acetone, and finally chloroform. After this the column should be thoroughly dried. A frit (usually $\approx 2\ \mu m$) is placed on one end of the column, and the appropriate fittings are attached and tightened. The column is then ready for filling.

B. Dry Packing

Dry packing is recommended for particle diameters greater than 30 μm and is the most common technique employed for preparation of pellicular packed columns. A method that produces good, reproducible columns is the "rotate, bounce, and tap" method, which can be mechanized (*5*).

The column is mounted vertically with a small funnel attached at the top and is filled with small increments of the packing material. During this process the column is bounced (from a height of about 1 cm) on a hard surface about 100 times per minute, while at the same time being rotated about once per second. During this operation the column is tapped gently near the level of the top of the packing. This whole process takes about 10–30 min. Several analysts have evaluated this technique (*6–8*).

C. Slurry Packing

Particles that are less than 20 μm in diameter are best packed by slurry techniques. These small particles possess high surface energy and tend to agglomerate in the dry state and stick to the walls of the column. Thus, by dispersing the particles in a suitable liquid (usually with ultrasound), a more uniform column bed can be obtained. The main problem associated with slurry techniques is the sedimentation of the particles. Two techniques have proved to be satisfactory to minimize this effect: a balanced-density slurry and a viscosity technique.

Figure 4.3 illustrates a typical setup for slurry packing LC columns. The equipment consists of a constant-pressure pump such as the Haskel DSXHT-602-C which is capable of operating routinely at 10,000–15,000 psi. Constant-volume pumps may also be employed as long as high flow rates can be generated. The pump is connected to a pressure gauge (if not included in the pump itself) and an on–off valve. The slurry is placed in a reservoir attached to the column. Between the two is usually placed a column extension which is constructed of the same tubing as the column and is connected to the column by means of a union. The reason for the exten-

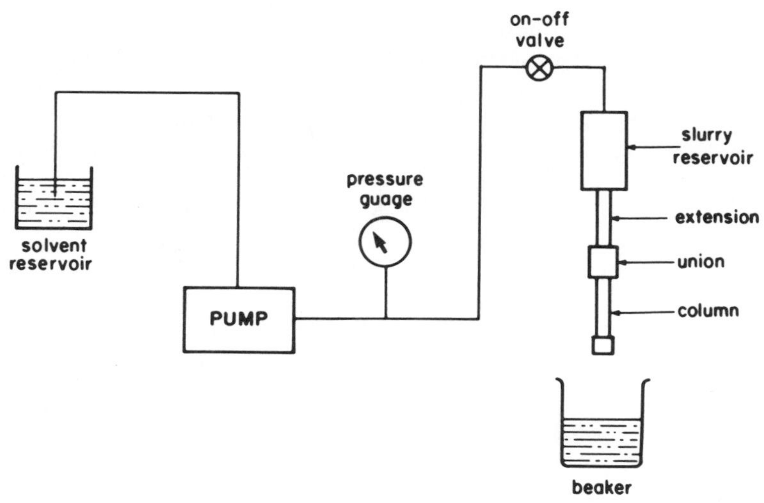

Fig. 4.3 Slurry-packing apparatus employed for microparticulate materials.

sion is that as the column becomes filled, the pressure drop across the top few centimeters becomes less; thus the material is not packed as densely as the rest of the column. Enough packing is used to fill both column and extension. When complete, the extension is discarded, leaving a uniformly packed column. Many variations in slurry-packing techniques have been reported with good success (*9–14*).

1. Balanced-Density Slurry Packing

This procedure employs a dispersion medium of high-density solvents to minimize sedimentation of the particles. For example, 50 ml of a solvent mixture consisting of 60.6% (wt) tetrabromoethane and 39.4% (wt) tetrachloroethylene may be used for 5 gm of silica particles, which will pack a 25 cm × 4.6 mm (id) column plus 5 cm of extension. The mixture is first degassed and homogenized in an ultrasonic bath and carefully added to the reservoir, filling it. The column and extension are filled previously with the same balanced-density solvent. The reservoir is connected to the pump and the slurry is forced at maximum flow rate (or pressure; e.g., 10,000 psi) into the column using hexane or similar solvent from the pump. When this latter solvent appears at the end of the column, the flow may be stopped since packing is complete. Before removing the column from the apparatus it is important to let the pressure of the system fall completely to ambient. A frit is placed on top of the column and the appropriate fittings are attached. The column is then ready for use.

2. Slurry Packing With Viscous Solvents

The apparatus illustrated in Fig. 4.3 may also be used for packing with viscous dispersion solvents. To pack a 25 cm × 4.6 mm (id) column, 5 gm of silica (for example) are mixed with 50 ml of a 20% solution of glycerol in methanol. This solution is degassed and treated ultrasonically in the same manner as described above for the preparation of balanced-density slurries. The column and extension are filled with the viscous dispersion solution; then the slurry is added to the reservoir. The pump is connected and hexane or a similar solvent is pumped at a high flow rate, generating pressures up to 10,000 psi or more. The faster the column is packed, the higher the efficiencies obtained.

V. STATIONARY-PHASE MATERIALS

Table 4.1 lists suppliers of column-packing materials and prepacked columns for LC. These include pellicular and microparticulate materials, ion-exchangers, silicas, aluminas, gel permeation materials, and a number

TABLE 4.1

Suppliers of Column Packing Materials

Supplier	Address
Applied Science Division, Milton Roy Co.	P.O. Box 440, State College, Pennsylvania 16801
Becker-Packard B.V.	P.O. Box 519, Vulcanusweg 259, Delft, The Netherlands
Bio-Rad Laboratories	Chemical Division, 32nd and Griffin Ave., Richmond, California 94804
Corning Glass Works	Houghton Park Corning, New York, 14830
E.I. Du Pont de Nemours and Co.	Instrument Products, Division, Wilmington, Delaware, 19898
Durrum Instruments Corp.	3950 Fabian Way, Palo Alto, California 94303
Hamilton Co.	12440 E. Lambert Road, Whittier, California 90608
L.K. B. Productur A.B.	Fredforsstigen, 22–24 Bromma, Stockholm, Sweden
Mackery, Nagel and Co.	Werkstrasse 6–8, D5160 Dueren, West Germany
E. Merck	Frankfurter Strasse 250, D-61 Darmstadt, West Germany
Perkin-Elmer Corp.	Instrument Division, Main Ave., Norwalk, Connecticut 06856
Pharmacia Fine Chemicals A.B.	Box 604, 751-25 Uppsala, 1, Sweden
Phase Separations Ltd.	Deeside Industrial Estate, Queensferry CH5 2LR, Clwyd, United Kingdom
H. Reeve Angel	9 Bridewell Place, Clifton, New Jersey 07014
The Separations Group	8738 Oakwood Ave., Hesperia, California 92345
Spectra-Physics	2905 Stender Way, Santa Clara, California 95051
Varian Associates Ltd.	Instrument Division, 611 Hansen Way, Palo Alto, California 94303
Whatman-Reeve Angel	Gaurt Street, London, England, SE1 6BD
Waters Associates, Inc.	183 Maple St., Milford, Massachusetts 01757

of different bonded phases. In the literature, reviews and research articles on many different packing materials used for LC have appeared (*15–22*).

REFERENCES

1. W. McReynolds, *J. Chromatogr. Sci.* **8,** 685 (1970).
2. J. J. deStefano and H. C. Beachell, *J. Chromatogr. Sci.* **8,** 434 (1970).
3. J. J. deStefano and H. C. Beachell, *J. Chromatogr. Sci.* **10,** 655 (1972).
4. B. L. Karger, K. Conroe, and H. Engelhardt, *J. Chromatogr. Sci.* **8,** 242 (1970).
5. J. H. Knox, *Lab. Pract.* **22,** 55 (1975).
6. L. R. Snyder, *Anal. Chem.* **39,** 698 (1967).
7. H. N. M. Stewart, R. Amos, and S. G. Perry, *J. Chromatogr.* **38,** 209 (1968).
8. D. Randau and W. Schnell, *J. Chromatogr.* **57,** 373 (1971).
9. R. E. Majors, *Anal. Chem.* **44,** 1722 (1972).
10. J. J. Kirkland, *J. Chromatogr. Sci.* **10,** 593 (1972).
11. R. M. Cassidy, D. S. LeGay, and R. W. Frei, *Anal. Chem.* **46,** 340 (1974).
12. J. Asshauer and I. Halasz, *Anal. Chem.* **46,** 139 (1974).
13. B. Coq, G. Gonnet, and J. L. Rocca, *J. Chromatogr.* **106,** 249 (1975).
14. I. S. Krull, M. H. Wolf, and R. B. Ashworth, *Am. Lab.* **10,** 45 (1978).
15. A. Pryde, *J. Chromatogr. Sci.* **12,** 486 (1974).
16. C. Horvath and W. Melander, *J. Chromatogr. Sci.* **15,** 393 (1977).
17. N. H. C. Cooke and K. Olsen, *Am. Lab.* **11,** 45 (1979).
18. F. M. Rabel, *Am. Lab.* **7,** 53 (1975).
19. R. E. Majors, *Am. Lab.* **8,** 13 (1975).
20. R. P. W. Scott and P. Kucera, *J. Chromatogr. Sci.* **12,** 473 (1974).
21. M. Gurkin, *Am. Lab.* **9,** 29 (1977).
22. A. P. Shroff, *Am. Lab.* **8,** 13 (1976).

Chapter 5
Detectors

I. INTRODUCTION

Detectors for modern LC are of several distinct types. There are those that rely on the difference in a bulk property of the solute-plus-mobile phase and the mobile phase itself. Thus a physical characteristic of the effluent is monitored, and the change that occurs in that characteristic during the elution of a solute produces the detector response. The most commonly used detector of this type is the refractive index detector, although others such as the electrolytic conductivity and dielectric constant detectors are available. They are often referred to as mass detectors or bulk-property detectors. In general, these are not suited to organic trace analysis because of the lack of both sensitivity and selectivity.

A second type of detector is defined as the solute detector. In this case a physical property of the solute is utilized for detection. This same property, of course, must be absent or minimal in the mobile phase itself. The UV absorption detector is by far the most common detector of this type. Others include fluorescence, radiochemical, and electrochemical detectors. The use of these detectors necessitates the choice of solvent systems that will not interfere with detection of the solute. This is normally of no great concern since the variety of solvents for LC is very large.

A third type of LC detector is the phase-transformation detector. At present, this class of detectors is not widely used in LC. These operate on the principle of removing the mobile phase, usually by evaporation, before measuring the solute. This is normally carried out by allowing the column effluent to deposit on a moving wire, chain, or ribbon, which passes through a heater where the volatile mobile phase is evaporated. The solute molecules remain on the wire and continue into a detection system, normally associated with flame-analytical techniques such as flame ionization or flame photometric detection. Technical problems such as evaporation of the solute from the moving wire and difficulty in obtaining uniform coating of the wire with solute have limited the desirability of such a system. However, research is continuing in this area and newer commercial detectors of this type should be much improved.

II. GENERAL DETECTOR REQUIREMENTS

No matter what detector one chooses for an LC system, there are certain minimum requirements that the detector must possess in order to function satisfactorily. The level of detector performance needed can vary greatly, depending upon the application of the LC system. Three general headings one might use for comparing detector requirements are sensitivity, noise, and linearity of response. These are the same parameters which are normally used to characterize gas chromatography detectors. For trace analysis, sensitivity is of prime importance. A sensitive detector permits the use of small sample sizes, thus maintaining longer column life and freedom from overloading, which can adversely affect chromatographic peak shape. High sensitivity is required for most analytical problems concerned with trace (parts per million or less) compounds in complex matrices, such as foods and soils. It is of less importance in areas such as quality control monitoring of products or for preparations where the components are in the percent range of concentration.

Practical considerations of sensitivity must include noise. Detector sensitivities are best described in terms of quantities required to produce a certain signal-to-noise ratio. Often detectors that are described as being very sensitive in an absolute sense are not useful at maximum sensitivity because of the associated noise, be it high-frequency noise, short-term noise, or baseline drift.

The linearity of response is certainly of importance where quantitation of peaks is required regardless of the level of the detector sensitivity. A detection system that provides a linear relationship with peak height or peak area is the simplest to use. However, logarithmic plots are sometimes required, but may be less satisfactory because of the nature of the mathematical conversions involved.

Design features play an important role in the successful operation of a detector for LC. Detector volume is of major importance. The cell size should be kept as small as possible while maintaining good sensitivity. Large-volume cells permit unnecessary peak broadening due to longitudinal mixing within the cell. Normally the volume of the detector cell should be smaller than 0.2 of the volume containing the eluted peak. This is important for early eluting peaks with k' values of less than 2, since in this area components are eluted in small volumes. For peaks having

$k' \geq 2$, cell volumes of 5–10 μl are satisfactory. Cells having volumes of 1–20 μl are the most common for LC detectors.

The shape of the cell is also important. Eluting solvents must be swept through the system cleanly with as little mixing as possible. Mixing often occurs at inlet connectors that are not fitted properly to the detector cell. This permits "dead" volume and results in solute band broadening.

The sensitivity of the detector to temperature changes, either of the mobile phase or of the ambient conditions, can have a pronounced effect on detector response. Usually, such effects appear as baseline changes, and where there is a frequent but perhaps small temperature change, such as from a nearby open window, significant short-term noise may result. Also, long-term drifting may result when the room temperature fluctuates slowly because of air conditioner off-and-on cycles, or from other equipment that may generate heat. This noise is particularly evident with mass detectors such as the refractive index detectors, and occurs to a very minor extent with some UV detectors at high sensitivity.

Vapor bubbles in detector cells are frequently the cause of many problems. It is advantageous for cells to be designed so that bubbles are swept through the system with no accumulation within the cell. Usually, degassing the mobile phase by boiling, applying vacuum, or by ultrasonification solves this problem. However, it may be necessary in some cases to apply a back-pressure to the cell outlet, which helps keep bubbles dissolved. This must be within the maximum operating pressure for the cell, which is generally in the range of 50–100 psi or greater.

The influence of mobile-phase flow rate on the detector is usually significant with the mass detectors such as the refractive index or conductivity detectors. UV detectors can also be affected by flow-rate change, but in this case, since the change is normally small, the effect may be of concern only when operating at high detector sensitivities.

One particular advantage of flow-through type detectors is that the solute molecules are not destroyed. This is unlike many detectors for gas–liquid chromatography (GC), which utilize flames or furnaces for detection purposes. This nondestructive nature of most LC detectors permits easy collection of solutes for confirmation purposes, if desired, or allows LC to be used as an efficient cleanup method for quantitation of the intact solute by other techniques.

There is a wide variety of detectors available that have been created specifically for LC. These often differ in design features and operating characteristics. A detailed discussion of the more common LC detectors including operation, advantages, and limitations is presented in the following paragraphs. A list of manufacturers of LC detectors is given in

TABLE 5.1

Manufacturers of Liquid Chromatography Detectors

Detector	Manufacturer
Absorbance	Aminco, 8030 Georgia Ave., Silver Spring, Maryland 20910
	Analabs, 80 Republic Dr., North Haven, Connecticut 06473
	Applied Automation, Pawhuska Rd., Bartlesville, Oklahoma 74004
	Beckman Instruments, Inc., P.O. Box C-19600 Irvine, California 92713
	Buchler, 1327 16th St., Fort Lee, New Jersey 07024
	DuPont, 1007 Market St., Wilmington, Delaware 19898
	Gilson Medical Electronics, P.O. Box 27, Middleton, Wisconsin 53562
	Hewlett-Packard, Route 41, Avondale, Pennsylvania 19311
	Instrumentation Specialties, 4700 Superior Ave., Lincoln, Nebraska 68503
	JEOL, 477 Riverside Ave., Medford, Massachusetts 02155 1418 Nakgami Akishima, Tokyo 196, Japan
	Jobin-Yvon Optical Systems, 173 Essex Ave., Metuchen, New Jersey 08840
	Laboratory Data Control, P.O. Box 10235, Riveria Beach, Florida 33404
	LKB, 12221 Parklawn Dr., Rockville, Maryland 20852
	Micromeritics Instrument Corp., 5680 Goshen Springs Rd., Norcross, Georgia 30071
	Perkin-Elmer, Main Ave., Norwalk, Connecticut 06856
	Pye Unicam, York St., Cambridge, CBI 2PX, Great Britain
	Serva, Heidelberg, Federal Republic of Germany
	Schoeffel Instrument Corp., 24 Booker St., Westwood, New Jersey 07675
	Siemens, Karlsruhe, Federal Republic of Germany
	Spectra Physics, 2905 Stender Way, Santa Clara, California 95051
	Tracor, 6500 Tracor Lane, Austin, Texas 78721
	Varian Associates, 611 Hansen Way, Palo Alto, California 94303
	Waters Associates, 165 Maple St., Milford, Massachusetts 01757
	Carl Zeiss, Inc., Oberkochen/Württemberg, Federal Rebublic of Germany
Refractive index	Analabs
	Applied Automation
	Cargille Labs, 55 Commerce Rd., Cedar Grove, New Jersey 07009
	DuPont
	Gow-Mac, 100 Kings Rd., Madison, New Jersey 07940
	Hewlett-Packard

(*continued*)

TABLE 5.1 *(cont'd)*

Detector	Manufacturer
	Laboratory Data Control
	Micromeritics
	Perkin-Elmer
	Siemens
	Tracor
	Varian Associates
	Waters Associates
Conductivity	Chromalytics Corp., Route 82, Unionville, Pennsylvania 19375
	Laboratory Data Control
	LKB
	Varian Associates
Flame ionization	Chromalytics Corp.
	Pye Unicam
	Tracor
	Varian Associates
Microadsorption	Gow-Mac
	Schoeffel
Fluorescence	Aminco
	DuPont
	G.K. Turner Associates, 2524 Pulgas Ave., Palo Alto, California 94303
	Laboratory Data Control
Radiochemical	Berthold, D-7547 Wildbad, Federal Republic of Germany
Electron capture	Pye Unicam

Table 5.1 and some examples of a number of LC detectors and important associated parameters appear in Table 5.2.

III. PHOTOMETRIC DETECTORS

A. UV Absorbance Detectors

1. Fixed Wavelength Filter Photometers

The most commonly used UV absorption detectors are single-wavelength types based on absorption of many mercury lines, 254 nm being most frequently used because it is generally applicable and also the most intense line. Provision is usually made in these detectors for wavelength changes by replacing filters.

TABLE 5.2

Characteristics of Some Common Detectors

Detector	Cell volume (μl)	Sensitivity (full scale)	Noise at maximum sensitivity	Range (full scale)
Absorption				
Beckman 151	20–100	5×10^{-3} AU	$\pm 5 \times 10^{-5}$ AU	0.005–2.56 AU
DuPont	8	1×10^{-2} AU	$\pm 5 \times 10^{-5}$ AU	0.01–2.56 AU
Gilson 263	8	5×10^{-3} AU	$\pm 5 \times 10^{-4}$ AU	0.005–2.56 AU
Gow-Mac	20	1×10^{-2} AU	$\pm 1 \times 10^{-4}$ AU	0.01–1.0 AU
LDC UV-Monitor	8	1×10^{-2} AU	$\pm 1 \times 10^{-3}$ AU	0.01–0.64 AU
Perkin-Elmer LC55	8	2×10^{-2} AU	$\pm 5 \times 10^{-5}$ AU (280 nm)	0.02–3.0 AU
Schoeffel	8	1×10^{-2} AU	$\pm 5 \times 10^{-4}$ AU	0.01–2.0 AU
Spectra-Physics AP8200	20	2.5×10^{-3} AU	$\pm 2 \times 10^{-5}$ AU	0.0025–1.28 AU
Varian Varichrom	8	5×10^{-3} AU	$\pm 3 \times 10^{-4}$ AU	0.005–5.4 AU
Waters 440	12.5	5×10^{-3} AU	$\pm 5 \times 10^{-5}$ AU	0.005–2.0 AU
Refractive Index				
DuPont 845	3	5×10^{-6} RIU	$\pm 5 \times 10^{-8}$ RIU	5×10^{-6}–2.5×10^{-3} RIU
Tracor	5	5×10^{-5} RIU	$\pm 5 \times 10^{-8}$ RIU	1×10^{-5}–1.28×10^{-3} RIU
Varian	5	1×10^{-5} RIU	$\pm 5 \times 10^{-8}$ RIU	1×10^{-5}–1.28×10^{-3} RIU
Waters 401	10	6×10^{-6} RIU	$\pm 5 \times 10^{-8}$ RIU	6×10^{-6}–3×10^{-3} RIU
Fluorescence				
Aminco	10	0.2 mg/ml[a]	0.2%	1000-fold
DuPont	16	0.3 mg/ml[a]	10%	6552-fold
Gilson	5–65	—	3%	1000-fold
LDC	10	2 mg/ml[a]	10% drift/hr	64-fold
Conductivity				
LDC	2.5	0.01 μmho/cm	5×10^{-4} μmho	0.01–100K μmho/cm

[a] Nanograms per milliliter of quinine sulfate detected at a signal-to-noise ratio of 2.

A schematic of the Beckman Model 151 is shown in Fig. 5.1. This is a double-beam dual-wavelength UV detector utilizing a low-pressure mercury light source for 254-nm wavelength and a phosphor screen for conversion to 280 nm. Both light sources pass through the sample and reference cells as illustrated in Fig. 5.1. Each wavelength has its own photocell to monitor continuously the transmitted light from both reference and sample cells. The electrical output from each set of photocells is transmitted to a low-noise logratiometer circuit which produces an output signal directly related to sample concentration.

The detector is capable of operating in four modes; it can monitor at 254 nm, at 280 nm, or in a differential mode that monitors the difference in absorption of one wavelength from the other or vice versa. The advantage of the differential modes has not been proved, since little work in this area has appeared yet in the literature. For trace analysis this aspect should not be seriously considered when purchasing any UV photometer, unless there is some reasonable proof of its value. Other wavelengths may be chosen (up to 700 nm), available as factory-installed options. Absorbance ranges from 0.005 to 2.56 absorbance units (AU) full scale are offered, with a typical high frequency noise level 0.0001 AU, peak-to-peak, and a 0.0002 AU/hr drift. The 20-μl flow cell (10-mm path length) comes equipped with a heat exchanger. The complete system can be used with 1-MV, 10-MV, or 1-MA recorders by means of a selector switch.

The DuPont Model 835 fixed-wavelength filter detector is similar in some features to the Beckman 151. It operates with a stabilized low-pressure mercury lamp. Maximum sensitivity is 0.01 AU full scale with a noise level less than 1.0% (peak-to-peak). The linearity of the detector is kept within 1% up to 2.56 AU full scale by using a differential logarithmic amplifier circuit. The cell has a smaller volume (8 μl, standard) than the Beckman system. The selection of filters includes 280 nm, 313 nm, 334 nm and 365 nm with a medium-pressure mercury source, and selected wavelengths between 380 nm and 650 nm are available with a quartz–iodine light source. For the latter lamp, a 24-μl cell is standard. This might be a source of undesirable band broadening with fast-eluting peaks because of solute mixing effects.

Gilson Medical Electronics, Inc., offers a UV detector (Model 263) identical to the Beckman detector, with the exception that it has an operating range of 0.005–1.0 AU full scale. Noise, drift, cell volume, and electronics are virtually the same.

The Laboratory Data Control (LDC) DuoMonitor is also the same as the Beckman dual-wavelength detector. The two standard wavelengths are 254 nm and 280 nm, set up in the same manner as the Beckman 151. These detectors consist of two modules, an optical unit and an electrom-

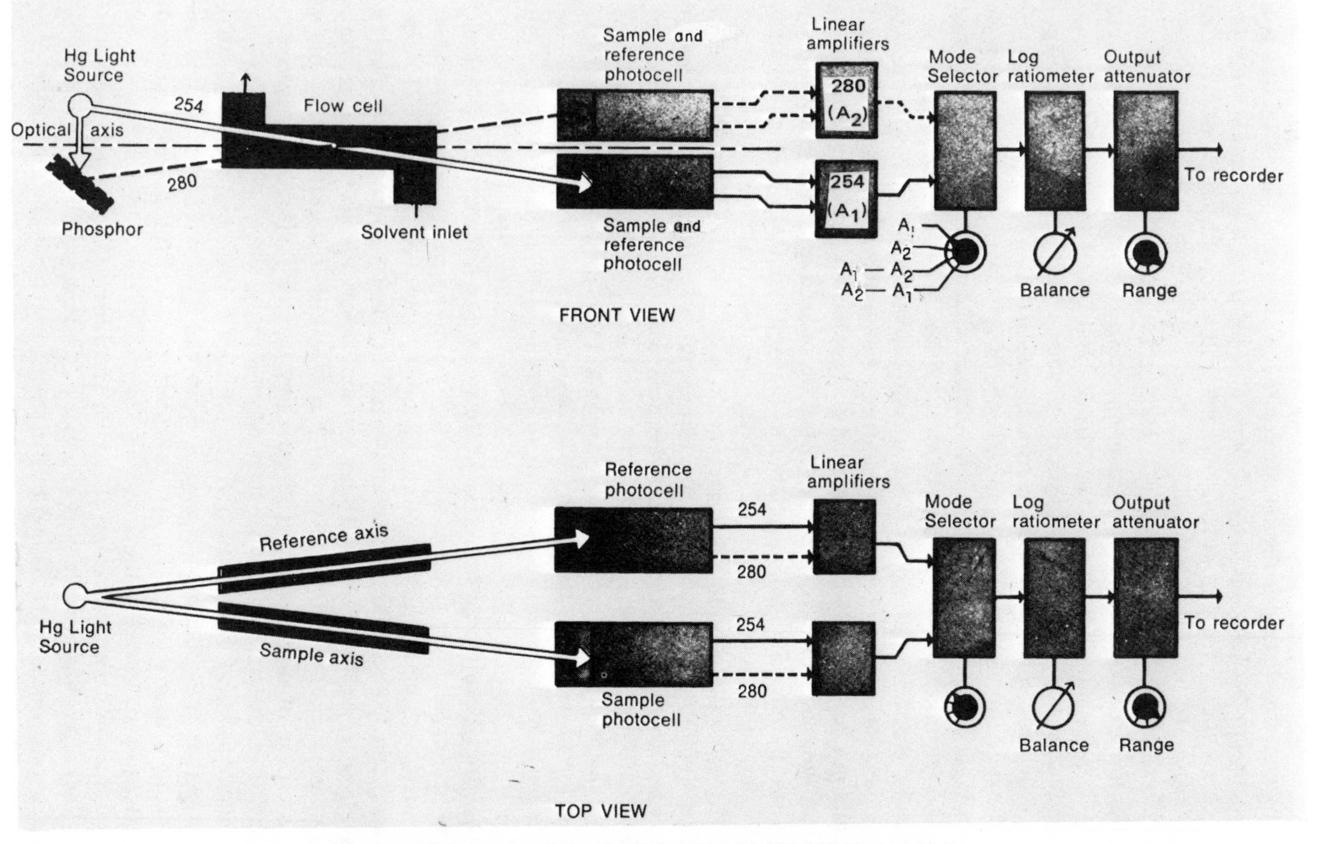

Fig. 5.1 Schematic of the Beckman Model 151 UV photometer.

eter. The LDC detector has an absorption range of 0.02–0.64 AU full scale with 1% noise at maximum sensitivity and photometer drift of less than 0.002 AU/hr. The cell volume is 25 μl.

The Spectra-Physics SP8200 UV-visible liquid chromatography detector utilizes a standard mercury vapor lamp which provides wavelengths of 254, 312, 365, 436, and 536 nm, when used with the appropriate filter. A phosphor-coated lamp is available to provide 280 nm radiation. The cell volume can either be 8 or 20 μl. The latter is suggested for most operations. Operation is similar to the others described and is based on comparison of the sample cell to a reference cell. The sensitivity is 0.0025 AU full scale which makes it the most sensitive detector available at a high-frequency noise (dry cell) of 2×10^{-5} AU at 254 nm.

The Varian fixed-wavelength dual-beam detector is operated on the same principles as those described above: that is, it involves a differential monitoring of a reference cell and the sample stream. The light source used is a hot-cathode mercury lamp which emits more than 90% of its energy as a resonance radiation of 254 nm. A 280-nm phosphor converter is also available for this unit. The small cell volume (8 μl) is a favorable feature of this detector, although the full-scale sensitivities range from 0.02 to 0.64 AU, the same as the LDC DuoMonitor. The noise level is good at $\pm 5 \times 10^{-5}$ AU.

The Waters Associates Model 440 absorbance detector can be operated in a single- or dual-wavelength mode. Figure 5.2 illustrates the setup for dual-wavelength monitoring. Each wavelength is monitored with its own independent sample and reference cells, and electronics. Thus, with the dual-wavelength accessory, two different solvent streams can be independently analyzed and recorded, making it in fact two complete UV absorbance detectors in one. For dual-wavelength monitoring of the same solvent stream, the cells are connected in series.

A novel feature of the Waters detector is their Taper-Cell design for the sample and reference cells (Fig. 5.3). This cell apparently keeps false signals due to refractive index changes to a minimum. Refractive index changes within the cell cause some of the light passing through to bend and strike the cell wall, where they may be absorbed. Thus the photodetector records the loss in energy as absorption. This refractive index effect changes with solvent composition, flow-rate changes, and even because of the solute itself. Figure 5.3 illustrates the refractive index changes within the cell due to the "liquid lens" effect caused by any of the factors described above. The Taper-Cell design permits less light to be absorbed by the cell walls than the conventional cell. This is especially important when working at high sensitivities.

The general specifications of the Model 440 are as follows: The cell vol-

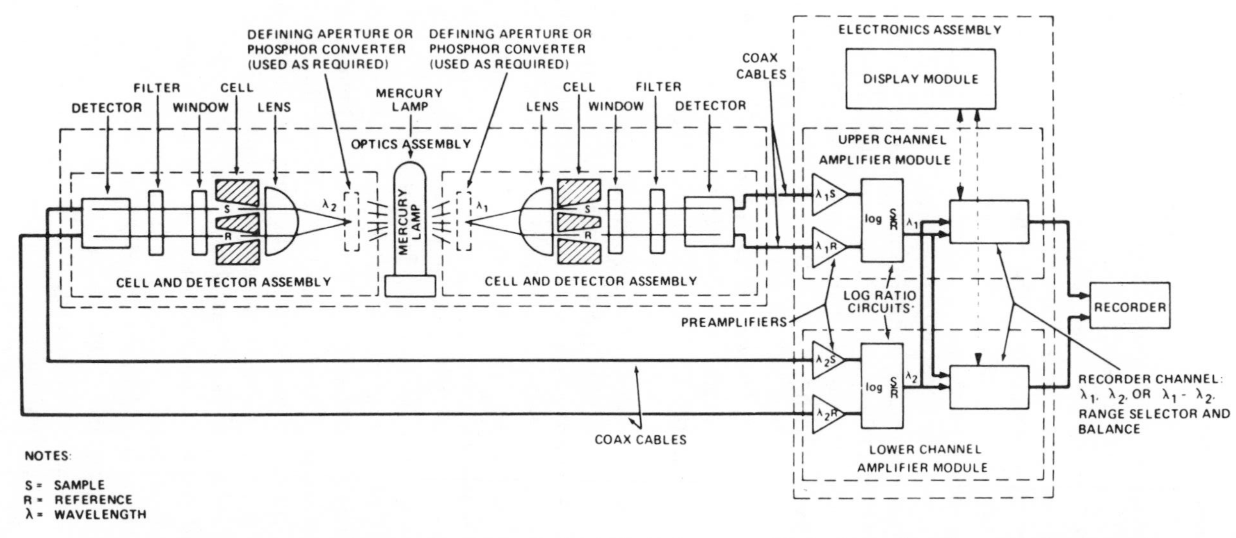

Fig. 5.2 Diagram of the Waters Associates Model 440 UV detector with dual-wavelength monitoring capacity.

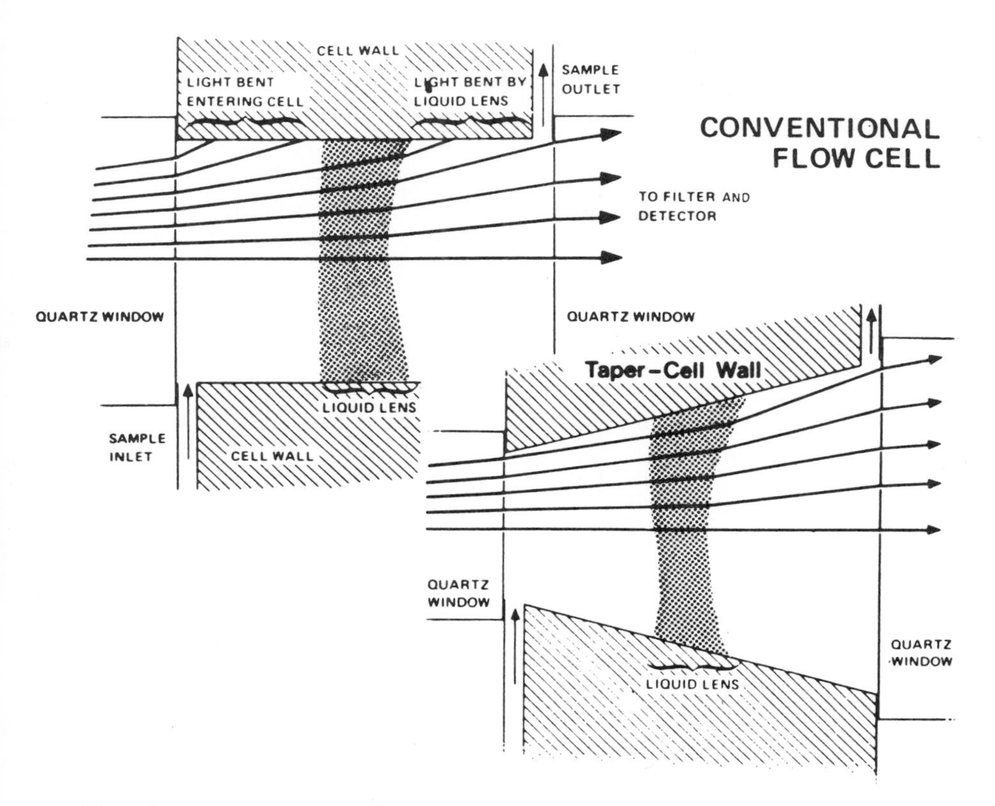

Fig. 5.3 Water Associates Taper-Cell design compared with a conventional cylindrical flow cell.

ume is 12.5 μl. Wavelengths may be selected by various filter and aperture combinations, including 254 nm (standard), 280 nm, 313 nm, 340, 365 nm, 405 nm, 436 nm, and 546 nm. Full-scale sensitivity ranges from 0.005 to 2.0 AU with a high-frequency noise of 5×10^{-5} AU (254 nm) with 1 ml/min methanol passing through the cell. In terms of signal-to-noise ratio, this detector appears comparable to the Spectra-Physics Model 8200, which has a noise level of 2×10^{-5} AU for a dry cell, thus does not take into account the effects of a moving solvent stream. The Model 440 also has a differential absorption mode which measures the difference in absorption at two wavelengths in the same manner as does the Beckman or Gilson detector. As stated earlier, the practicality of such monitoring has not yet been proved for application to trace analysis.

2. Applications of Filter Detectors

The successful use of a filter photometeric LC detector often depends upon the proper choice of wavelength. Although 254 nm is the most intense mercury line, this wavelength is generally the least selective of those offered. Where detector selectivity is not of great concern (for example, in clean samples where background interferences are at a minimum), the high intensity of this line offers excellent sensitivity. Also, since most aromatics absorb energy in this region, at least to some degree, 254 nm can be used in many applications and at relatively low solute concentrations.

For trace analysis, however, selectivity is often a very important con-

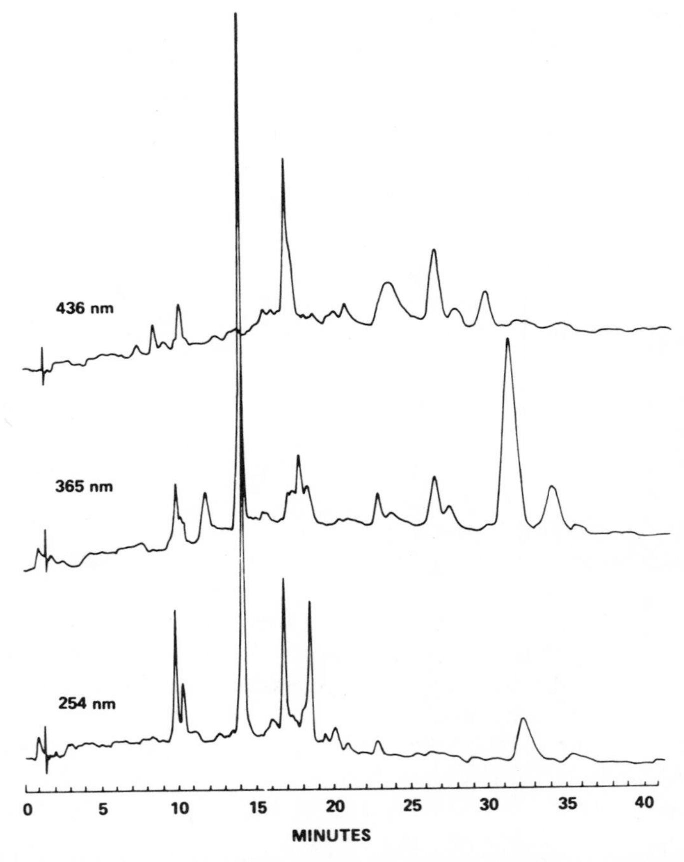

Fig. 5.4 Comparison of chromatograms of a vitamin extract at 254 nm, 365 nm, and 436 nm. Courtesy of Spectra-Physics, Stender Way, Santa Clara, California.

sideration. The differences in absorption characteristics of solute and sample coextractives must be exploited as much as possible for best analytical conditions. Thus, the UV spectrum of the compound of interest should be known so that, in the first instance, a filter passing UV light as close to the absorption maximum as possible is used. However, if background interferences become a problem at the selected wavelength, a different filter should be chosen for an area where the difference between sample and background absorption is as large as possible. The effect of choosing different filters for a particular analysis can often be very dramatic. Figure 5.4 shows a chromatogram of a vitamin extract monitored at three different wavelengths. The chromatograms show the significant effect changing wavelength has on the response. Each wavelength provides different information about the components of the extract.

Proper wavelength selection is of great importance where the compounds of interest are in extremely low concentrations, often in complex sample matrices. Figure 5.5 compares sensitivities of carbamate insecticides, carbofuran (2,3-dihydro-2,2-dimethylbenzofuranyl-7-*N*-methylcarbamate), 3-hydroxycarbofuran (2,3-dihydro-2,2-dimethyl-3-hydroxybenzofuranyl-7-*N*-methylcarbamate), and 3-ketocarbofuran (2,3-dihy-

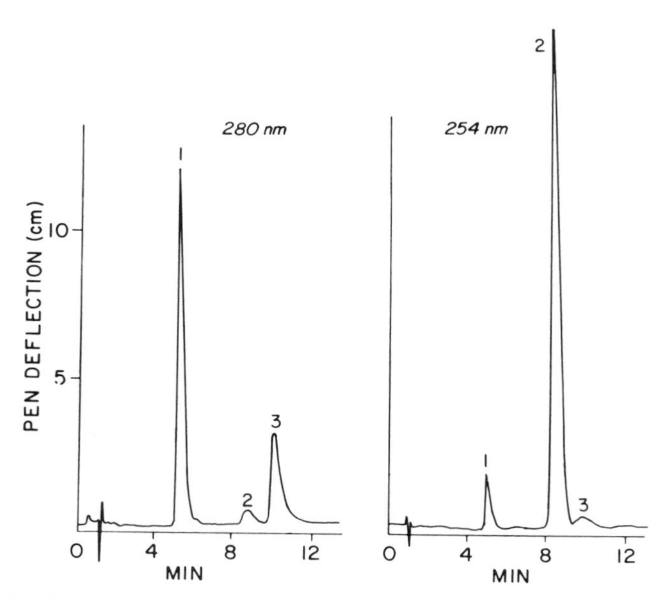

Fig. 5.5 Comparison of responses of 200 ng carbofuran (1); 135 ng 3-ketocarbofuran (2); and 200 ng 3-hydroxycarbofuran (3), at 280 nm and 254 nm detection, 0.01 AU full scale. Chromatography conditions: mobile phase, 6% isopropanol in isooctane (v/v), 1.0 ml/min; column, 25 cm × 2.8 mm i.d., 5 μm LiChrosorb Si 60.

dro-2,2-dimethyl-3-ketobenzofuranyl-7-*N*-methylcarbamate) at 254 nm and 280 nm. It can be seen that 3-ketocarbofuran is almost 20-fold more sensitive than carbofuran at 254 nm, but is about 20-fold less sensitive at 280 nm. This is due to a combination of the increase in sensitivity of the carbofuran and 3-hydroxycarbofuran at 280 nm, as well as a corresponding decrease in sensitivity of the 3-ketocarbofuran.

Figure 5.6 shows the results of a turnip sample analysis for carbofuran and 3-ketocarbofuran by LC with simultaneous detection at 254 and 280 nm. The advantage of 280 nm can be readily seen for carbofuran detection. The sample coextractives produce a much greater response at 254 nm, making quantitation of the pesticide much more reliable at the longer wavelength. However, 3-ketocarbofuran is not detected at all at 280 nm under the conditions used, whereas it is plainly visible at 254 nm. These results indicate the need for optimizing detector parameters for

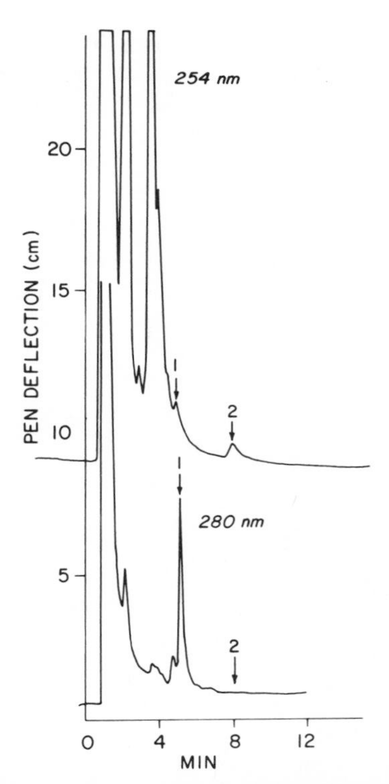

Fig. 5.6 Analysis of a turnip extract containing 1.0 ppm carbofuran (1) and 0.1 ppm 3-ketocarbofuran (2) at 280 nm and 254 nm, 0.01 AU full scale. Chromatography conditions as described in Fig. 5.5; 75 mg equivalent turnip injected. (1) carbofuran; (2) 3-ketocarbofuran.

maximum selectivity and sensitivity. Also, as shown above, a set of conditions that are optimal for one compound may not be satisfactory for another.

Optimization of detectors might include some minor modifications. Callmer and Nilson (*1*) modified a UV detector by rebuilding the detector mounting assembly to include cooling coils to maintain a constant detector temperature. This, coupled with a thermostatted analytical column, reduced detector noise by tenfold to 0.002 AU at a flow rate of 80 ml/hr isooctane. A constant temperature greatly improves the baseline of most filter photometric detectors and is especially beneficial when working at high sensitivites. Keeping the detector cell and photodetector at a constant as well as a cool temperature also reduces background noise. As a detector is operating, heat from the lamp can slowly raise the temperature of the cell and electronic components, causing an increase in noise. Increased temperature also promotes bubble formation within the cells. Cooling the detector assembly even by wrapping $\frac{1}{8}$-in copper tubing around the mounting assembly and passing tap water through it can reduce these effects. High-frequency electrical noise can be eliminated by inserting an electrical resistance–capacitance filter at the detector output (*2*).

3. Variable Wavelength Detectors

Variable wavelength (or spectrophotometric) detectors are not as widely used as the filter photometers mainly because of the difference in cost. Although more expensive, the spectrophotometric detectors offer several advantages. These include the availability of a wide range of wavelengths and the facility of changing wavelengths without changing filters or lamps, as is necessary in filter photometers. Also, since most of these variable wavelength detectors operate from a continuum source and not from the discreet lines of a mercury lamp, it is possible to use wavelengths as low as 190 nm. Wavelength selection is normally carried out with a grating monochromator. The number of variable wavelength detectors designed especially for liquid chromatography is limited at present. Some of these are discussed below.

The Schoeffel Model SF 770 Spectroflow Monitor, the Spectra-Physics Model 770, and the Waters Model 450 are essentially the same detector. The wavelength range is 190–400 nm with a deuterium lamp (standard), whereas for 350–630 nm a tungsten–halogen lamp is required (accessory). Interchange between the two ranges is a simple operation. The bandwidth is 5 nm for each, using a grating monochromator. Figure 5.7 shows a schematic of the detectors. The cells are 8 μl in volume. The operational range is 0.01–2.0 AUFS with a drift of 5×10^{-4} AU/hr. The only significant difference between these detectors is the reported high-frequency

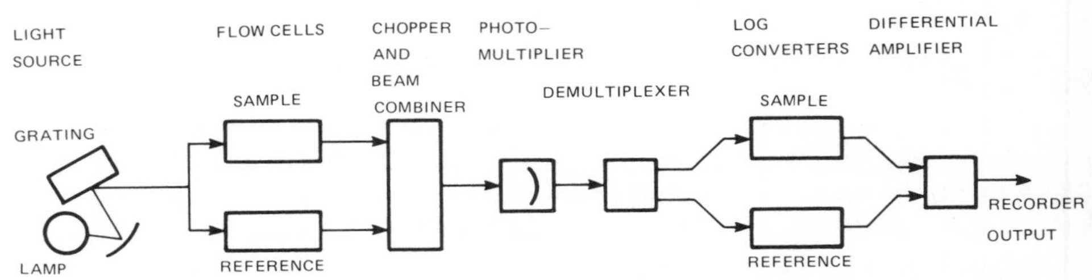

Fig. 5.7 Schematic diagram of the Spectra-Physics and Schoeffel variable wavelength detectors.

noise level. Schoeffel and Waters report this to be about $\pm 2.5 \times 10^{-4}$ AU (280 nm), and Spectra-Physics claims theirs is $\pm 1.0 \times 10^{-4}$ AU.

The Varian Vari-chrom LC detector is a double-beam, UV-visible spectrophotometer with capabilities similar to the Schoeffel or Spectra-Physics detectors. However, the Vari-chrom has a standard wavelength range from 200 to 720 nm with an option of 200–900 nm by changing the photodetector. It also offers variable wavelength band-passes from 2 to 16 nm. The full-scale sensitivity is 0.005–2.0 AU with a "photometric accuracy" of $\pm 1.0\%$ AU full scale, which is equivalent to a high-frequency noise level of 0.0001 AU. Some temperature control is provided by the water-jacketed inlet and outlet tubing to the detector cell. Wavelength selection is made by turning a large circular dial (attached to a grating monochromator), as opposed to digital selection in the two previous detectors discussed.

The Zeiss Model PM2 DLC spectrophotometric detector operates over a range of 200–850 nm, a deuterium source with separate power supply being used for 200–450 nm and a halogen filament lamp with built-in power supply for 290–850 nm. The wavelength accuracy (digital selection) is ± 1 nm (which is comparable to the others discussed) at a fixed spectral bandwidth of 10 nm. The operating ranges are 0.02–2.56 AU full scale with a high-frequency noise of ± 0.0001 AU (250 nm) and drift of 0.005 AU/hr. The flow cell is 8 μl in volume.

Other variable wavelength LC detectors are marketed by Beckman, Du Pont, Perkin-Elmer, Laboratory Data Control (LDC), Hewlett-Packard, and Tracor. The above variable wavelength detectors all have similar major features, including the noise levels and the absorbance ranges. While the main advantage of these detectors over the filter photometers is their selection of wavelengths, especially in the 190- to 250-nm range, their biggest drawback is the lower sensitivity compared to the filter photometers. This is mainly due to the fact that a monochromator and associated mirrors are used to select a wavelength from a continuum source, whereas the filter photometer isolates a single line from a low- to medium-pressure mercury lamp. Because of this, the variable wavelength detectors are 5- to 10-fold noisier, resulting in a proportional decrease in sensitivity (as signal-to-noise ratio).

The use of wavelengths in the 200- to 250-nm region is probably of definite value where samples are relatively clean and the solutes absorb poorly at 254 nm. Figure 5.8 compares UV detection of some fatty acid methyl esters at 215 nm and 254 nm. The longer wavelength is useless for analysis of these compounds. However, for trace analysis of complex samples such as soils or foods, the utility of this lower region is questionable because of the absorption of coextractives. Even the number of choices of solvents for the mobile phase greatly diminishes as wavelengths close to 200 nm are used for monitoring.

In the selection of an absorbance detector for LC, the fixed wavelength photometers should be considered first because of their lower price and

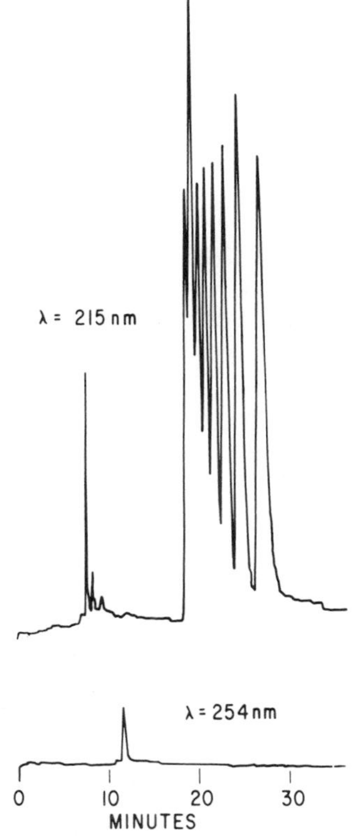

Fig. 5.8 Comparison of 215-nm and 254-nm detection of some fatty acid methyl esters separated by LC.

simpler optics and cell assemblies. However, if this type of detector does not satisfy the requirements, the variable wavelength detectors should be considered.

4. Recording Spectrophotometric Detectors

These LC detectors are modifications to existing recording spectrophotometers. Usually the conversion is made by replacing the cell holder assembly with a microflow cell. Many companies are now offering this capability as an accessory to their classic equipment. This can be a definite advantage to those who already have a spectrophotometer, since the only cost for application to LC is for the appropriate accessory. However, the cost of a complete spectrophotometer with microflow cell attachment for use only as an LC detector would unlikely be justified since the sensitivity and noise of such detectors for LC are at best comparable to the variable wavelength detectors mentioned in the previous section. The Perkin-Elmer LC 55 detector for LC and the Varian Variscan detector are examples of classic spectophotometers equipped with a microflow cell and marketed as LC detectors.

B. Fluorometric Detectors

1. General

The fluorometric detector is one of the most selective detectors available for LC. This is because of the nature of the fluorescence process and is obtained through structural restrictions of chemical species; that is, only certain types of molecules have the ability to fluoresce and even then only under specified conditions. The selectivity is also the main reason why this detector has only limited use at present. However, the formation of fluorescent derivatives of nonfluorescent compounds enables fluorescence to be useful in most areas of concern to the analytical chemist. Methods in fluorometric derivatization have been described in the literature (*3*) and are discussed in Chapter 7.

The fundamental difference between fluorometric and UV-visible absorption detection is that in fluorescence the wavelength of emitted light that reaches the photodetector is different from the incident light being directed onto the sample stream. Only the fluorescence emission from the sample reaches the detector, since the incident light is blocked by an appropriate filter. Thus, errors associated with transmission or reflection of the incident radiation are greatly reduced. The sensitivity of photometric detectors is determined by the background noise level, be it from the electronics of the measuring instrument itself or from fluctuations in the sample flow due to cell geometry. Also, UV absorption measurements are indirect; that is, detector output is determined by measuring the decrease in intensity of the incident beam as it is transmitted through the sample cell.

Fluorescence techniques are direct. The measured light values are positive and increase with quantity of emitting species present. If no fluorescent components are in the sample stream, then no emission signal can be obtained. However, in practice the mobile phase often contains traces of fluorescent impurities and thus a low background signal is obtained. The detector noise of most fluorometers results from the signal amplification circuitry.

2. Commercial Detectors

The Aminco Fluoro-Microphotometer is a filter photometer that can be equipped with a microflow cell for use with LC. The detector is designed so that emitted light is monitored at 90° to the incident light. The light first passes through an aperture plate (which controls the quantity of light incident on the sample stream), and then through a primary (excitation) filter to illuminate the sample. The fluorescence emission passes out of the cell and through the secondary (emission) filter to the photomultiplier tube, where the electrical signal is amplified for output. The excitation light originates from a 4-W cold source which emits a phosphor band spectrum at 360–370 nm and mercury lines at 365, 404, and 436 and 546 nm. The detector has a seven-step attenuation range from 0.1 to 100 × whereas the smallest flow cell available is 10 μl. The sensitivity of this detector is comparable to that described below. A wide range of primary and secondary filters is available. The best combination of filters for maximum sensitivity must be chosen empirically.

The LDC FluoroMonitor is a fixed-wavelength fluorescence detector. The excitation is at 360 nm with emission being monitored between 400 and 700 nm. Figure 5.9 shows a diagram of the optics and cell assembly. The detector can be operated in a direct mode, which simply monitors the fluorescence in one of the cells only, or it can function in a differential mode by measuring the difference in fluorescence of two streams. The 360-nm band is isolated by the excitation filter before it passes into the cone condenser (with a highly reflecting internal surface) and is directed onto the two cells. The emitted light then travels through the emission filter which only passes light >400 nm. This light is then detected by the photocell. The light source, cells, and detector are all on the same line. The cells have a 10-μl capacity. The minimum concentration detectability is 1 ppb (part per billion) of quinine sulfate (equivalent to noise). Drift is typically 10%/hr at maximum sensitivity. The main disadvantage of this detector is its inflexibility. The excitation and emission wavelengths

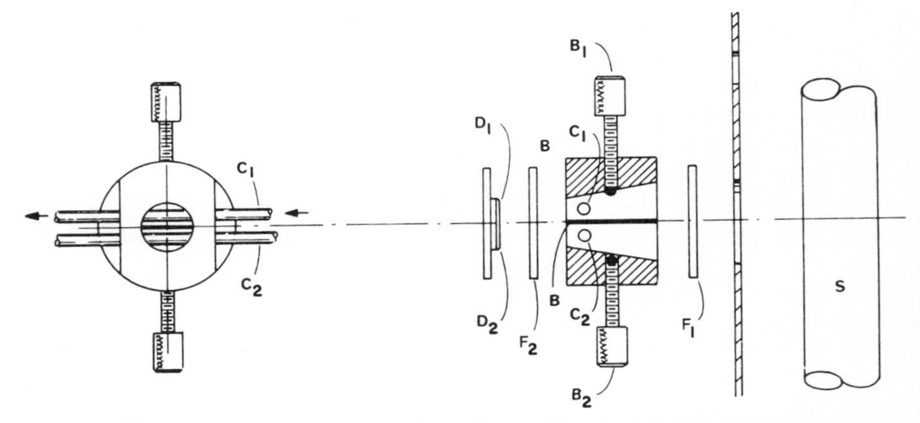

Fig. 5.9 Diagram of the FluoroMonitor optics. S is a low-pressure hot cathode mercury lamp with phosphor coating. F_1 and F_2 are the primary and secondary filters, respectively. B is the cone condenser. C_1 and C_2 are the cell chambers. B_1 and B_2 are background adjustment screws. D_1 and D_2 are photosensitive elements of the photocell.

cannot be easily changed as is the case with other available fluorescence detectors.

The DuPont Model 836 fluorescence detector is a dual UV absorption fluorescence monitor in one instrument. It is a modification of the Model 835 UV detector with the addition of a dual-purpose flow cell, additional photomultiplier, and linear amplifier. The wavelength for excitation is selected by one or two filters, one of which passes 250–390 nm light while the second transmits light in the range of 325–385 nm. The emission filters allow the passage of narrow bands which have peak transmission at 310, 357, 377, 408, 457, 502, and 566 nm. Unlike the LDC fluorescence detector, the emitted light is monitored perpendicular to the incident radiation. The Z-shaped cell (20-mm path length) has a 16-μl volume (larger than the LDC cell) and highly polished inner walls for maximum reflection of emitted light to the photomultiplier tube.

Steichen (*4*) evaluated this detector. A block diagram is depicted in Fig. 5.10, showing how simultaneous UV absorption and fluorescence measurements are made. The quinine sulfate detectability is near 0.3 ppb at a signal-to-noise ratio of 2:1. These data indicate that the DuPont detector appears significantly more sensitive than the LDC fluorescence monitor.

The Schoeffel FS 970 fluorescence flow monitor has a variable wavelength excitation capability. A continuous range of wavelengths is provided by a deuterium lamp (190–400 nm) and a tungsten–halogen lamp (350–700 nm) with wavelength selection by a digitally controlled monochromator accurate to ± 0.5 nm with a band-pass of 5 nm. The align-

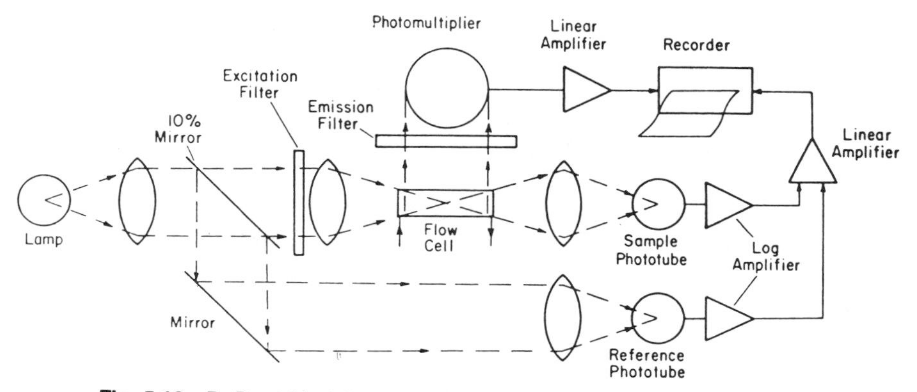

Fig. 5.10 DuPont Model 836 dual absorbance–fluorescence detector.

ment is the same as the LDC fluorescence detector (straight line). The cell volume is 10 μl. Subpicogram quantities of quinine sulfate can be detected. Because of the variable wavelength capability of this detector, it appears to be the most versatile of the detectors discussed.

The Varian Associates Fluorichrom detector is a compact filter photometer similar in design and operation to the Aminco detector. Picogram quantities of fluorescent solutes can be detected in some cases.

3. Modifications and New Designs

Manufacturers of filter fluorometers and spectrofluorometers, which are normally used for classical fluorescence measurements (in cuvettes), are designing microflow cells for adaptation to LC in the same manner as are the manufacturers of absorbance spectrophotometers. Perkin-Elmer, Turner, and Aminco are three such companies that market microflow accessories for their fluorometers. The spectrofluorometers have the added capability of stopped-flow scanning, which can be more accurate in peak identification because of the selectivity of fluorescence spectra (*5*).

Several research groups have worked on fluorescence detector design. Cassidy and Frei (*6*) designed a simple flow cell for the Turner Model III fluorometer. The design utilized the thin-layer scanning door and a Lucite light pipe. The detector was suitable for LC and could detect low-nanogram quantities of fluorescent compounds. Greenhalgh and Marshall (*7*) modified a similar Turner fluorometer without the thin-layer scanning door by building a microflow cell directly where the normal curvettes were placed. A Lucite light pipe was designed to direct the emitted light to the photocell. This detector was found to be slightly more sensitive than that described above (*6*).

Perchalski *et al.* (*8*) compared a 150-W Eimac lamp (a miniature high-pressure xenon arc lamp with collimating mirror) with a standard xenon lamp for use in an Aminco–Bowman spectrophotofluorometer modified with a 16-μl flowcell. They found that the high-pressure source provided two to three times the photon flux, which resulted in an almost tenfold increase in sensitivity for their particular applications. The major disadvantage was that light-sensitive solutes have a tendency to be degraded to a greater degree because of the high intensity source. However, the degree of decomposition depends upon the compound. Also, for LC analy-

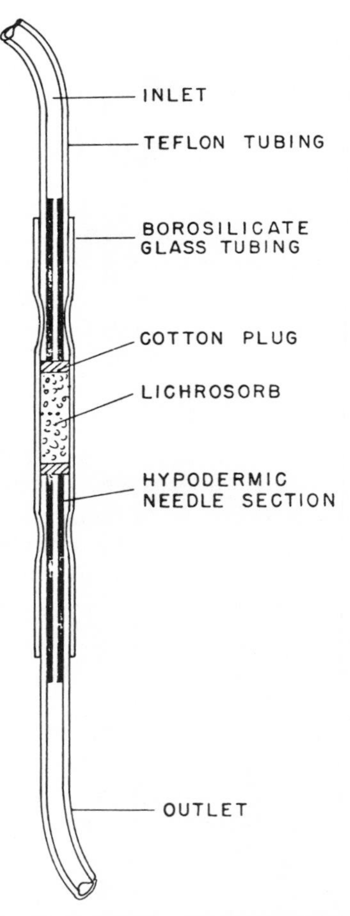

Fig. 5.11 Adsorbent-packed flow cell for the LDC FluoroMonitor, designed to fit directly in place of the existing cells. From Panalaks and Scott (9), with permission from the Association of Official Analytical Chemists.

sis this would be expected to be reduced because of the short residence time of the solute in the light path due to the flowing stream.

Panalaks and Scott (*9*) modified the LDC FluoroMonitor to improve the fluorometric analysis of aflatoxins B_1 and B_2. They found that when the sample cell is packed with small-particle silica gel such as LiChrosorb Si 60 (5 μm), (Fig. 5.11), the fluorescence intensity of the B aflatoxins is greatly enhanced. For the four aflatoxins they found relative increases in intensity for B_1, B_2, G_1, and G_2 to be 250-, 1900-, 5-, and 12-fold, respectively, compared to the results obtained with the unpacked cell. Figure 5.12 compares results obtained with this modified cell to the normal flow cell. Both sample and reference cells required packing to equalize the transmission of background fluorescence to the photodetectors. The great influence of adsorption on fluorescence yield has been noted earlier by Lloyd (*10*) who studied this same effect on polynuclear aromatic hydrocarbons.

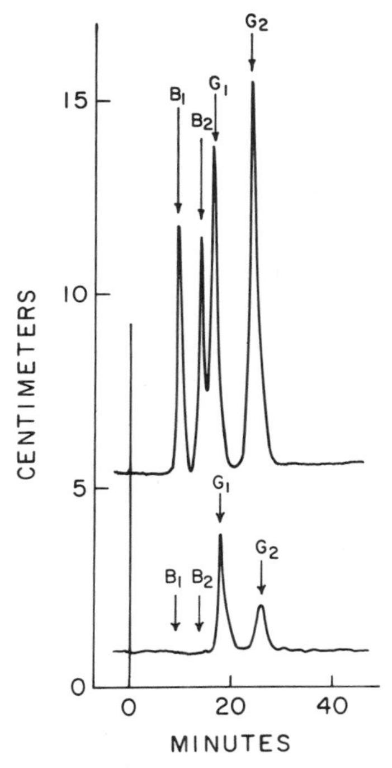

Fig. 5.12 Comparison of packed (top chromatogram) and unpacked flow cells for the fluorometric determination of aflatoxins B_1, B_2, G_1, and G_2; 20 ng each B_1 and G_1, 10 ng each B_2 and G_2 for the packed cell, and 240 ng each B_1 and G_1, 120 ng each B_2 and G_2 for the unpacked cell. Attenuation 4× for each. Courtesy T. Panalks, unpublished results, Health Protection Branch, Ottawa, Canada, 1977.

Thacker (*11*) described a miniature flow fluorometer for LC that was machined from a block of aluminum. It contained a low-pressure mercury lamp along with excitation and emission filters as well as a quartz flow cell and photocell. A photoconductor was installed to compensate for fluctuations in lamp intensity. The operational characteristics appear to be similar to commercially available flow fluorometers for LC.

A free-falling drop detector has been described in the literature (*12*). The eluant from an LC column enters a spherical chamber with a highly reflecting surface, through a 22-gauge Teflon needle. The solvent forms a drop at the end of the needle which falls through the excitation light path. The emitted light is then detected by a photodetector after passing through an emission filter. Thus the detector signal is recorded in pulses that are related to the size and frequency of drops. This detector has not been extensively evaluated or compared to flow-cell type detectors. However, it appears to offer few advantages over commercially available flow-cell fluorometers.

Asmus *et al*. (*13*) have developed a general detection technique for liquid chromatography based on fluorescence enhancement. The effluent from a chromatography column is mixed with a solution of a fluorescent dye. The fluorescence of the dye is enhanced in the presence of polar solutes. However, problems of short-term noise due to noninstantaneous mixing and dilution of the eluant with the dye solution (10-fold in this work) make it questionable for routine operation. Sensitivity appears to be better than the refractive index detector. Because sensitivity is based on polarity, care must be taken in solvent selection. Also, in adsorption chromatography the sensitivity would probably increase with the retention time of a solute (the more polar compounds would have a longer retention time), and vice versa with reversed-phase chromatography. With some improvements in design and more evaluation of a large number of dyes, this detector could be much preferred to the RI detector in terms of sensitivity for compounds such as polar aliphatic hydrocarbons, which cannot be detected by natural fluorescence or UV absorption.

C. Radiochemical Detectors

1. General

Radiochemical detection of LC effluents offers much promise for the analysis of radioactive species, either naturally occurring or as the result of radiolabeling of compounds used in metabolism or distribution studies in animals or the environment. Two principles can be used for radiochemical detection; scintillation counting and direct monitoring of radiation with a counter of the Geiger-Müller type.

The Geiger-Müller counter is one of the oldest and most often used instruments for the detection of radioactivity. Its operation is based on the production of ions in a chamber when an ionizing species passes between two electrodes which have a high potential across them. The electrons produced by the interaction of the nuclear particle with the chamber gas (usually argon, helium, or some polyatomic gas such as methane) are accelerated to the appropriate electrode, thus causing an electric current to flow, which is converted into a recorder response.

Scintillation counters operate on the principle that light emission occurs in certain compounds when they are exposed to radiation (usually β-particles). When a particle strikes a scintillator molecule, a photon of light is released. This light is then detected by a photomultiplier tube, where it is converted into an electrical pulse and is amplified. The detector response is normally proportional to the quantity of radiation coming in contact with the scintillator. Because β-particles vary greatly in their energies, the radioactive substance should be placed as closely as possible to the scintillator substance so that even the lowest-energy β-particles can be detected. For most intimate contact, liquid scintillators that dissolve the radioactive compound are frequently used. However, for flow detectors, this requires a secondary pumping system and a mixing "T" to add the scintillator to the column effluent before it enters the detector. In many instances solid scintillators can be used, although these are much less efficient than their liquid counterparts. In the latter case, the detector flow cell is packed with the solid scintillator (compounds such as anthracene, stilbene, terphenyl, sodium iodide, and calcium fluoride), and responses are generated when radioactive substances pass over these crystals. The advantage of these is that the radioactive species can be easily recovered since they are not dissolved in the scintillator.

Sieswerda *et al.* (*14*) mathematically treated flow versus batch counting of radioactive effluents in column liquid chromatography. They found that although flow counting can be successfully applied to LC, the sensitivity and accuracy of batch counting are superior because of the greater counting time involved.

Radiochemical detection is one of the most selective monitoring approaches available for liquid chromatography. It is probably the system least affected by sample type of any of the detectors available. However, it can monitor only for radioactive species; for application to general analytical techniques, this requires the formation of labeled derivatives in the same manner, for example, that fluorescent derivatives must be made for nonfluorescent compounds for detection by fluorometry. Many radiolabeled reagents such as acid chlorides or anhydrides are commercially available at reasonable prices for derivatization.

2. Detector Design

A radiochemical detection system has been designed by Berthold (Karlsruhe, West Germany) for monitoring chromatographic effluents. The system is based on scintillation counting for which two cells are available (Fig. 5.13). The homogeneous scintillation flow cell is used for liquid scintillators, which are premixed with the column effluent; it has a volume of 0.16 ml. This large cell volume makes it impractical for high-efficiency LC, although selectivity is such that in many instances sharp peaks are not required. The heterogeneous cell is packed with a solid scintillator such as cerium-activated lithium glass, and has a void volume of 0.7 ml, which makes it of little analytical use in LC but useful for monitoring columns for fraction collection for preparative purposes. The efficiency of the homogeneous cell is about 95%, whereas only 12% is attained by the heterogeneous cell. Background noise for the two are 10 counts per minute (cpm) for the former and 30 cpm for the latter. A 4000-fold linear range is provided by each cell.

Most other flow detectors for scintillation counting of chromatographic effluents are available only as accessories to classical scintillation counters. These cells normally have volumes in the range of 0.5–4.0 ml and are thus unsuitable for analytical but satisfactory for preparative work. Several companies such as Nuclear-Chicago (Des Moines, Iowa), Picker Nuclear (White Plains, New York), Packard (Downers Grove, Illinois), Beckman (Fullerton, California) and Intertechnique Instruments (Dover, New Jersey) offer flow-cell accessories to their scintillation equipment.

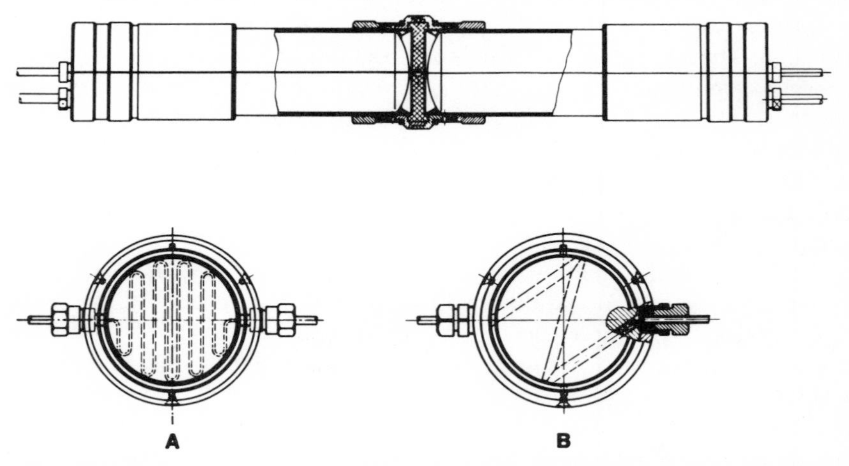

Fig. 5.13 Flow cells for the Berthold scintillation detector. A, design for homogenous scintillation; B, Z-shaped heterogeneous scintillation cell. Top figure shows arrangements of photodetectors.

McGuinness and Cullen (*15*) have reviewed the uses of these continuous-flow scintillation detectors.

Schutte (*16*) compared homogeneous and heterogeneous counting for the detection of liquid chromatographic effluents. Homogeneous scintillation was found to be more sensitive, especially for weak β-emitters such as tritium-labeled compounds. The cell volume ranged from 0.16 to 1.4 ml; detection limits of about 2 nanocuries (nCi) for ^{14}C and 5 nCi for ^{3}H were obtained with the homogeneous system, and 10 nCi for ^{14}C and 1 microcurie (μCi) for ^{3}H were obtained with the heterogeneous cell. Sieswerda and Polak (*17*) described a combined scintillation flow-cell fraction collector that automatically recorded peaks of radioactive solutes, as well as collected the peaks for further counting over longer periods of time, giving more accurate results.

Direct radiometric counting of effluents is in principle preferred to scintillation counting since there is no need for solid or liquid scintillators. Bangan (*18*) described a flow cell designed to direct the column effluent over the mica window of a thin-window-tube counter. The detector was sensitive enough to detect low-energy β-emitting particles at levels of 2.5 nCi/ml. Van Urk-Schoen and Huber (*19*) designed and evaluated a microradiometric cell with a volume in the range of 4–30 μl. The design was essentially the same as that described above by Bangan (*18*) with the exception that their cell was very small, which enabled it to be used directly with LC. This detector could detect low-nanocurie levels of ^{86}Rb or ^{137}Cs.

There has been little reported use of radiochemical detectors in LC, mainly because of the lack of technology in this area, combined with the very specialized nature of this type of detector. However, radiochemical detectors in association with radiochemical derivatization techniques have much potential for general application to LC.

D. Atomic Absorption Detector

Although no atomic absorption (AA) detectors have yet been designed especially for LC, a number of investigators have evaluated this approach for a variety of problems. The major advantage AA has as an LC detector is its selectivity. It is an element-selective instrument that makes it especially well suited for application to trace metallic components in complex matrices. The AA unit is also easily connected to the LC system since the column exit can be directly linked to the AA nebulization chamber. The choice of solvents for the mobile phase, however, does have an effect on the flame characteristics and can affect detector operation.

Jones and Manahan (*20–23*) carried out detailed studies on the application of AA to the detection of organometallic compounds separated by LC. They evaluated a Perkin-Elmer Model 403 AA spectrophotometer as the detector. When using water as the mobile phase they found the detector response to be influenced by several variables, including eluant flow rate into the AA burner. Maximum sensitivity was found at a flow rate of 2–3 ml/min, which was just equivalent to the drawing rate of the aspirator. However, the nebulization efficiency (quantity of water drawn in to the nebulization chamber that actually reaches the flame) is only about 25% at these flow rates. Organic solvents are much more efficient, approaching 100% in many instances. Table 5.3 lists the nebulization efficiencies of several common organic solvents at different flow rates.

For the organic solvents, the increase in nebulization rate results in a corresponding increase in sensitivity. However, when metal chelates were analyzed in organic solvents, the sensitivity was much greater than that expected solely on the basis of nebulization efficiency. Other contributing factors included the quenching or enriching effect of the organic solvent on the flame and the flow rate into the flame. Table 5.4 lists the detection limits of several metals in both aqueous and organic mobile phases.

Cassidy *et al.* (*24*) used AA detection for the analysis of organosilicon compounds in industrial process waters. A Varian Techtron AA-5 spectrophotometer with a variable-flow nebulizer was used as the detector, and chromatographic separation was achieved by either reversed-phase or gel permeation (molecular sieve) LC. The detection limits for organo-

TABLE 5.3

Nebulization Efficiencies for Some Common Solvents[a]

Solvent	Flow rate (ml/min)	Percent nebulized
Methanol	2.0	86
	1.0	100
Ethanol	2.0	83
	1.0	100
Chloroform	2.0	100
	1.0	100
Benzene	2.0	100
	1.0	100
Water	2.0	26
	1.0	32
	0.5	38
	0.3	74

[a] From Jones *et al.* (*23*), with permission from the American Chemical Society.

TABLE 5.4

Detection Limits of Some Metals[a]

Metal	Minimum concentration (μg/ml)	Detection limit (ng)
	Aqueous mobile phase	
Fe	0.7	7.2
Cr	2.14	21.4
Zn	0.5	5.0
Cu	0.13	1.3
	Organic mobile phase	
Fe	0.79	7.9
Cr	0.67	6.7
Co	2.60	26.0

[a] Based on a signal twice the standard deviation of the background noise. From Jones *et al.* (*23*), with permission from the American Chemical Society.

silicons as chromatographically separated peaks were in the range of 0.5–5 μg. Funasaka *et al.* (*25*) used an Hitachi 303 AA for the detection of some organomercury compounds, using *n*-hexane as the mobile phase on a Corasil I adsorption column. Detection limits were in the range of 10 ng for some organomercurials.

E. Flame Photometric Detector

Freed (*26*) reported on the combination of LC and flame photometric detection of a number of metal ions. Separation was carried out by ion-exchange chromatography. The flame photometric detector was constructed from a Beckman total-consumption burner with a hydrogen–oxygen flame. A Schoeffel quartz prism monochromator in conjunction with an Aminco blank-subtract photometer was used for detection of emission lines. The chromatographic effluent directly entered the aspirator and then into the flame. Flow rate had a significant effect on the detector response. Sensitivity diminished substantially at faster flow rates. Normally, less than 1.0 ml/min was found most useful. The linear range for the elements sodium and calcium was 2–10 ppm, while the precision was found to be 1.2 and 1.4%, respectively. The author found LC–flame photometry very useful in the separation and quantitation of rare earth elements in the low ppm range. The sensitivity of this detector does not appear to be as good as atomic adsorption detection, although selectivities are similar.

F. Laser Detector

Photometric detectors that use lasers as the light source offer much promise as LC detectors because of their great photon flux. This is especially applicable to fluorescence where background signal (background emission) is low and response is linear to concentration over a wide range. The application of tunable lasers in analytical chemistry has been reviewed (*27*). However, application to LC has been limited. Freeman *et al.* (*28*) used an infrared laser as an LC detection system and adjusted the wavelength to monitor carbon–hydrogen stretching vibrations at 2950 cm^{-1}. This precluded the use of many hydrocarbon solvents, although mobile phases of carbon tetrachloride or chloroform could be used. Minimum detectability was usually in the low microgram range.

IV. ELECTROCHEMICAL DETECTORS

A. General

Electrochemical detectors show much promise for use in LC, although at present few types are available commercially. All are dependent upon some electroactivity of the solute in the mobile phase. This is normally evident through oxidation or reduction, or the ability to act as a conductor of an electrical charge. For this, the solvents normally required are aqueous-based systems that might extend from aqueous alcohol mixtures to strong buffer solutions. The application of electrochemical detectors to organic solutes in organic solvents has not yet been extensively investigated, although these detectors have much promise in this area.

The electrochemical detectors are generally simpler in design and operational characteristics than the optical detectors since light sources, lenses, monochromators, filters, or photodetectors are not required.

The sensitivity, in general, is extremely good by present standards, and is capable of detecting picograms of many electroactive species. Noise levels arise from several sources, including electronics, fluctuation in electrical standing current, flow irregularities due to cell geometry, temperature fluctuations, and anomalous reactions at the electrode surfaces. Changes in flow rate can also affect electrochemical response. The cleanliness of the electrodes in fixed-electrode systems as opposed to the dropping mercury electrode (DME) will also affect operation and thus the utility of the detection system.

The detectors are considered to be selective since only electroactive species will provide a response. The selectivity can often be varied by altering voltages (for oxidation–reduction reactions). For example, easily oxidizable solutes could be selectively detected in the presence of solutes that are more difficult to oxidize, by maintaining the potential below that necessary for oxidation of the latter but high enough to cause a reaction with the former. Much work is now being carried out in this area by independent researchers, and new sensitive electrochemical detectors should become available commercially.

B. Conductivity Detectors

Conductivity (or conductometric) detectors have not enjoyed wide use in LC up to the present because of their selective nature. Thus only a few are commercially available for LC, of which the Laboratory Data Control (LDC) ConductoMonitor Model 701 appears to be the most popular. Others are produced by Chromalytics, LKB, and Varian. All of these are similar in principle of operation to that described below for the LDC detector.

The ConductoMonitor continuously and quantitatively measures the electrical conductivity of solutions passing through the flow cell. The readings are in specific conductivity units (K_{sp}) which may result from detector operation in either an absolute or a differential mode. Conductivity differences as small as 0.1% K_{sp} can be detected in the differential mode. The cell volume is only 2.5 μl, but the inlet connective tubing is 25 μl. However, if the inlet tubing volume could be reduced significantly, detector efficiency would be much improved. The full scale range is from 0.01 to 10^5 μmho/cm, which makes the detector suitable for use with mobile phases varying from distilled water to concentrated salt solutions. Polarization effects are prevented by the high-frequency electronic design.

The cell assembly (Fig. 5.14) consists of two separate cells sharing a common electrode. Each cell has its own inlet and outlet tubing, which are internally grounded to the cell housing. These cells are electrically connected as part of the ratio arms of a bridge amplifier. The voltage across the cell electrodes is less than 1.0 V at a frequency of 2000 Hz. The output signal is proportional to the change in cell resistance. The high-frequency noise is less than 0.2% full scale. Since the detector monitors conducting species in solution, most applications of this detector would be for ion-exchange chromatography or for gel permeation chromatography with elution by aqueous solvents.

Tesarik and Kalab (*29*) reported the design of a conductometric detector for use with liquid chromatographic systems based on an aqueous mobile phase. Their detector was designed in a block of organic glass and was used primarily with glass chromatography columns which fitted

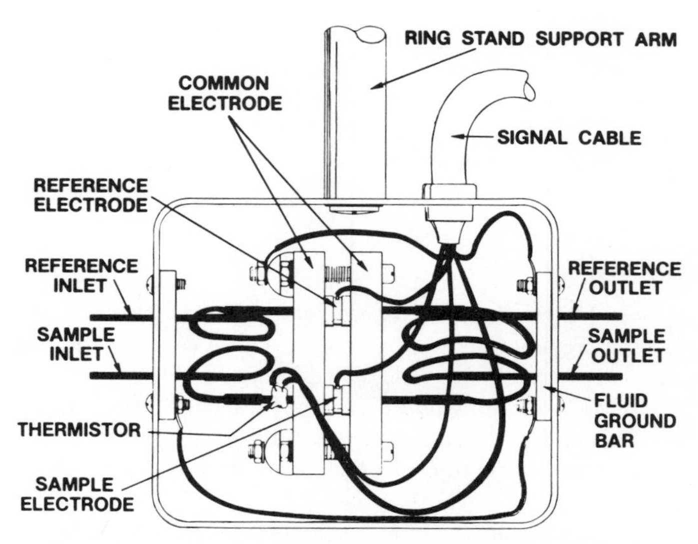

Fig. 5.14 Diagram of the LDC ConductoMonitor Model 701.

directly into the detector block. The distance between the electrodes was 0.2 mm, which could be varied by means of adjusting screws. The volume between the electrodes was about 0.18 μl and the total volume was about 0.5 μl, making it suitable for application to high-efficiency LC. Minimum detectable quantities of potassium chloride, eluted as a peak from a column of glass beads, were 5 μl of a 1×10^{-5} *M* KCl solution or about 0.4 ng KCl per injection at a noise level of less than one tenth of that reported by others (*30*). These results were under optimized conditions for pure standards of KCl, which were essentially unretained by the glass beads. However, this does give some indication of the very good sensitivity that can be obtained with this system. Conductometric detection has much promise for the analysis of water-soluble components such as amino acids and carbohydrates in biological fluids.

C. Coulometric Detectors

Several electrochemical detectors based on current measurements at fixed electrodes have been designed (*31–33*), and some of these have recently become commercially available. They are suitable for solutes that exhibit electroactivity (oxidation or reduction) at predetermined potentials. A model designed by Blank (*31*) appears to be extremely sensitive to easily oxidizable compounds such as norepinephrine or dopamine.

Figure 5.15 shows a diagram of the detector, designed for dual simulta-

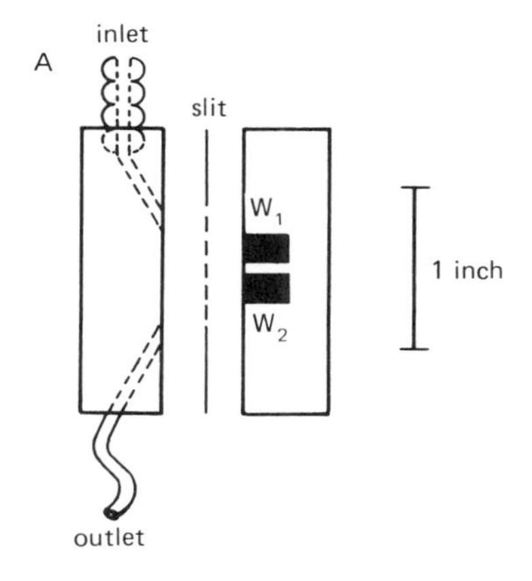

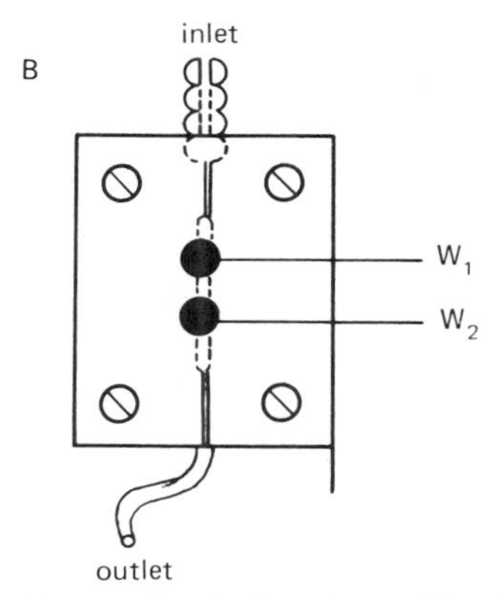

Fig. 5.15 Coulometric detector for dual analyses. The slit is a spacer made from a 0.005-inch PTFE sheet with a rectangular hole cut into it, which acts as the cell compartment. W_1 and W_2 are two carbon-paste electrodes. A, side view disassembled; B, front view assembled. From Blank (*31*), with permission from Elsevier Scientific Publishing Co., Amsterdam.

neous analyses at two different electrode potentials. The electrical connections to the carbon-paste electrodes are provided by platinum wires. The effluent from the detector flows into a small container which contains the reference (calomel) electrode and the auxiliary platinum wire electrodes. An example of the selective nature of this detector is shown in Fig. 5.16 for the selective detection of 6-hydroxydopamine (6-OHDA) in a mixture of norepinephrine (NE) and dopamine (DA). 3,4-Dihydroxybenzylamine (DHBA) was included as an internal standard. Since 6-OHDA is exclusively oxidized at +0.45 V whereas the other compounds are not, only a response for this compound is observed at that

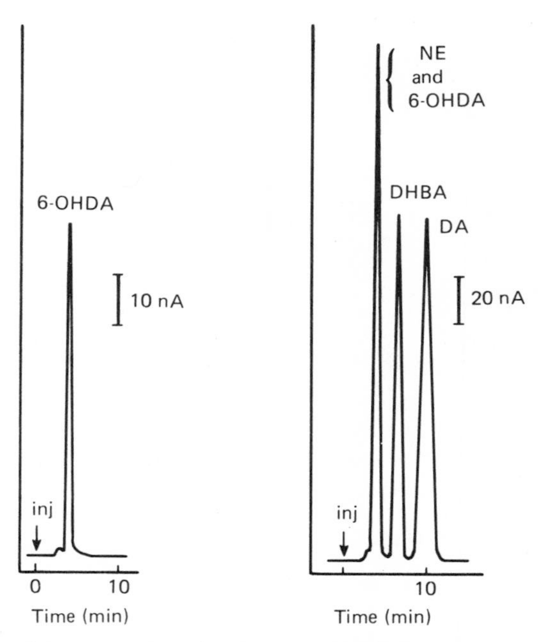

Fig. 5.16 Simultaneous detection of components at +0.45 V, first electrode (A) and at +0.8 V, second electrode (B); 20 μl of a solution containing 7.2×10^{-5} *M* NE, 1.03×10^{-4} *M* DHBA, 2.22×10^{-4} *M* DA, and 3.7×10^{-5} *M* 6-OHOA. For details of chromatography, see Blank (*31*). From Blank (*31*), with permission from Elsevier Scientific Publishing Co., Amsterdam.

voltage. However, at +0.8 V electrode potential, all four compounds are reduced and thus produce a response. The advantage of selective detection is shown here where NE elutes at the same retention time as 6-OHDA. This system shows excellent sensitivity for the above and related compounds. Between 0.03 and 0.08 pmoles of the four compounds can be detected at a 2:1 signal-to-noise ratio. It is also interesting to note that only 1% of the solute produces the detector response because of the electrode efficiency (Coulometric yield). Further evaluation of this type of detector for a variety of chemicals in a number of solvent systems would be useful in determining its promise for analytical applications.

A wall-jet electrode detector has been designed and evaluated for application to LC (*34,35*). The electrochemical principle is the same as that described above, although cell geometry and electrode materials were different and the detector was not of a dual nature as was the former. However, it was designed so that the working cell volume could be adjusted by varying the distance between the electrode and the inlet nozzel tip. Minimum detectable quantities (equivalent to high-frequency noise) for chromatographic peaks were about 10 pg for ferrocyanide or 100 pg for keto or hydroxy steroids.

MacDonald and Duke (*36*) constructed and evaluated a small-volume solid electrode flow-through cell for use with pulse polarographic techniques. They found that the pulsed method reduced noise and increased detector response for an overall increase in sensitivity of 5- to 20-fold. Also, the pulsed mode was much less affected by variation in effluent flow rate. The routine use of the pulsed mode extended operation time between electrode cleanings.

The use of glassy carbon electrodes in the design of a coulometric detector for LC has been studied by Lankelma and Poppe (*37*). Glassy carbon electrodes show much promise for detecting electrochemically oxidizable compounds with high current yield. The design of their system is shown in Fig. 5.17. The construction was basically similar to that

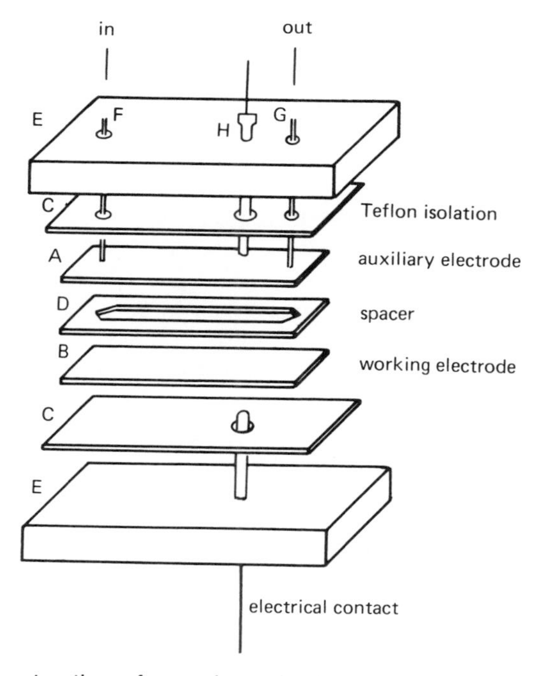

Fig. 5.17 Construction of a coulometric detector cell with glassy carbon electrodes. A and B are glassy carbon plates that act as auxiliary and working electrodes respectively; C, PTFE isolation sheets; D, PTFE isolation (spacer) sheet; E, steel plates; F and G, inlet and outlet tubes (outlet tube being the reference electrode). From Lankelma and Poppe (*37*), with permission from Elsevier Scientific Publishing Co., Amsterdam.

described by Blank (*31*) in that the cell was formed using PTFE as a spacer sandwiched between two opposing walls. However, rather than having carbon-paste electrodes implanted in the cell walls, the walls themselves were constructed of glassy carbon which acted as the working and auxiliary electrode surfaces. This provided a much larger surface area for electrochemical reactions to occur. The very thin nature of the cell (50–100 μm) coupled with the large electrode surface greatly enhanced the Coulometric yield (percent of solute reaching the electrode surface) to nearly 100%. This resulted in a significant improvement of sensitivity which enabled the detection of picogram quantities of solutes such as fluphenazine (a pharmaceutical compound) eluted from chromatographic columns. The linear range was calculated to be $> 10^6$. The application of this detection system to organic or organic–aqueous solvent systems has the same excellent possibilities for routine applications as that described for the cell based on carbon-paste electrodes (*31*).

D. Polarographic Detectors

Micropolarographic detectors have been designed for LC by several groups. One of the first uses was reported in 1952 by Kemula (*38*), who used polarography as a detector for classic column chromatography. Koen *et al.* (*39*) reported on the design of a micropolarographic detector for use with LC based on the dropping mercury electrode (DME). It was constructed of a solid body with a small cylindrical space (1 × 1 mm diameter) which acted as the detection cell. In order to keep dead volume to a minimum, the cell was designed so that the column effluent flowed in a thin film around the mercury drop. The drop time was 1 sec as opposed to 4 sec in classical polarography. This fast time was essential for accurate measurement of very narrow chromatographic peaks. The drops fell into a pool of mercury, which acted as the unpolarized electrode.

This detector was applied to the analysis of the organophosphate insecticides parathion and methylparathion in crops (*40*). Minimum detectable levels were about 0.03 ppm. A similar DME polarographic detector was designed and evaluated by Fleet and Little (*34*). Their design had a cell volume of 2–5 μl and used a silver–silver chloride element as the reference electrode. The drop rate was very rapid ($\ll 1$ sec) in the pulsed polarographic measuring mode. The minimum detectable concentration in the cell was found to be 10^{-7} *M* *p*-nitrophenol.

Stillman and Ma (*41*) designed a micropolarographic detector based on the design of Blaedel and Strohl (*42*). Figure 5.18 shows a diagram of the

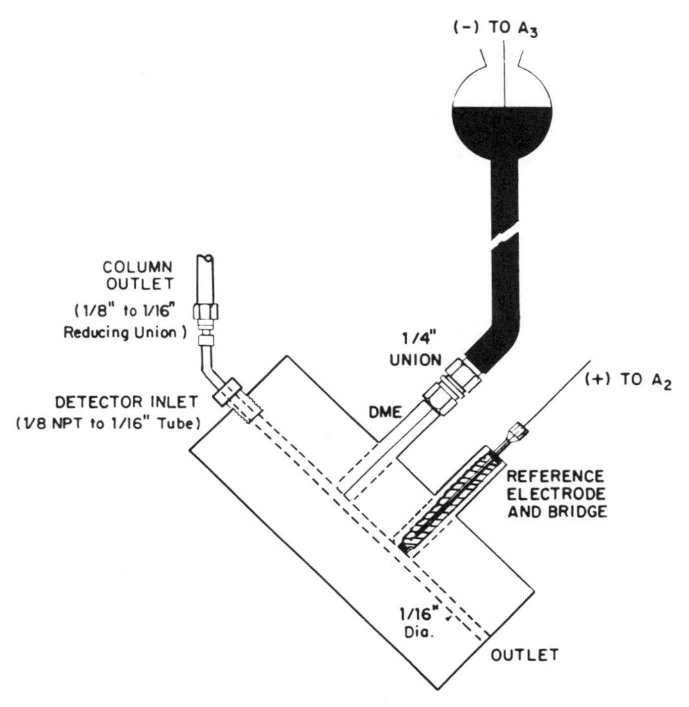

Fig. 5.18 Polarographic cell for LC. From Stillman and Ma (*41*), with permission from Springer-Verlag, Berlin and New York.

apparatus. The detector body was machined from a block of PTFE. The DME was a glass capillary tube 3 in. long with a 0.5-sec drop rate capable of accurate measurement of peaks with a bandwidth of 10 sec or less. The minimum detectable quantities of parathion were about 10 ng (lower end of linear range) with a linear response (peak area) over a 10^3-fold concentration range. Reproducibility of replicate injections of 10 μg of parathion was $\pm 1.8\%$ (coefficient of variation).

Joynes and Maggs (*43*) constructed a polarographic detector for LC based on a commercially available carbon-impregnated silicone rubber membrane as the electrode. They compared this with a micro-DME for monitoring LC eluants and found that the membrane electrode had several advantages, including (1) low standing current and noise; (2) was less prone to oxide films; (3) was 100 times less affected by oxygen; and (4) was simpler in operation. Detection limits were in the range of 10^{-9} *M* of some organo–nitro compounds.

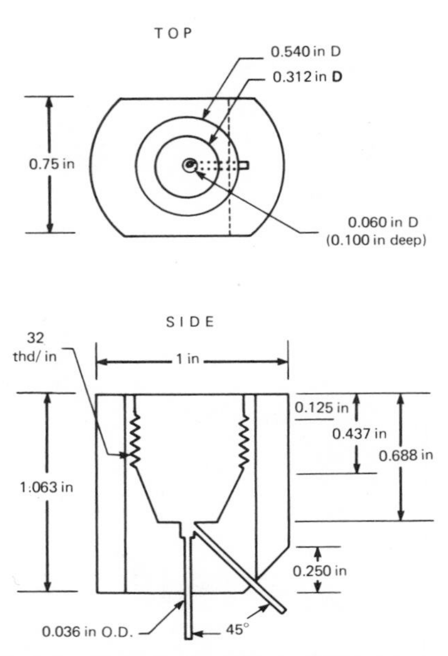

Fig. 5.19 Flow-through cap for the liquid membrane electrode. From Shultz and Mathis (*44*), with permission from the American Chemical Society.

E. Ion-Selective Electrodes

Shultz and Mathis (*44*) reported on the design of a microflow adapter for a liquid-membrane ion-specific electrode. Figure 5.19 shows a diagram of the detector. The cap was machined from a 1-inch diameter nylon cylinder that contained a small (5-μl volume) cavity for the detector cell. This was then screwed snugly to the electrode which was operated in the normal manner. The detector head and reference Ag–AgCl electrode were immersed in a small container of eluant solution and electrode response was measured with a pH meter and recorder. Low nanomole quantities of NO_2^- and NO_3^- could be detected with this system with the particular electrode used. Figure 5.20 shows the selective detection of nitrate, nitrite, and some phthalates after separation by LC. This detection system has some promise as a selective detector for LC since there are a variety of ion-selective electrodes available that could be used. Also, sensitivity might be improved by increasing the surface area of the electrode contact with the solvent stream by using a wide flat cell design rather than the small cavity shown in Fig. 5.19. For fast-eluting peaks, the mass transfer between electrode and eluant may be of some importance. Further work in this promising area should be carried out.

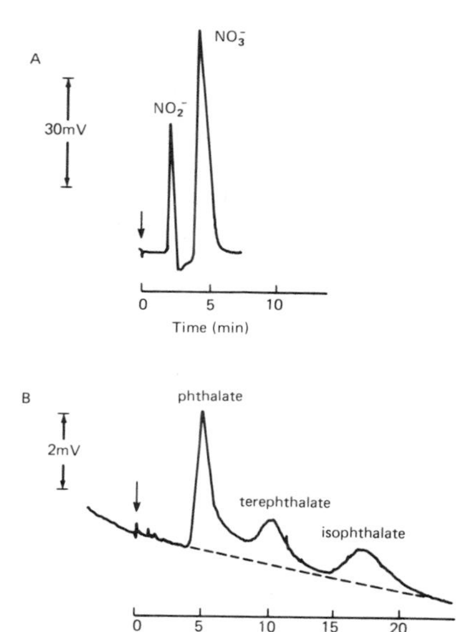

Fig. 5.20 Chromatogram (A) 1 μl of 0.1 *M* NO_3^- and 0.1 *M* NO_2^-; eluant: 0.01 *M* KH_2PO_4, 0.01 *M* Na_2SO_4, pH 7.0; flow rate 0.92 ml/min. (B) 1.0 μl of 0.04 *M* of each phthalate ion; other conditions as in (A). From Shultz and Mathis (*44*), with permission from the American Chemical Society.

V. SOLUTE TRANSPORT DETECTORS

A. General

Solute transport detectors represent an entirely independent group of liquid chromatographic detectors. These instruments, also referred to as phase-transformation detectors, function by first separating the solute from the mobile phase. This is invariably done by evaporating the solvent as the eluant is deposited on a moving wire or chain. The dry solute is then transported to a detector, usually based on flame techniques such as flame ionization (FID) or alkali flame ionization detection (AFID). The wire passes through the flame, where the solute is burned and detected.

The Pye Unicam LCM2 FID system adds an intermediate step of converting all carbon-containing compounds to CO_2 by first burning the solute in an excess of air. The CO_2 then passes through a molecular entrainer with the aid of a mixture of hydrogen and nitrogen, and into a pyrolysis furnace where the CO_2 is converted to methane in a reducing atmosphere with the aid of a nickel catalyst. The methane is then detected by the FID system. Apparently, the conversion of the solute to gaseous material for FID detection greatly increased the sensitivity of this system, compared to direct combustion of the solute on the wire. However, the minimum detectable quantities are at best in the low microgram range.

Other commercial transport detectors have been made available (e.g., by Chromalytics and Tracor), but to date the detector has experienced limited use. The major reasons for this lie in the design of the detector and in the inherent shortcomings of the technique. The modifications and new designs described below attempt to remedy the shortcomings of this type of detector. However, it must be kept in mind that none of these have been rigorously tested for application to routine LC analyses and thus there still remains the question of the ultimate suitability of this detection approach for LC.

B. Modifications and Improvements

Pretorius and van Rensburg (*45*) modified the Pye System 2 flame ionization detector (FID) by coating the solute transport wire with a porous layer (such as ceramic material). They studied several layers which permitted a larger quantity of eluant to be deposited, and thus increased the detectability of the solute. The layers were thermally stable (*ca* 750°C) and inert to both solvent and solute. For best results the inert material had to be uniformly coated on the wire and wetted by the solvent. In order to ensure uniform layer thickness, a simple apparatus was designed to coat the wires, the thickness of which was dependent upon the composition of the ceramic slurry. A layer consisting of kaolin increased detector sensitivity by about 300-fold compared to the uncoated wire when 190 μg of squalene was determined with hexane as the mobile phase. The coated wires were particularly suited to aqueous solvents, which do not evenly wet the uncoated stainless steel wire. However, some problems were encountered with flaking of the brittle layers when they were wrapped on the collector spool. This was not considered to be serious.

Stolyhwo *et al*. (*46*) used a transport system consisting of a twisted steel core with a fine steel wire wrapped around it. This provided a greater degree of entrapment of the liquid than a single-strand wire, which is normally used. The detector provided a detection limit of 100 ng of triolein eluted from a silicic acid column with chloroform.

Karmen and co-workers (*47,48*) developed a "buffer-storage" system to store the solute-coated transport wires after the eluant was evaporated. They found that whereas chromatography required 30 min per analysis, FID detection only required 3 min. Thus, one FID was capable of detecting effluents of up to 10 LC systems in an "off-line" mode. Sensitivities to carbon compounds were in the low microgram range with reproducibility of ± 10%, which is about twice the variation for the FID when operated with gas chromatography. Thus, the transport system (without storage) apparently contributed a significant error in this particular system because of changes in flame characteristics as the wire passed through the flame, as well as the poor reproducibility of applying the solute to the wire.

van Dijk (*49*) studied the method of application of the chromatographic effluent onto the conveyor wire and found that spraying the liquid increased detector sensitivity by a factor of 20–50. Also, because the wire was more evenly coated, irregular response was much reduced and the troublesome "spike effect" eliminated. Also, the spray process removed the influence of mobile-phase flow rate on the fraction of solute deposited on the wire. Figure 5.21 is an illustration of the nozzle. Compressed air at 0–3 atm was forced through the nozzle, atomizing the eluant, and

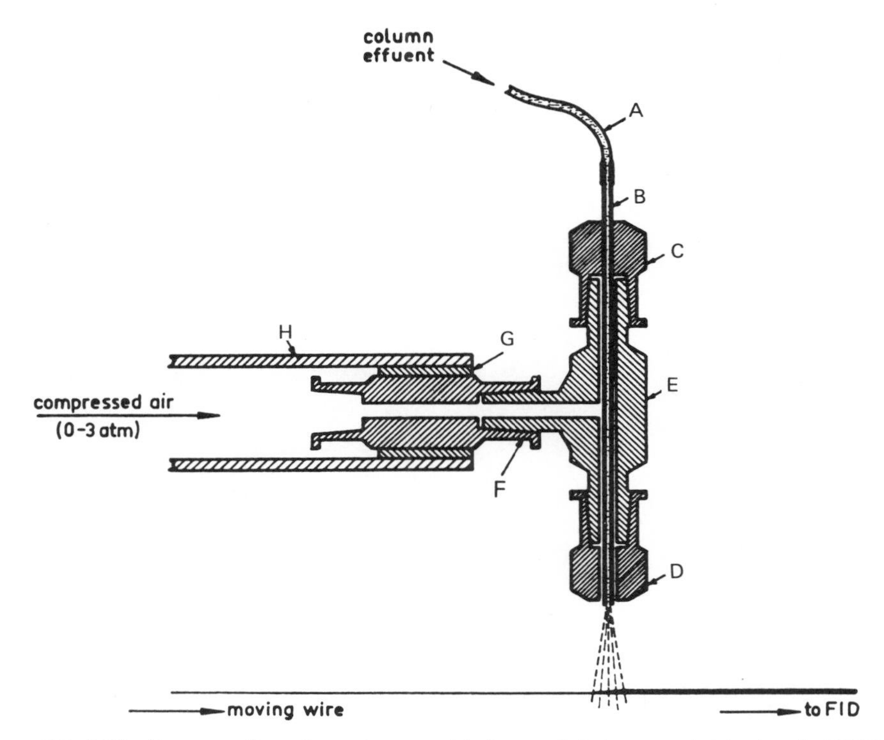

Fig. 5.21 Cross section of nozzle assembly for a solute transport detector. A, PTFE tubing; B, stainless steel needle tubing; C and D, Kel-F luer caps; E, Kel-F T-connector; F, Kel-F union; G, rubber tubing; H, 0.5-inch o.d. stainless steel tubing. From van Dijk (*49*), with permission from Preston Publications, Inc.

directing it toward the wire. The distance between nozzle tip and wire could be varied for optimum sensitivity.

Dubsky (*50*) studied the use of a horizontal disk for collection of the column effluent. The disk consisted of a rotating circular net which continuously received the solute and entered an FID after evaporation of the solvent. Detectability was similar to the above solute transport systems, being in the low-microgram range. Reproducibility was about ±4%. No significant spiking was observed in the chromatograms, and noise was much reduced by incorporation of a resistance–capacitance filter circuit into the detector outputs (*51*).

A vertical rotating disk constructed from a bar of ceramic alumina was successfully used by Szakasits and Robinson (*52*) in their design of an FID detector for LC (Fig. 5.22). They found improved baseline stability compared to the metal conveyors, and since the disk was selfsupporting there was no need for pulleys. For replicate analyses of *n*-eicosane the standard deviation was found to be 1.02%, which was much better than that reported for most other transport detectors. Sensitivity was reported to be of the order 3–4 ng/sec for many compounds, and linearity was excellent over 4 orders of magnitude. This transport system appears to be the simplest one yet designed and the one that has the most promise for practical use.

The use of an alkali flame ionization detector (AFID) in place of an FID increased the selectivity of the solute transport detector. Slais and Krejci (*53*) converted an FID system to an AFID for selective detection of ther-

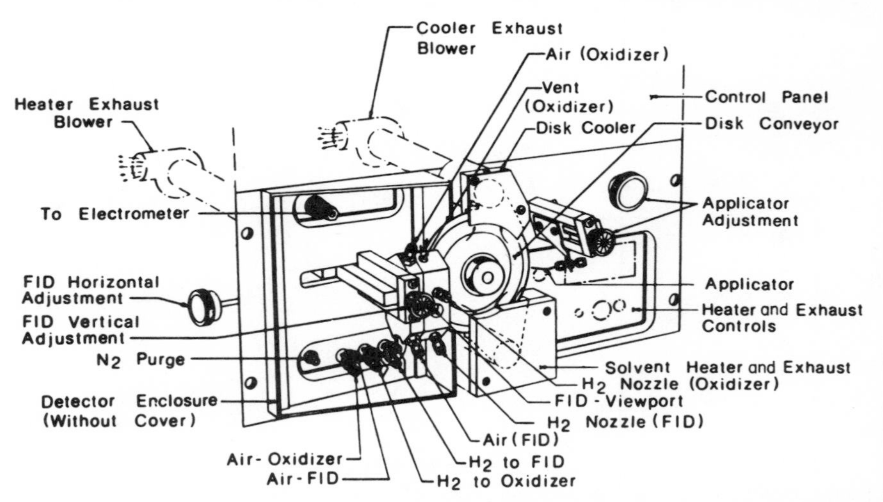

Fig. 5.22 Vertical rotating disk transport detector. From Szakasits and Robinson (52), with permission from the American Chemical Society.

mally labile halogen derivatives of some tetrahydrofurans. The transport system consisted of a stainless steel wire which was reeled off a spool in a manner similar to some of those described above. The major difference in the detector was the inclusion of a salt-tip consisting of sodium sulfate. The sensitivity of the detector to 2-(tert-butoxy)-3-chlorotetrahydrofuran was 300 ng/sec at a 2:1 signal-to-noise ratio, and a selectivity ratio of greater than 2000:1 for halogens versus hydrocarbons was obtained using squalene as the hydrocarbon.

Boshoff *et al.* (*54*) designed a thin-layer transport detector for LC. A quantity of the effluent from the column (depending upon a splitting ratio) is deposited on a moving chromatoplate and subsequently detected according to established TLC methods. In this particular study the chromatoplate was treated by a general fluorescent derivatization technique (*25*) and then recorded with a thin-layer chromatography (TLC) scanner. Standard 20 × 5 cm silica gel plates were used for deposition of the solute. Solvent was removed by heating the plate to 90°C, and was drawn off by vacuum. Figure 5.23 shows a diagram of the apparatus and Fig. 5.24 shows a comparison of the TLC detection system with normal UV detection for four steroids. Only three are detected by direct UV, whereas all four are detected by the TLC detection method. The advantage of this approach is that much selectivity can be gained through the use of different TLC sprays. It is possible for example to have the solutes deposited simultaneously on several different plates for analysis with a number of different sprays for a more positive identification of the peaks. However, the use of this system requires additional time for treatment of the plates, as well as equipment for scanning the spots. This might not be desirable for many applications.

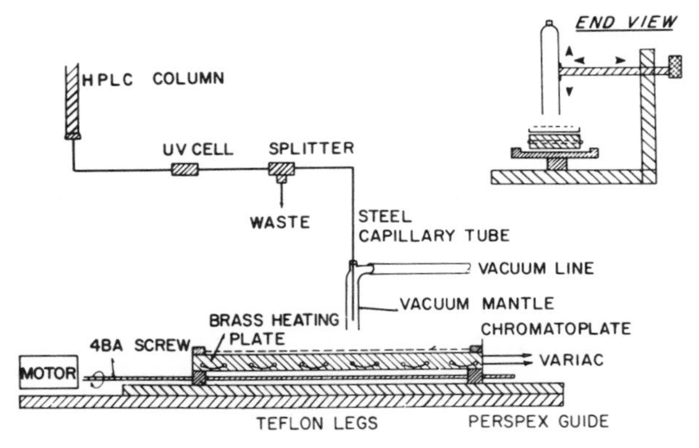

Fig. 5.23 Diagram of TLC detector apparatus. From Boshoff *et al.* (*54*), with permission from Elsevier Scientific Publishing Co., Amsterdam.

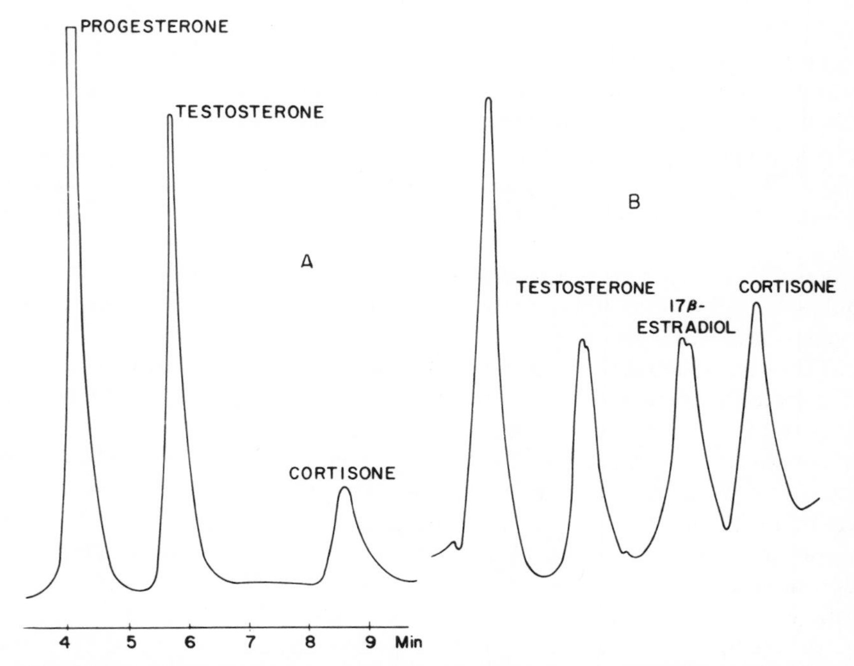

Fig. 5.24 Comparison of: (A) absorption and (B) TLC (fluorescence) detection for the steroids progesterone, testosterone, 17β-estradiol, and cortisone: 200 ng of each passed through the UV detector while 66 ng of each was deposited on the TLC plate because of the split ratio. From Boshoff *et al*. (*54*), with permission from Elsevier Scientific Publishing Co., Amsterdam.

In addition to the studies described above, other similar reports on the design and evaluation of transport detectors have appeared in the literature (*55–58*).

VI. REFRACTIVE INDEX DETECTORS

A. General

Refractometric detectors are based on the principle of interfacial refraction and reflection of light between two different media. The deflected or refracted light is measured continuously with a photocell which produces an output signal relative to the change in refractive index of the mobile phase.

Because change in refractive index caused by a solute is a bulk property not specific to that particular component, the refractive index (RI) detector is considered to be a general detector, the response of which is based essentially on the total mass of the solute in the sample stream, and not on a specific property of the solute. The detector can be used for a wide variety of solvent streams. The RI detector based on reflection rather than refraction of light has the advantage that dark solutions can be readily monitored, although dark solutions are not often encountered in LC since sample sizes are normally quite small and are well diluted by the time the components reach the detector.

Temperature stability is of major concern with the RI detector. Very small fluctuations can cause significant detector noise. Thus, most commercial detectors are especially thermostatted to minimize this problem. Also, all detectors now operate in a differential mode where the difference in refractive index of mobile phase and mobile phase-plus-solute is measured. This is accomplished by directing the mobile phase from the pump through the reference cell before entering the column. The mobile phase that leaves the column then passes through the sample cell. Since RI detectors are sensitive to flow changes, flow programming as well as gradient elution is almost impossible.

RI detectors are not particularly suited to trace analysis because of lack of both sensitivity and selectivity. The detector is also limited by the above-mentioned sensitivity to flow and solvent composition. The discussion included herein serves to inform the analyst of the principles and operation of RI detection.

B. Commercial Detectors

The DuPont Model 845 differential refractometer is a Fresnel type of instrument based on the change in transmittance of a glass–liquid interface with refractive index change in mobile phase when the interface is illuminated near the critical angle. Figure 5.25 shows an optical diagram of the detector. The refractive index of the sample cell is monitored relative to the refractive index of the reference cell. A lamp illuminates both cells (which are the thin areas between a prism and a stainless steel plate). While some light is reflected at the glass–liquid interface, some passes through the cell area and is reflected back by the stainless steel plate. It is this latter diffuse light that is directed to the dual photodetectors, the intensity of which is proportional to the refractive index change of the flowing stream passing through the sample cell. The Model 845 cells are thermostatted such that the difference is less than 0.01°C. The cell volume is 3 μl, which permits little band broadening within the cell and makes it

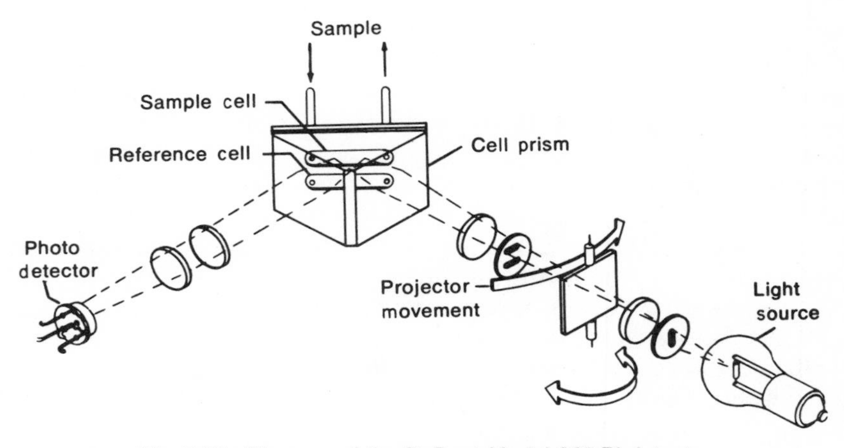

Fig. 5.25 Diagram of the DuPont Model 845 RI detector.

particularly suitable for monitoring fast-eluting peaks (at $k' \leq 2$) with high-efficiency columns. For mobile phase changes, the detector requires about 30 min to reequilibrate. The full-scale sensitivity ranges from 5×10^{-6} to 1.29×10^{-4} refractive index (RI) units with a high-frequency noise level of $\pm 5 \times 10^{-8}$ RI units and a drift of 1×10^{-6} units/hr at 25°C. The detector is capable of detecting low microgram quantities and has a 500-fold linear range.

The Varian RI detector is the same in principle as the DuPont model. It is a dual-beam Fresnel type of detector with essentially the same optical design as shown in Fig. 5.25 and offers the same capabilities.

The Waters Associates Model 401 differential refractometer measures the deflection of a beam of light due to the difference in refractive index between a sample and reference liquids in a single compact sample cell. A schematic is shown in Fig. 5.26. A light beam from an incandescent lamp passes through an optical mask which confines the beam to within the cell

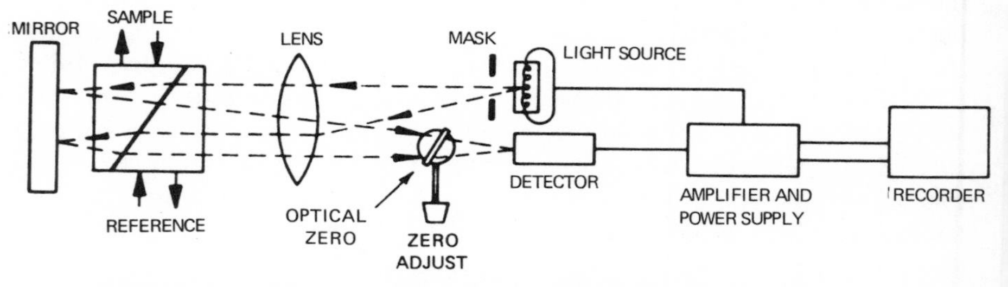

Fig. 5.26 Schematic diagram of the Waters Model 401 RI detector.

region. The lens collimates the light beam, which then passes through the analytical cell (containing the sample and reference liquids in separate chambers) to the mirror. The beam is then reflected back through sample and reference compartments to the lens, which focuses on the photodetector. It is the location of the light beam and not the intensity that is altered by a change in refractive index of the liquid streams. (This is the major difference between the design of this detector and the Fresnel type of instrument.) As the beam changes location on the photodetector, an output signal is generated proportional to the degree of light-beam deflection. The volumes of the sample and reference cells are 10 μl each. Full-scale sensitivities range from 6×10^{-6} to 3×10^{-3} RI units.

A third type of RI detector is marketed by Gow-Mac. This detector is based on the Christiansen effect and consists of a sample flow cell that is packed with a solid material having the same refractive index as the mobile phase. Thus, a beam of light passes directly through the cell to the detector. However, when a solute passes through the cell, the refractive index changes, causing a decrease in amount of light that strikes the photodetector. This is known as the Christiansen effect. A number of solid materials are available for use with a large variety of solvents over a wide refractive index range. However, the nature of this detector makes it impossible to use with gradient elution. It has not been evaluated enough yet to fully assess its capabilities.

C. RI Detector Modifications

Deininger and Halasz (*59*) modified a Waters Model R-4 RI detector by altering the heat exchanger and inlet tubing. They found excessive peak broadening in the heat exchanger of the original model. This was greatly reduced by using a geometrically deformed heat exchanger 20 cm in length. Figure 5.27 shows a sketch of the altered heat exchanger tubing. The tubing was formed into a coil about 8 mm in diameter and cast into an aluminum heat exchanger block that fitted into the detector. This design greatly reduced the band broadening to an acceptable level for use with LC. However, the design did not improve the heat exchanger capability and in fact it was somewhat less efficient than the original detector. No detectable effect could be noticed with either detector when ambient temperature changed from 20° to 25°C. Generally, heat exchangers should be constructed of tubing with an internal diameter as small as possible. This keeps band broadening to a minimum. The length of the tubing should be only long enough to ensure temperature equilibration of the two streams.

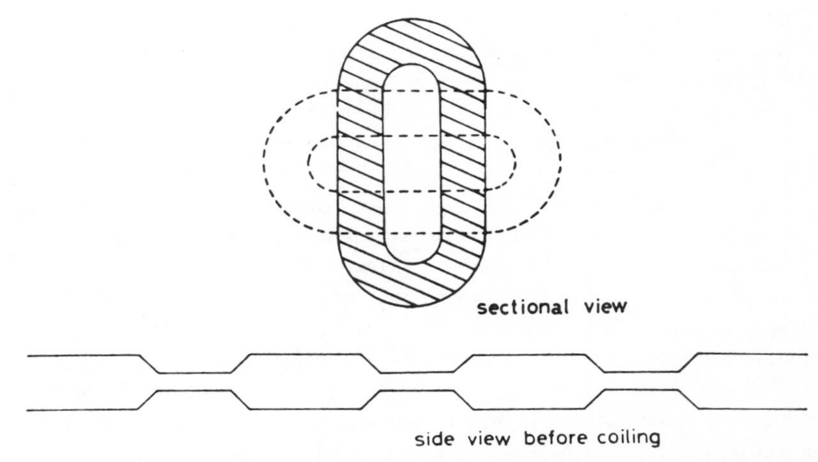

Fig. 5.27 Sketch of geometrically deformed heat exchanger tubing for RI detector. The tubing is 1.5 mm o.d. × 0.5 mm i.d.

VII. MASS SPECTROMETER

There have been several reports in the literature on coupling LC with mass spectrometric (MS) detection, either in a continuous or discontinuous sampling manner. Emphasis in most cases has been placed on the design of suitable interfaces (*60–66*). Such designs included split-stream analysis, where a small fraction of the LC eluant was directed into the mass spectrometer either in a chemical ionization source (*61*) or by sampling of an ionization chamber (*60*); sample collection on a probe where the solvent was evaporated before MS analysis (*63,64*); the use of a silicone rubber membrane molecular separator for removing a large quantity of eluting solvent (*62*); wire transport of the solute after evaporation of the mobile phase (*66*); and post-column extraction into a volatile organic solvent followed by transport via a wire to the mass spectrometer. However, because of the complexity and cost of MS equipment, it is unlikely that it would ever be used solely as an LC detector. A detailed state-of-the-art review of LC–MS has appeared in the literature (*67*).

VIII. OTHER DETECTORS

A. Heat-of-Adsorption Detector

The microadsorption detector is designed to detect very small increases in temperature when a solute adsorbs on, or interacts with, a selected adsorbent. The reference and sample thermistors can be arranged in several ways, but both are normally required to reduce the influence of system temperature fluctuations. Several heat-of-adsorption detectors have been evaluated for use with LC (*68–71*). Scott (*72*) gave a theoretical treatment of the heat-of-adsorption detector and concluded that it is not a viable detection system for LC because of the relationship of detector temperature change with the true concentration profile of the solute. The peaks provided by the detector are usually initially positive as the solute adsorbs onto the adsorbent, but then drop to a negative value below the baseline as the solute desorbs form the adsorbent. Best estimation of sample concentration is obtained by integrating the response to obtain a better peak shape. The sensitivity of this detector is similar to a refractive index detector.

B. Electron-Capture Detector

Application of electron-capture (EC) detection to LC has been studied by several investigators. Maggs (*73*) used an electron-capture detector in conjunction with a solute-transport system where the solvent was removed before the solutes were volatilized into the detector. However, the inherent problems of the transport system made high-sensitivity measurements difficult.

Nota and Palombardi (*74*) developed an electron-capture detection system based on the nebulization of a fraction of the chromatographic eluant, which was then swept into an EC detector (tritium source). Figure 5.28 shows a schematic of the detector setup. A stainless steel capillary

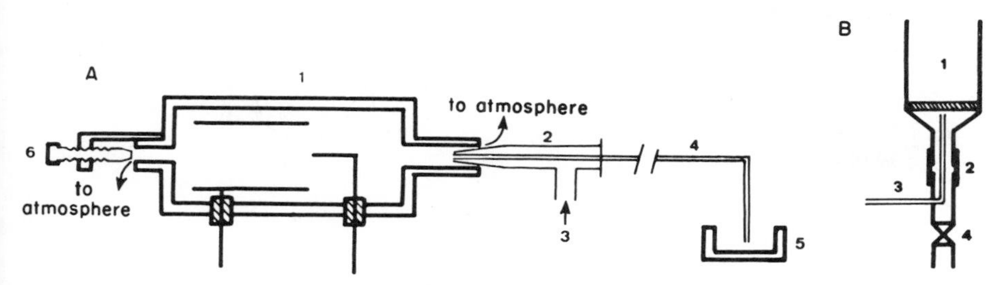

Fig. 5.28 Schematic of LC electron capture detector. (A) 1, EC detector; 2, nebulizer; 3, gas inlet; 4, Teflon capillary; 5, sample reservoir; 6, needle valve. (B) Chromatography column–detector coupling unit: 1, column; 2, silicone rubber joint; 3, stainless steel capillary, which inserts into the Teflon capillary of the detector; 4, tap. From Nota and Palombardi (*74*), with permission from Elsevier Scientific Publishing Co., Amsterdam.

inserted into the column drain removes a small portion of the eluant for EC analysis. Gas flow rates were optimized for best sensitivity with minimum noise. The sensitivity of this detection system was found to be similar to the same detector when used with gas chromatography. Thus, picogram quantities of substances like chloroform, lindane, or nitrobenzene could be detected. Chromatography solvents such as hexane, benzene, pyridine, and methanol were satisfactorily tested with this detector.

Wilmott and Dophin (*75*) described an EC detector in which the total LC effluent was completely volatilized by means of an oven and the vapors passed through the detector cell housed in a separate oven. A ^{63}Ni source was used. The detector could operate effectively with solvent flow rates of up to 3.5 ml/min with optimum results normally at <1 ml/min, which is satisfactory for most separations. Detection limits were also in the picogram range for many compounds, including organochlorine pesticides. A detector based on this concept is currently available from Pye Unicam Ltd. (Fig. 5.29). This detector appears to be the most sensitive yet described for the LC analysis of organochlorine compounds (although it is about 100-fold less sensitive than detection of the same compounds by GC-EC). Figure 5.30 shows an LC chromatogram of two dinitrotoluene isomers by EC and UV detection. The strong EC sensitivity of the $—NO_2$ makes EC detection very suitable for such compounds.

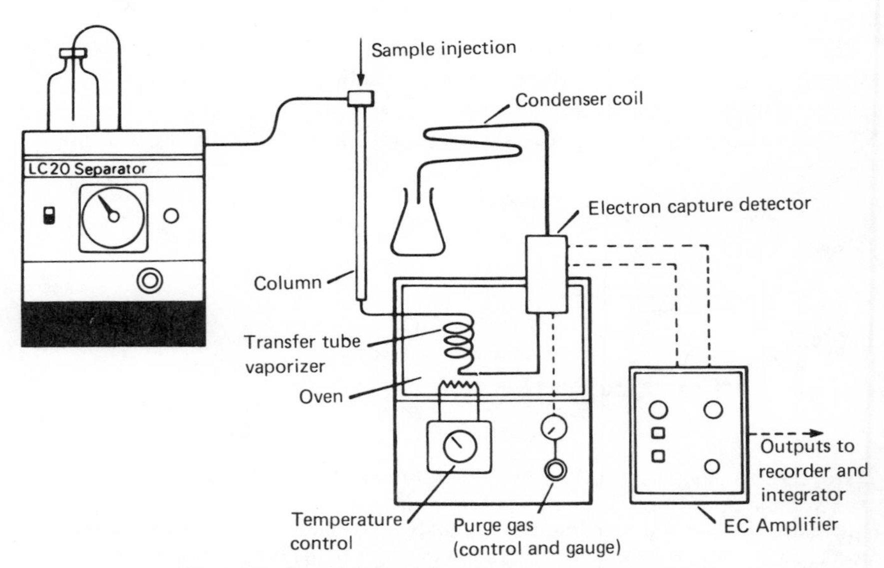

Fig. 5.29 Pye Unicam LC electron-capture detector.

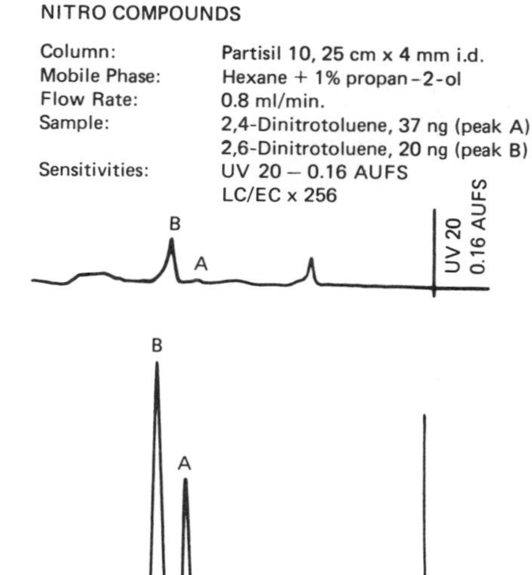

Fig. 5.30 Comparison of UV and electron-capture detection of 2,4-dinitrotoluene (peak A) and 2,6-dinitrotoluene (peak B) separated by LC.

C. Piezoelectric Detector

A general detection system for LC has been described based on the piezoelectric effect (*76*). This detector makes use of a quartz resonator which vibrates at a constant frequency when placed in an electric potential. The frequency of vibration is altered with a change in mass of the crystal. For LC, the effluent is sprayed on the surface of a piezoelectric quartz crystal, the solvent evaporated, and the mass of the remaining solute determined by the change in frequency of the crystal. Evaporation of the solvent is necessary since this effect operates only in the gas phase or in a vacuum. The detection limits appear to be similar to the UV detector, being in the low nanogram range for many compounds. However, the discontinuous nature of the detection makes its routine application questionable.

D. Dielectric-Constant Detector

Klatt *(77)* has evaluated an LC detector based on the measurement of dielectric constant of the effluent. The response is determined by the shift in frequency resulting from dielectric-constant changes in the flowing stream due to the elution of a solute. The detector is not affected much by flow-rate change and has good stability and reproducibility. Solvents with widely different dielectric constants can be used as mobile phases. This has proved to be superior to several capacitance detectors described *(78–80)* where the detectors had to be altered to operate with solvents of widely different dielectric constants. The sensitivities of these detectors are in the same range as the sensitivity of refractive index detectors. More work is required on the evaluation of these detectors before they can be considered useful as general-application LC detectors.

E. Spray-Impact Detector

Mowrey and Juvet *(81)* described a novel LC detector which they have termed the "spray-impact detector." When a stream of solvent from an LC column is sprayed onto a conducting target electrode, the droplets become electrically charged and a potential of over 2000 V is produced at the electrode surface. When a solute is eluted from the column, the electrode potential (and current) changes. These changes are monitored and related to solute concentration. The actual mechanism of charge formation and solute influence are not yet completely understood.

The design of the detector (Fig. 5.31) is simple, and it is easily constructed in the laboratory. The detection limits vary greatly for organic compounds, depending on their nature. They range from subnanograms per second up to micrograms per second with linearity extending from 2 to 4 orders of magnitude.

Unfortunately, one of the major disadvantages of this detector is that because of the nature of the measurements, some solutes give negative peaks. This could result in difficulty in interpretation of chromatograms when mixtures of positive- and negative-producing solutes are present together. Also, peak shape can change at higher concentrations, often becoming flat-topped or even negative. The frequent non-Gaussian peak shape attributed to the detector might make quantitation or qualitative identification of the peaks difficult, since the peak maxima change. More work on this detection system is required before it can be considered suitable for routine use with LC.

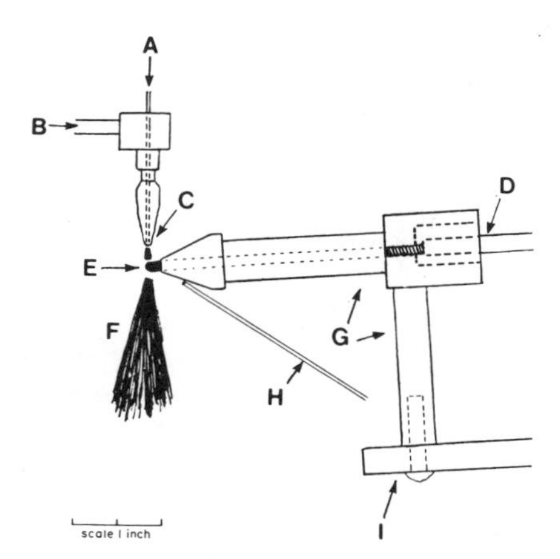

Fig. 5.31 Spray-impact detector. A, column outlet; B, air inlet; C, stainless steel aspirator jet; D, coaxial cable to electrometer; E, glassy carbon (or gold) target electrode; F, spent spray; G, PTFE body; H, glass capillary to aspirator vacuum for removal of charged droplets; I, laminated plastic mount. From Mowrey and Juvet (*81*), with permission from Preston Publications, Inc.

F. Photoconductivity Detector

This detector has been introduced by Tracor, Inc., and is a combination of photometric and electrochemical detection (*82,83*). The detection mechanism is based upon post-column photochemical reactions which produce ionic species that may be detected electrochemically. Thus, the effluent from the LC is split into two streams, one of which is the reference and the other for analysis. The analysis stream passes through a quartz reaction coil where it is subjected to intense UV irradiation. The stream then passes into the conductivity cell. The reference stream passes through a delay coil (without irradiation), then into the reference side of the conductivity cell. The response is measured as a difference in conductivity of the two cells. This detector system has been applied to halogenated and some sulfur- or nitrogen-containing compounds. Subnanogram quantities of some compounds may be detected. The theory of operation has been discussed (*82*).

REFERENCES

1. K. Callmer and O. Nilson, *Chromatographia* **6,** 517 (1973).
2. G. Brooker, *Anal. Chem.* **43,** 1095 (1971).

3. J. F. Lawrence and R. W. Frei, "Chemical Derivatization in Liquid Chromatography." Elsevier, Amsterdam, 1976.
4. J. C. Steichen, *J. Chromatogr.* **104,** 39 (1975).
5. E. D. Pellizzari and C. M. Sparacino, *Anal. Chem.* **45,** 378 (1973).
6. R. M. Cassidy and R. W. Frei, *J. Chromatogr.* **72,** 293 (1972).
7. R. Greenhalgh and W. Marshall, unpublished work (1974).
8. R. J. Perchalski, J. D. Winefordner, and B. J. Wilder, *Anal. Chem.* **47,** 1993 (1975).
9. T. Panalaks and P. M. Scott, *J. Assoc. Off. Anal. Chem.* **60,** 583 (1977).
10. J. B. Lloyd, *Analyst (London)* **100,** 529 (1975).
11. L. H. Thacker, *J. Chromatogr.* **73,** 117 (1972).
12. F. Martin, J. Maine, C. C. Sweeley, and J. F. Holland, *Clin. Chem.* **22,** 1434 (1976).
13. P. A. Asmus, J. W. Jorgenson, and M. Novotny, *J. Chromatogr.* **126,** 317 (1976).
14. G. B. Sieswerda, H. Poppe, and J. F. K. Huber, *Anal. Chim. Acta* **78,** 343 (1975).
15. E. T. McGuinness and M. C. Cullen, *J. Chem. Educ.* **47,** A9 (1970).
16. L. Schutte, *J. Chromatogr.* **72,** 303 (1972).
17. G. B. Sieswerda and H. L. Polak, *in* "Liquid Scintillation Counting" (M. A. Crook, P. Johnson, and B. Scales eds.), Vol. 2, p. 49. Heyden, London, 1972.
18. D. R. Bangan, *Biochem. J.* **62,** 552 (1956).
19. A. M. Van Urk-Schoen and J. F. K. Huber, *Anal. Chim. Acta* **52,** 519 (1970).
20. D. R. Jones and S. E. Manahan, *Anal Chem.* **48,** 1879 (1976).
21. D. R. Jones and S. E. Manahan, *Anal. Chem.* **48,** 502 (1976).
22. D. R. Jones and S. E. Manahan, *Anal. Lett.* **8,** 569 (1975).
23. D. R. Jones, H. C. Tung, and S. E. Manahan, *Anal. Chem.* **48,** 7 (1976).
24. R. M. Cassidy, M. T. Hurteau, J. P. Mislan, and R. W. Ashley, *J. Chromatogr. Sci.* **14,** 446 (1976).
25. W. Funasaka, T. Hanai, and K. Fiyimura, *J. Chromatogr. Sci.* **12,** 517 (1975).
26. D. J. Freed, *Anal. Chem.* **47,** 187 (1975).
27. J. I. Steinfield, *CRC Crit. Rev. Anal. Chem.* **5,** 225 (1975).
28. N. K. Freeman, F. T. Upham, and A. A. Windsor, *Anal. Lett.* **6,** 943 (1973).
29. K. Tesarik and P. Kalab, *J. Chromatogr.* **78,** 357 (1973).
30. R. L. Pecsok and D. Saunders, *Anal. Chem.* **40,** 1756 (1968).
31. C. L. Blank, *J. Chromatogr.* **117,** 35 (1976).
32. P. T. Kissinger, C. Refshauge, R. Dreiling, and R. N. Adams, *Anal. Lett.* **6,** 465 (1973).
33. C. Refshauge, P. T. Kissinger, R. Dreiling, C. L. Blank, R. Freeman, and R. N. Adams, *Life Sci.* **14,** 311 (1974).
34. B. Fleet and C. J. Little, *J. Chromatogr. Sci.* **12,** 747 (1974).
35. J. Yamada and H. Matsuda, *J. Electroanal. Chem.* **44,** 189 (1973).
36. A. MacDonald and P. D. Duke, *J. Chromatogr.* **83,** 331 (1973).
37. J. Lankelma and H. Poppe, *J. Chromatogr.* **125,** 375 (1976).
38. W. Kemula, *Rocz. Chem.* **26,** 281 (1952).
39. J. G. Koen, J. F. K. Huber, H. Poppe, and G. den Boef, *J. Chromatogr. Sci.* **8,** 192 (1970).
40. J. G. Koen and J. F. K. Huber, *Anal. Chim. Acta* **51,** 303 (1970).
41. R. Stillman and T. S. Ma, *Mikrochim. Acta* p. 491 (1973).
42. W. J. Blaedel and J. H. Strohl, *Anal. Chem.* **36,** 445 (1964).
43. P. L. Joynes and R. J. Maggs, *J. Chromagtor. Sci.* **8,** 427 (1970).
44. F. A. Shultz and D. E. Mathis, *Anal. Chem.* **46,** 2253 (1974).
45. V. Pretorius and J. F. J. van Rensburg, *J. Chromatogr. Sci.* **11,** 355 (1973).
46. A. Stolyhwo, O. S. Privett, and W. L. Erdahl, *J. Chromatogr. Sci.* **11,** 263 (1973).

47. A. Karmen, M. L. Karasek, L. D. Lane, and B. M. Lapidus, *J. Chromatogr. Sci.* **8,** 438 (1970).
48. B. M. Lapidus and A. Karmen, *J. Chromatogr. Sci.* **10,** 103 (1972).
49. J. H. van Dijk, *J. Chromatogr. Sci.* **10,** 31 (1972).
50. H. Dubsky, *J. Chromatogr.* **71,** 395 (1972).
51. M. Dressler and M. Deml, *J. Chromatogr.* **56,** 23 (1971).
52. J. J. Szakasits and R. E. Robinson, *Anal. Chem.* **46,** 1648 (1974).
53. K. Slais and M. Krejci, *J. Chromatogr.* **91,** 181 (1974).
54. P. R. Boshoff, B. J. Hopkins, and V. Pretorius, *J. Chromatogr.* **126,** 35 (1976).
55. R. P. W. Scott and J. G. Lawrence, *J. Chromatogr. Sci.* **8,** 65 (1970).
56. E. P. Foster and A. H. Weiss, *J. Chromatogr. Sci.* **9,** 266 (1971).
57. H. Coll, H. W. Johnson, A. G. Polgar, E. E. Siebert, and K. H. Stross, *J. Chromatogr. Sci.* **7,** 30 (1970).
58. M. Krejci and M. Dressler, *Chem. Listy* **65,** 574 (1971).
59. G. Deininger and I. Halasz, *J. Chromatogr. Sci.* **8,** 499 (1970).
60. E. C. Horning, D. I. Carroll, I. Dzidic, K. D. Haegele, M. G. Horning, and R. N. Stillwell, *J. Chromatogr.* **99,** 13 (1974).
61. P. J. Arpino, B. G. Dawkins, and F. W. McLafferty, *J. Chromatogr. Sci.* **12,** 574 (1974).
62. P. R. Jones and S. K. Yang, *Anal. Chem.* **47,** 1000 (1975).
63. R. E. Lovins, S. R. Ellis, G. D. Tolbert, and C. R. McKinney, *Anal. Chem.* **45,** 1553 (1973).
64. H. R. Schulton and H. D. Beckey, *J. Chromatogr.* **83,** 315 (1973).
65. B. L. Karger, D. P. Kirby, P. Vouros, R. L. Foltz, and B. Hildy, *Anal. Chem.* **51,** 2324 (1979).
66. R. P. W. Scott, C. G. Scott, M. Munroe, and J. Hess, Sr., *J. Chromatogr.* **99,** 395 (1974).
67. P. Arpino and G. Guiochon, *Anal. Chem.* **51,** 682A (1979).
68. M. N. Munk, *J. Chromatogr. Sci.* **8,** 491 (1970).
69. K. P. Hupe and E. Bayer, *J. Gas Chromatogr.* **5,** 197 (1967).
70. M. N. Munk and D. N. Raval, *J. Chromatogr. Sci.* **7,** 48 (1969).
71. J. L. Cashaw, R. Segura, and A. Zlatkis, *J. Chromatogr. Sci.* **8,** 363 (1970).
72. R. P. W. Scott, *J. Chromatogr. Sci.* **11,** 349 (1973).
73. R. J. Maggs, *Column* **2,** 5 (1968).
74. G. Nota and R. Palombardi, *J. Chromatogr.* **62,** 153 (1971).
75. F. W. Wilmott and R. J. Dophin, *J. Chromatogr. Sci.* **12,** 695 (1974).
76. W. W. Shultz and W. H. King, *J. Chromatogr. Sci.* **11,** 343 (1973).
77. L. N. Klatt, *Anal. Chem.* **48,** 1845 (1976).
78. R. Vespalec and K. Hana, *J. Chromatogr.* **63,** 53 (1972).
79. R. Vespalec, *J. Chromatogr.* **108,** 243 (1975).
80. W. F. Erbelding, *Anal. Chem.* **47,** 1983 (1975).
81. R. A. Mowrey and R. S. Juvet, *J. Chromatogr. Sci.* **12,** 689 (1974).
82. D. J. Popovich, J. B. Dixon, and B. J. Ehrlich, *J. Chromatogr. Sci.* **17,** 643 (1979).
83. D. H. Rodgers and R. C. Hall, *Pittsburgh Conf. Anal. Chem.* Pap. No. 8 (1977).

Chapter 6
Chromatography Theory

I. FUNDAMENTAL PRINCIPLES

A. Introduction

The following is a brief discussion of how an analyst can evaluate the performance of a chromatographic system in order that it may be optimized for a particular analysis. Several books have appeared that rigorously treat chromatography theory (*1–4*), and the reader is directed to these for additional in-depth discussion.

When a sample mixture is injected onto an LC column it begins to migrate down the column under the influence of the mobile phase. During this process the various components of the sample mixture will begin to separate, depending upon their affinity for the stationary phase in the presence of the mobile phase. The components that are only weakly retained by the stationary phase (for example, nonpolar compounds in adsorption chromatography and very polar compounds in reversed-phase chromatography) will pass through the column and be eluted first. Thus, they will be the first peaks to appear in the resulting chromatogram. The strongly retained components will elute later, the relative separation being dependent upon the degree of retention by the stationary phase of each sample component. Thus, the components pass down the column at different speeds which can be related to the distribution of each component in the stationary and mobile phases.

During the passage down the column the component bands will spread longitudinally from the initial discrete zone of the injection. This band spreading or band broadening is related to column efficiency. Figure 6.1 shows a chromatogram of the separation of a two-component mixture. The two bands are separated on their relative differences in distribution between the mobile and stationary phase. This may be referred to as differences in distribution isotherm. The distribution isotherm is defined as an isothermal plot of the equilibrium quantity of solute taken up per unit

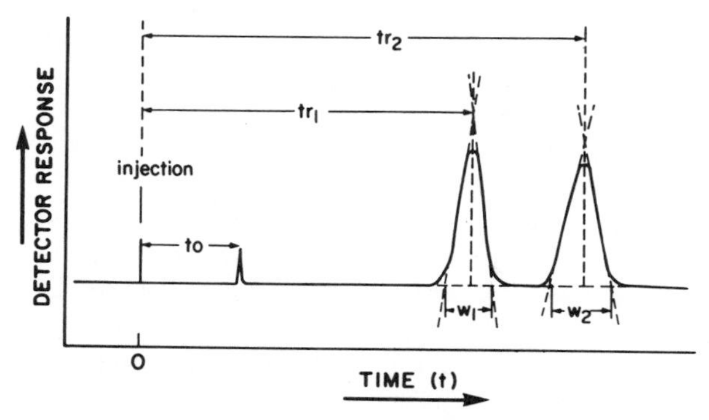

Fig. 6.1 Typical chromatographic separation of a two-component mixture indicating important parameters always measured in the same units (time or distance).

weight of the stationary phase versus the concentration of the compound in the mobile phase. As the equilibrium quantity of solute retained by the stationary phase increases for a given concentration in the mobile phase, the migration rate of the solute through the column becomes slower.

Figure 6.2 shows three types of isotherms and the resulting peak shape. If the isotherm is linear, as in Fig. 6.2A (i.e., an increase in the concentra-

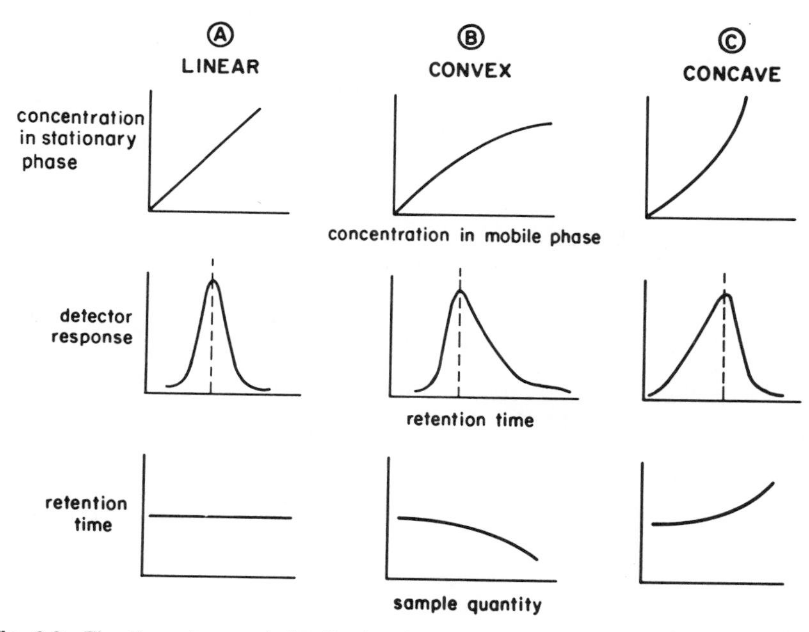

Fig. 6.2 The three types of distribution isotherms and their effect on peak shape and retention time. A, linear (ideal) isotherm; B, convex isotherm; C, concave isotherm.

tion of the solute in the mobile phase causes a corresponding linear increase of solute in the stationary phase), then a Gaussian distribution of the solute molecules will occur as the molecules pass down the column. This is an ideal situation, and most chromatographic peaks exhibit at least some degree of tailing, as shown in Fig. 6.2B. Concave isotherms, shown in Fig. 6.2C, are rarely observed in LC. Another consequence of a non-ideal isotherm is that retention time measured from the top of the peak (tr_1 or tr_2 in Fig. 6.1) will change with increasing quantity of sample injected. Usually, under conditions of a convex isotherm, this means that the peaks will elute slightly faster with increased quantity of solute injected. The actual magnitude of this change depends upon the degree of peak tailing, which is related to the extent of deviation of the convex isotherm from the ideal.

B. Retention

The retention of the two peaks in Fig. 6.1 is best characterized by the elution volume, V_r, which is the volume of mobile phase required to elute solute molecule at its maximum concentration (peak height) from the column. This can be calculated by multiplying the mobile phase flow rate F_c (ml/min), by the retention time in minutes, t_r. Thus for the first peak of Fig. 6.1

$$V_{r1} = t_{r1} \times F_c \tag{1}$$

and for the second peak

$$V_{r2} = t_{r2} \times F_c \tag{2}$$

The time required to elute an unretained solute—that is, one that elutes completely in the mobile phase with no retention by the stationary phase—is called t_0. The volume required to elute this is

$$V_0 = t_0 \times F_c \tag{3}$$

This is a measure of the volume of mobile phase in the column and is referred to as the void volume of the column or the "dead" volume. This value remains constant for any given column, regardless of the type of mobile phase or flow rate.

A very important term in LC is k', referred to as the capacity factor, which is a measure of retention of a retained solute compared to an unretained solute.

$$k' = (t_{r1} - t_0)/t_0 \tag{4}$$

The k' value for a solute is constant for any given chromatography system and does not vary with mobile-phase flow rate, column length, or column diameter. Because of this, k' is a very useful value since an analyst can calculate the new retention time of a solute peak if he replaces a column with one that contains the same packing material but which may be of different dimensions. To do this, it is necessary first to calculate the t_0 value of the new column by rearranging Eq. (4); substituting in the t_0 value for the new column—say, t_{0_x}, the following expression is obtained

$$t_{r_x} = t_{0_x}(1 + k') \tag{5}$$

where t_{r_x} is the new retention time.

The true t_{r_x} may not be exactly as calculated, since the selectivity of the new column may not be identical to the old, thus causing a change in k' value. However, it does serve as a very good approximation.

C. Column Efficiency

Efficiency is a measure of the broadening of a solute band as it migrates down the column. A highly efficient column will produce well-defined sharp peaks, a feature desired by the analyst. If band broadening is significant, the peaks will be broad, resulting in both poorer separation from other sample components and poorer sensitivity (see Chapter 4).

The sample is usually applied to the top of the column as a discrete narrow plug. As the sample moves down the column, the individual molecules follow many different pathways, as illustrated in Fig. 6.3. Because

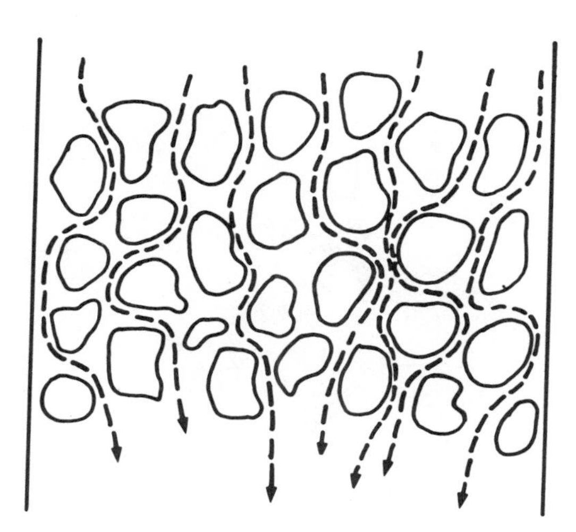

Fig. 6.3 Illustration of the pathways taken by solute molecules as they pass down a chromatography column.

the molecules move at random, some paths will be shorter than others; thus some molecules of the sample band will migrate down the column more rapidly, causing dispersion. The discrete plug of sample soon becomes distributed in a Gaussian manner (assuming a linear isotherm). A nonhomogeneously packed column will have pathways that vary significantly in length, resulting in much band broadening. The problem of "channeling" due to nonhomogeneous packing is well known in classical column chromatography, and the distorted bands are easily observed if colored substances are chromatographed. This same effect results in poorly packed LC columns.

Another factor involved in band broadening is the particle size. Large porous particles offer a wider range of path lengths and flow velocities, as shown in Fig. 6.4. Molecules may be trapped in minute "voids" within the particles where mobile-phase flow is very slow. By reducing the particle diameter, one reduces the distance a molecule can penetrate the particle as well as reducing the differences in path length and the total time spent within the particle. This serves to increase mass transfer between mobile and stationary phases, thus reducing band broadening (see Chapter 4, Fig. 4.2).

A numerical measurement of column efficiency can be obtained in terms of "theoretical plates." This theory was first put forward by Martin and Synge (*5*) to treat the processes involved in chromatography mathematically, enabling the analyst to understand and evaluate them in order that they can be optimized.

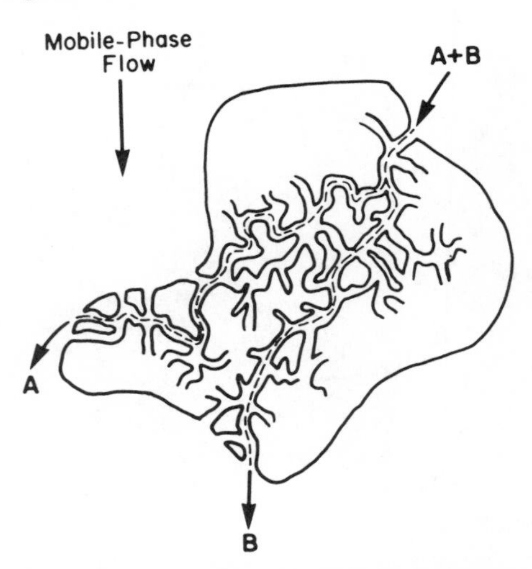

Fig. 6.4 Porous particle illustrating the differing pathways for two identical molecules, A and B.

The number of theoretical plates, N, of a given column may be calculated from Fig. 6.1 by the following expression

$$N = 16\left(\frac{t_{r_1}}{w_1}\right)^2 \quad (6)$$

where t_{r_1} and w_1 are measured in the same units of time or distance. From this equation, it can be seen that for any given t_{r_1}, the value N will increase as the peak width w_1 becomes smaller. Thus, large numbers of theoretical plates indicate relatively narrow peaks. It should be pointed out that w_1 is not related to N^{-1} but to $N^{-0.5}$. Thus, doubling the number of plates only decreases the peak width by $2^{0.5}$ (i.e., 1.41). For example, if a column had 1000 plates, and this were increased to 4000, the peak width would not be $\frac{1}{4}$ as wide as the original but only $\frac{1}{2}$ as wide.

Another useful parameter is the height equivalent to a theoretical plate, H, defined as

$$H = L/N \quad (7)$$

where L = column length, usually in millimeters. Maximum efficiency is obtained when H is as small as possible.

Column efficiency is also affected by mobile-phase flow rate. Figure 6.5 shows a plot of H versus mobile phase linear velocity v, which may be calculated according to the equation

$$v = L/t_0 \quad (8)$$

It can be seen from this plot that chromatographic efficiency increases (i.e., H decreases) with a decrease in linear velocity. This means that one can decrease the peak width w (measured in milliliters) when the

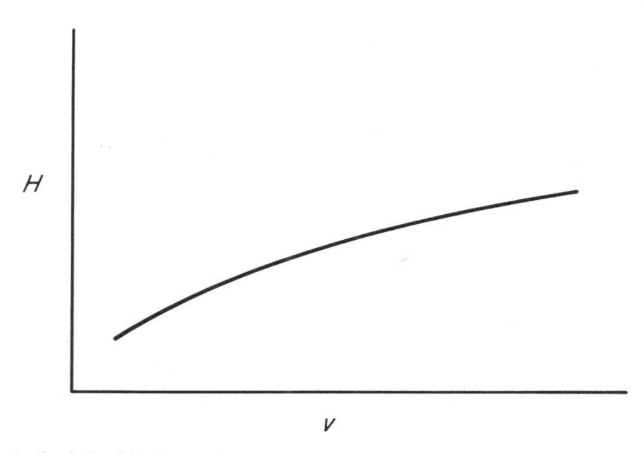

Fig. 6.5 Plate height H plotted versus the linear velocity v of the mobile phase.

mobile-phase flow rate is reduced, leading to a slight increase in peak height, depending upon the slope of the H versus v plot.

D. Resolution of Peaks

Resolution R_s or the degree of separation of two proximate peaks is an important term in chromatography. It is defined mathematically as the distance between peak centers divided by the average peak width measured either in time or volume units.

$$R_s = 2\left(\frac{t_{r2} - t_{r1}}{w_1 + w_2}\right) \tag{9}$$

A more fundamental equation for R_s was given by Purnell (*6*), where Eq. (9) can be transformed to one relating efficiency, N, and k'. Thus

$$R_s = \frac{N^{0.5}}{4}\left[1 - \left(\frac{1 + k_1'}{1 + k_2'}\right)\right] \tag{10}$$

If a selectivity factor, α, is introduced, where

$$\alpha = k_2'/k_1' \tag{11}$$

then Eq. (10) becomes

$$R_s = \frac{N^{0.5}}{4}\left(\frac{\alpha - 1}{\alpha}\right)\left(\frac{k_2'}{1 + k_2'}\right) \tag{12}$$

This equation relates three essentially independent parameters (k', N, and α) to resolution. By changing the mobile-phase composition, one can change k' and to some degree α, resulting in an improved separation of two peaks.

An improvement in resolution can also be obtained by increasing N. With a given column, one can only do this by increasing column length, i.e., adding a second column. This, of course, is not too useful since the separation time becomes longer and mobile-phase flow rate is limited because of the added back-pressure of the second column. Also, the improvement in resolution is only related to the square root of N. Thus, doubling the column length does not improve resolution by a factor of 2 but only by $2^{0.5}$, or 1.41. In cases where column efficiency is poor, it is preferable to use a better column.

II. ADSORPTION CHROMATOGRAPHY

A. Nature of the Process

Intermolecular forces play the major role in separating molecules in adsorption chromatography. These include van der Waals forces (London dispersion forces) that exist between the surface of the stationary phase and the adsorbed molecules, and which are dependent upon the relative masses of the solute and solvent (mobile phase) molecules. Electrostatic forces associated with both mobile phase and solute molecules also play an important role since these are related to molecular polarity. Adsorption chromatography, thus, is based on the balance of these forces between mobile and stationary phases (i.e., the equilibrium of the solute molecules in the free and adsorbed states). Separations are based on the difference in the distribution equilibria of different solutes. A strongly adsorbed molecule will be retained by the stationary phase for a longer period of time than one that has a weak interaction with the adsorbent. Thus the latter will be eluted from the column first.

It should be pointed out that physical adsorption (resulting from electrostatic and molecular dispersion forces) are not the only mechanisms that can be involved in an adsorption chromatographic system. Chemisorption (hydrogen bonding and charge-transfer forces between electron donors and electron acceptors) may also be involved. However, the rates of mass transfer associated with these are slow compared to physical adsorption and thus are to be avoided when employing adsorption chromatography. Rapid mass transfer is important if efficient separations are to be achieved. This is highly dependent upon the choice of materials used for both mobile and stationary phases.

By far the most common adsorbent used in LC is silica gel. It consists of noncrystalline polymers of SiO_2 and is a stable porous solid with a surface structure similar to that illustrated in Fig. 6.6. The surface consists of silanol (≡SiOH) and siloxane (Si—O—Si) groups. The silanol groups are

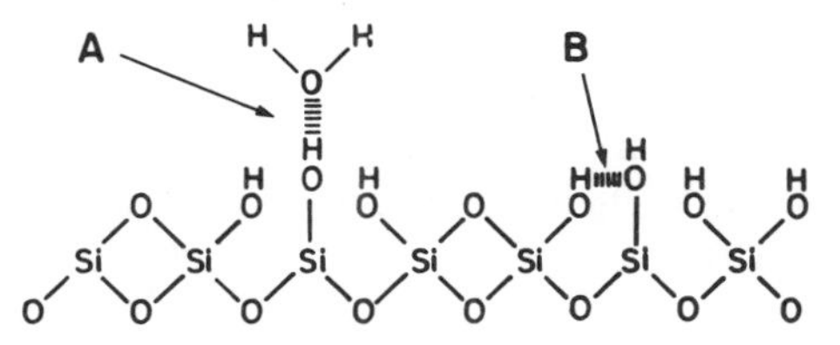

Fig. 6.6 A possible structure of the surface of silica gel. A, hydrogen bonding with water; B, intramolecular hydrogen bonding.

considered to be most important for separations. These groups are somewhat acidic, although not uniformly so, since the strength depends on the location (e.g., the most acidic sites are situated adjacent to silica atoms where intramolecular hydrogen bonding may occur; see "B" in Fig. 6.6). These are the sites that lead to chemisorption, which is an undesirable occurrence, as mentioned above. Usually such sites are blocked by the addition of a small percentage of a polar modifier such as H_2O to the mobile phase. Scott and Kucera (*7*) evaluated 10 commercially available silica gels in terms of efficiency, sample capacity, surface area, column permeability, etc.

Alumina (Al_2O_3) has properties similar to silica gel in that the retention of compounds is based on polarity. However, some differences in selectivity are to be expected since silica gel is weakly acidic and alumina weakly basic. The differences would be most noticeable with solutes such as amines, which would be retained longer on silica gel, or phenols, which might be retained longer on alumina. Alumina is commercially available for LC in bulk packing or as prepacked columns. However, this adsorbent is used far less than silica gel.

Several models to describe molecular adsorption have been proposed (*1,8–12*). Although no one model is satisfactory to describe all adsorption chromatographic behavior, these do shed much light on the molecular processes involved and relate these to actual experimental results. The recent work by Scott and Kucera (*10*) is particularly enlightening. When silica gel comes in contact with solvents or solvent mixtures, some molecules adsorb onto its surface to form a single or double immobile layer. The single (or monolayer) coverage may arise from contact with solvents that are nonpolar and that do not hydrogen bond. Experimental data indicate that nonpolar solvents do not form stable double layers. Polar solvents such as ethyl acetate or isopropanol form monolayers when present in concentration of 1% or less in nonpolar solvents such as heptane. As the concentration increases, say up to 3%, the second layer begins to form. This second layer is less tightly held and is formed more slowly with increasing concentration.

Further experimental results have shown that for compounds eluting at $k' < 10$, the solute molecules interact only with the primary (first monolayer) and do not replace it. Under conditions where a significant double layer exists, the solute molecules appear actually to displace the second layer and interact with the primary layer. In both cases, significant interaction with the adsorbent surface does not appear to exist. The above holds true for small concentrations of polar modifiers in very nonpolar solvents and not necessarily for other types of solvent mixtures.

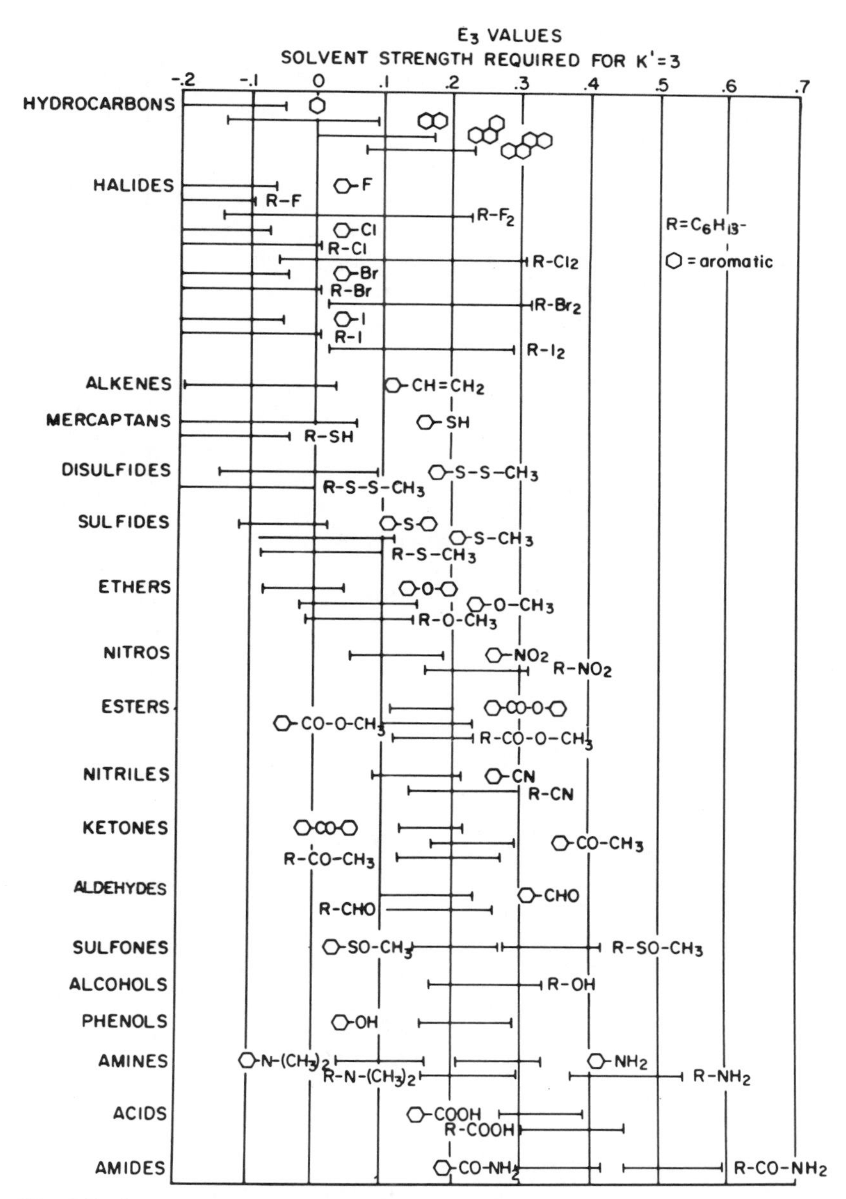

Fig. 6.7 E_3 values on silica gel. From Saunders (*13*), with permission from the American Chemical Society.

B. Mobile-Phase Selection

Saunders (*13*) devised a graphic system that enables an analyst to estimate the strength of the mobile phase to use with silica gel (300 m^2/gm surface area) in order to obtain a k' of 3, for a large series of compounds. This is presented in Fig. 6.7. This figure indicates the solvent strength (E_3) required to a produce a $k' = 3$ for the indicated compounds. Solvent strength (ϵ_0) is defined by Snyder (*1*) as the adsorption energy per unit area of a standard adsorbent. In practice, log k' varies linearly with ϵ_0. In Fig. 6.7 the horizontal bars indicate the range of values obtained for silica gel in a fully activated state (higher E_3 value) and when highly deactivated with water (lower E_3 value).

Figure 6.8 illustrates how a suitable solvent strength may be chosen from a selection of six solvents. These are particularly useful for LC since they have low viscosities, low UV absorption cutoffs, and they cover the whole range of solvent strength useful for adsorption chromatography. The figure plots solvent strength ϵ_0 versus binary solvent compositions. As an example, the dashed line indicates solvent compositions required for a solvent strength of 0.30. For pentane, this represents the addition of 75% methylene chloride, 49% ethyl ether, or 1.9% acetonitrile. If isopropyl chloride is used as the primary solvent, it would have to contain

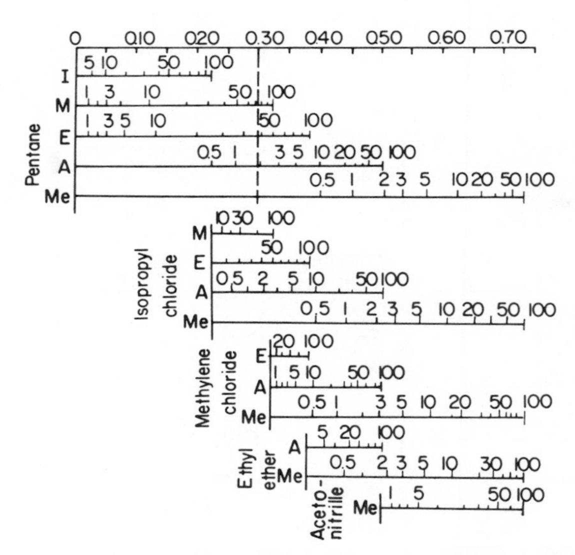

Fig. 6.8 Mixed solvent strengths on silica gel. M, methylene chloride; I, isopropyl chloride; E, ethyl ether; A, acetonitrile; and Me, methanol. From Saunders (*13*), with permission from the American Chemical Society.

TABLE 6.1

Properties of Chromatographic Solvents[a]

Solvent	Polarity ($\epsilon_0(Al_2O_3)$)	Viscosity (cP, 20°C)	UV cutoff (nm)
n-Pentane	0.00	0.23	210
iso-octane	0.01	0.54	210
Petroleum ether, Skellysolve B, etc.	0.01	0.30	210
n-Decane	0.04	0.92	210
Cyclohexane	0.04	1.00	210
Cyclopentane	0.05	0.07	210
Carbon disulfide	0.15	0.37	380
Carbon tetrachloride	0.18	0.97	265
Amyl chloride	0.26	0.43	225
Xylene	0.26	0.62–0.81	290
Iso-Propyl ether	0.28	0.37	220
Iso-Propyl chloride	0.29	0.33	225
Toluene	0.29	0.59	285
n-Propyl chloride	0.30	0.35	225
Benzene	0.32	0.65	280
Ethyl ether	0.38	0.23	220
Ethyl sulfide	0.38	0.45	290
Chloroform	0.40	0.57	245
Methylene chloride	0.42	0.44	245
Methyl-iso-butylketone	0.43	—	330
Tetrahydrofuran	0.45	—	220
Ethylene dichloride	0.49	0.79	230
Methylethylketone	0.51	0.43	330
1-Nitropropane	0.53	—	380
Acetone	0.56	0.32	330
Dioxane	0.56	1.54	220
Ethyl acetate	0.58	0.45	260
Methyl acetate	0.60	0.37	260
Amyl alcohol	0.61	4.10	210
Diethylamine	0.63	0.38	275
Nitromethane	0.64	0.67	380
Acetonitrile	0.65	0.37	210
Pyridine	0.71	0.94	305
Butyl cellosolve	0.74	—	220
2-Propanol, 1-propanol	0.82	0.23	210
Ethanol	0.88	1.20	210

(continued)

TABLE 6.1 *(cont'd)*

Solvent	Polarity ($\epsilon_0(Al_2O_3)$)	Viscosity (cP, 20°C)	UV cutoff (nm)
Methanol	0.95	0.60	210
Ethylene glycol	1.11	19.90	210
Acetic acid	Large	1.26	210
Water	Very large	1.00	210

[a] Data from Snyder (*1*).

50% methylene chloride, 37% ethyl ether, or 1.7% acetonitrile to produce a solvent strength of 0.30.

Although Figs. 6.7 and 6.8 serve to aid the analyst in selecting suitable mobile phases for a variety of compounds, the approach must be regarded as approximate, since the true k' values for any compound can only be determined experimentally. Also, only k' is of interest here, not selectivity. In this case Fig. 6.8 can be useful since the several solvent combinations described above for an $\epsilon_0 = 0.30$ will have some differences in selectivity, depending upon the solute and sample matrix analyzed.

Table 6.1 lists the ϵ_0 values for alumina along with viscosity and UV cutoff of a number of solvents useful for LC.

III. PARTITION CHROMATOGRAPHY

Partition chromatography is based on the separation of solutes by their differential partitioning between two immiscible phases. This usually involves a stationary phase coated (physically, or chemically bonded) on an inert solid support, normally silica gel and an immiscible mobile phase usually consisting of water or aqueous salt solutions modified with polar organic solvents such as acetonitrile or methanol (for reversed-phase chromatography).

Normal-phase partition chromatography involves the use of a polar substance coated on the support material as stationary phase and an immiscible nonpolar mobile phase. This type of chromatography is not much used since adsorption chromatography appears to work just as effectively, even though the chromatographic principles are different, and therefore the chromatographic selectivity. Separations in both systems are based on solute polarity; i.e., the least polar compounds are eluted first and the most polar solutes are retained the longest. Reversed-phase chromatography has found a far greater degree of popularity, especially chemically bonded reversed-phase chromatography. This system makes use of a nonpolar stationary phase and a highly polar immiscible mobile phase. Thus, compounds are separated by their relative hydrophobicity; i.e., the most polar compounds are usually eluted first while the nonpolar solutes are retained longer. Thus arises the term "reversed" phase partition chromatography [coined by Howard and Martin in 1950 (*14*)]. The separations are usually the reverse of adsorption or normal-phase partition chromatography.

A. Reversed-Phase Partition Chromatography on Chemically Bonded Phases

Probably the most spectacular advance in LC as an analytical technique in the 1970s was the development and use of chemically bonded stationary phase packing materials. Since Halasz and Sebestian (*15*) introduced the first chemically bonded phase (BP) in 1969, many bonding techniques have been investigated, and have resulted in the discovery of many useful products to aid the chromatographer. These phases range greatly in polarity and can be designed for selective separations. The most common BPs used at present are nonpolar hydrocarbonaceous phases such as octadecyl- or octyl-phases bonded to 5 or 10 μm porous silica gel.

Figure 6.9 illustrates a possible structure of a chemically bonded hydrocarbon phase. There are several types of bonding techniques that have been employed for the preparation of BPs. These are summarized in Table 6.2. The ester types (*15*) are not popular because of their instability toward hydrolysis by water and reaction with other solvents. Amino phases [as in (2) of Table 6.2] have been prepared and evaluated (*23–25*) but have not found much general application. Preparation of carbon-type packings has been described by Locke *et al.* (*26*). These are prepared using Grignard or organolithium reagents.

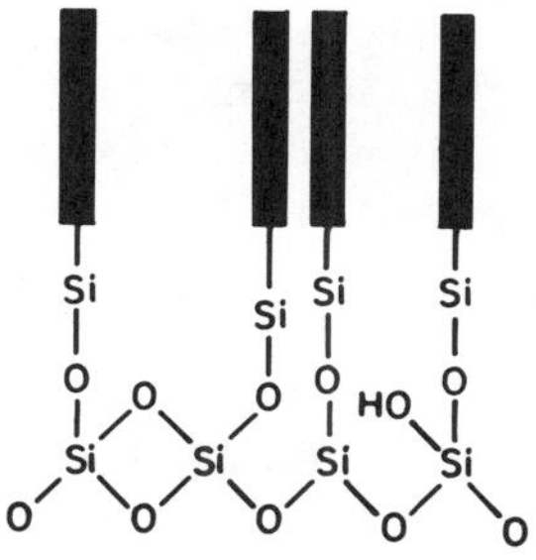

Fig. 6.9 Surface structure of a chemically bonded phase. The solid bar indicates the hydrocarbon chain.

TABLE 6.2

Types of Chemical Bonds in BP Materials[a]

Structure	Type
1. ≡Si—O—R	Ester
2. ≡Si—NR_2	Amino
3. ≡Si—OH → ≡Si—CR_3	Carbon
4. ≡Si—O—Si—CR_3	Siloxane

[a] R, polar or nonpolar.

Siloxane bonded phases are the most popular at present and are available commercially from many suppliers. When bonded in a monomeric fashion to the silica gel support, they form stable, efficient phases with practically all organic solvents used for LC, and over a pH range of about 2–9. The recent literature indicates that most applications in LC are in fact done by reversed-phase chromatography using siloxane BP materials. The reason for this is that these are very versatile, enabling the analyst to analyze very nonpolar compounds (such as polyaromatic hydrocarbons, organochlorine pesticides, or dioxins) to polar compounds (such as carbohydrates, amino acids or nucleosides), using one column and only requiring different mobile phases. It has been estimated that in the next few years 70% or more of LC separations will be carried out on BP materials (*16*). Several important reviews and papers have appeared on the subject of BP chromatography.

The actual separation mechanism in BP chromatography is currently the subject of much discussion. The research up to the present indicates that there are three possible mechanisms. First, the separation can be considered on the basis of partitioning between two liquid phases, the stationary and mobile phases. Another possible mechanism is adsorption onto the nonpolar phase, much like normal adsorption chromatography except that instead of having a liquid–solid interaction with mobile phases and silica gel (for example), the solute–stationary phase interactions would depend only upon the nonspecific Van der Waals dispersion forces, thus making the solute–mobile phase interactions predominant.

This view has found support in the "hydrophobic" effect presented by Horvath and Melander (*17,18*). The separation mechanism is related to the "repulsion" of the solute by the mobile phase, a concept based on the solvophobic theory (*21,22*). Figure 6.10 illustrates the suggested forces involved as a solute molecule is adsorbed onto the stationary-phase surface. The white arrows indicate that the binding of the solute to the hydrocarbon is facilitated by a decrease in the molecular surface area exposed to the solvent. The black arrows indicate forces of attraction between solute and mobile phase. The balance of these effects determines the retention of the solute by the stationary phase. As an example, polar substituents on the solute molecule increase interactions with the mobile phase, thus causing the solute to be eluted more rapidly.

The third type of mechanism has been proposed by Scott and Kucera (*19*) and Knox and Pryde (*20*). In this model [which has been experimentally verified under wetting conditions (*19*)], the organic modifier, e.g., methanol, in a mobile phase consisting of methanol–water, will preferentially be adsorbed onto the bonded phase, probably forming a monomolecular stationary layer different from the mobile phase. Solute molecules eluting with a k' value of 10 or less interact with this methanol layer but do not replace it under wetting conditions. Thus the solute molecules do not come in contact with the BP at all. In this case the BP serves as a support for the adsorbed layer of methanol. Wetting conditions have been established for several organic modifier concentrations, some of which are listed in Table 6.3 for aqueous mixtures of methanol, acetonitrile, and isopropanol (*17*).

These latter two theories might be combined since one deals only with the structure of the stationary phase whereas the second deals with the solute–mobile phase interactions leading to retention by the stationary phase. All of this helps us better understand the processes involved in chromatographic separations employing chemically bonded reversed-phase materials.

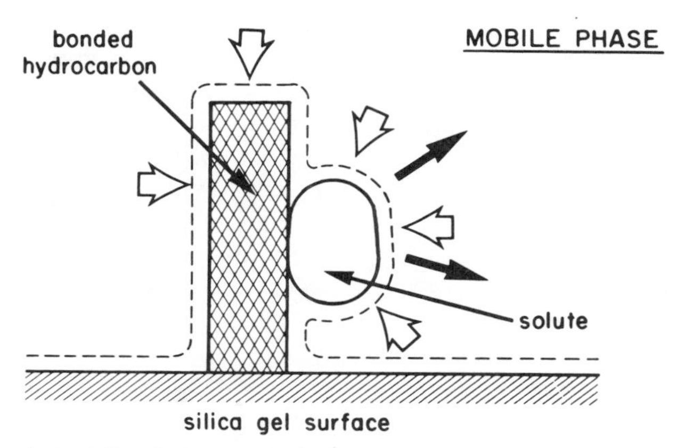

Fig. 6.10 Association between a solute and hydrocarbon bonded phase, which is primarily due to solvent effects in the mobile phase. See text for explanation.

TABLE 6.3

Wetting Characteristics of Some Reversed-Phase Materials[a]

Material	Carbon content (% w/w)	Chain length	Maximum % water for wetting for 3 solvents		
			Methanol	Aceto-nitrile	Isopro-panol
ODS (Whatman)	5.0	18	100	100	100
ODS2 (Whatman)	16.9	18	32	64	74
RP-2 (Merck)	5.0	2	50	76	84
RP-8 (Merck)	12.0	8	50	68	84
RP-18 (Merck)	19.8	18	55	58	82

[a] From Scott and Kucera (*19*), with permission from Elsevier Scientific Pub. Co., Amsterdam.

Solvent Selectivity and Temperature

Solvent selectivity is very important in reversed-phase chromatography, just as it is in adsorption chromatography. Usually, optimum separations of compounds are found empirically by testing various combinations of solvents for the mobile phase. It is important, however, that the analyst have some idea of how a change in mobile-phase modifier can change the selectivity of a chromatographic system. Figure 6.11 compares solvent selectivity versus functional groups attached to a benzene or naphthalene ring (*27*). A LiChrosorb RP-8 column was used with three mobile phases: 50% methanol, 40% acetonitrile, and 37% tetrahydrofuran, each in water. The mobile phase selectivity is given as percent change in k' relative to the 50% methanol–water mobile phase. Benzene eluted with the same k' value in each of the three systems studied; thus the percent change in k' for 40% acetonitrile and 37% tetrahydrofuran is zero. The graphs readily show the effect of the type of mobile-phase modifier on selectivity. For example, phenol elutes with a shorter k' with 40% acetonitrile than with 50% methanol, but with a much longer k' with 37% tetrahydrofuran. In some cases there is a positive change from one solvent to the next, whereas for others the k' change is the opposite, depending upon the functional group and size of the hydrocarbon.

The control of pH in reversed-phase chromatography is also important when the solutes have some acidic or basic character. The pH can be adjusted to eliminate ionization effects that can cause peak broadening. Table 6.4 lists some useful buffer systems as well as acids, bases, and salts that may be employed for ion suppression. Note that no halides are

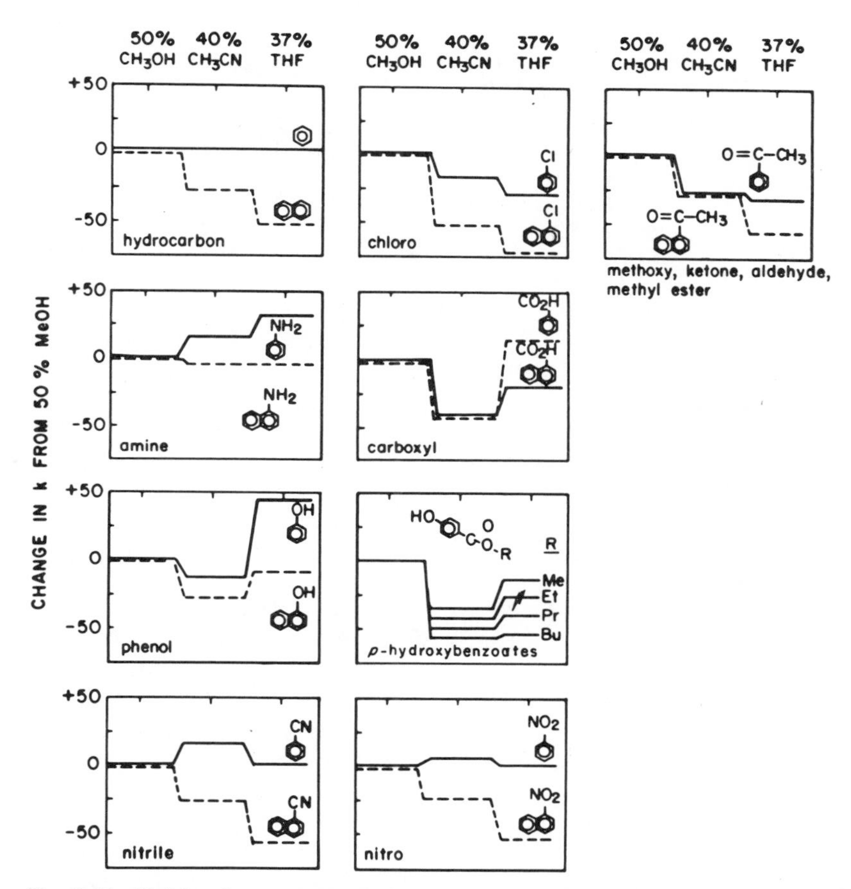

Fig. 6.11 Mobile-phase selectivity for some monosubstituted benzenes, naphthalenes, and *p*-hydroxybenzoates. From Bakalyar *et al.* (*27*), with permission from Elsevier Scientific Publishing Co., Amsterdam.

included; they must be avoided because of corrosion problems with stainless steel parts in the pump and elsewhere in the LC system.

Temperature is another factor that can influence chromatographic separations by altering efficiency and column selectivity (*28–31*). Normally, efficiency improves with temperature because of the increase in rate of mass transfer between mobile and stationary phases. However, in some systems (*29*) efficiency was found to decrease when the column temperature was raised from 25 to 60°C. This effect was experienced with 5-μm RP-8 and RP-18 columns, but not with a 10-μm RP-18 column.

Column selectivity for some compounds has been found to be signifi-

TABLE 6.4

Chemicals Used for pH Control in LC

Buffers	pH range
$H_3PO_4/KH_2PO_4/K_2HPO_4/KOH$	2–12
Acetic acid/sodium acetate	3–6
Acetic acid/NH_4 acetate/NH_4OH	5–9
$NH_4HCO_3/(NH_4)_2CO_3/NH_4OH$	8–10
$NaHCO_3/Na_2CO_3/NaOH$	9–11
$H_3BO_3/Na_3BO_3/NaOH$	7–11
Acids, bases, salts:	
Perchloric acid	
Phosphoric acid	
Acetic acid	
Ammonium acetate	
Ammonium carbonate	
Sodium bicarbonate	
Sodium carbonate	
Sodium hydroxide	

cantly influenced by temperature. Snyder (*28*) refers to "regular" temperature behavior when the retention order of a series of compounds is not altered with changes in temperature of the chromatographic system. Thus, changing temperature is not an effective means of altering selectivity for "regular" solutes. However, in cases when increased temperature significantly alters molecular shape, column selectivity can be altered. The change in shape of the molecules appears to be from a flat linear structure at a low temperature to a more spherical compact structure at a higher temperature. If this is indeed true, then as Snyder suggests, the following temperature-induced structural changes will result in increased retention: (a) increased branching of alkyl substituents, (b) increased cyclization, or (c) decreased overall length. Also, aryl-substituted aromatic rings will be more retained than fused aromatic rings, and out-of-plane substituents in molecules will cause greater retention at higher temperatures.

B. Liquid–Liquid Partition Chromatography

This type of chromatography may be carried out in both a normal and reversed-phase mode. It involves the use of stationary phases prepared by physically coating a liquid onto a support material such as silica gel, in a manner very much like that done in GC. The mobile phases must be immiscible with the stationary phase in order to prevent "bleeding" and to maintain a stable equilibrium so that chromatographic separations are reproducible. To ensure this, a precolumn consisting of a packing material heavily loaded with stationary phase is inserted between the pump and injection port to ensure that the mobile phase is saturated with the stationary-phase liquid, in order to prevent its removal from the analytical column.

Although liquid–liquid partition chromatography found some use as a technique is the early 1970s, there are several reasons why it has largely become replaced with chemically bonded phases. Saturation of the mobile phase with stationary phase is essential for consistent results. If for any reason this changes, stationary phase will be lost from the column, causing altered chromatography characteristics. This may occur by injecting a large volume of sample in a solvent that readily dissolves the stationary phase. The selection of solvents for mobile-phase preparation is very limited because of increased miscibility of the two phases as the solvent modifier concentration is increased. Gradient elution is almost impossible in this type of system because of the ever-changing nature of the mobile phase, either for normal or reversed-phase separations.

IV. MACRORETICULAR RESINS

Recently, Amberlite XAD-2 macroreticular resin has been employed as a stationary phase for LC (*32*). This is a macroporous styrene–divinyl benzene non-ionic copolymer that is capable of adsorbing both neutral and ionic species (*33–35*). Since it is commercially available only in hard white spheres of 20–50 mesh, it is necessary to grind, sieve, and then fractionate by solvent elutriation in order to obtain a small particle size with a range of 3.6–8.4 μm in diameter (*32*). Although not extensively tested, this material is advantageous because of its low cost and its useful pH range (0–14), which is far greater than any chemically bonded phases. The only limitation is that column permeability is low, thus limiting the mobile-phase flow rate.

The principle of chromatographic retention appears to be the same as provided by chemically bonded hydrocarbonaceous phases. It can also be used for ion-pair chromatography in the same manner as the bonded phases (*36*).

It is possible that with further work this type of stationary-phase material will be commercially available in a quality directly suitable for LC applications. Some work (*32*) has indicated that efficiencies can be as good as those obtained with chemically bonded phase packings. Table 6.5 is an eluotropic scale for various solvents on three resins. The solvents are ranked in order of increasing strength (*37*).

TABLE 6.5

Eluotropic Scale for Various Solvents on Amberlite XAD-2 and XAD-4, and Bio-Beads SM-2[a]

Solvent[b]	ϵ_0[c] (XAD-2)	ϵ_0[d] (SVB)
tert-Butanol	−0.072	−0.072
2,2,4-Trimethylpentane	−0.018	−0.022
Isopropanol	−0.016	−0.017
Isobutanol	−0.016	−0.018
1,1,2-Trichloro-1,2,2-trifluoroethane	−0.003	−0.023
Methanol	0	0
sec-Butanol	0.010	0.010
Ethanol	0.010	0.010
n-Propanol	0.015	0.014
n-Butanol	0.020	0.030
Propylene carbonate	0.052	0.049
Isohexanes	0.062	0.055
n-Nonane	0.067	0.068
n-Pentane	0.072	0.067
Acetonitrile	0.072	0.065
Light petroleum	0.075	0.066
n-Heptane	0.078	0.071
Hexanes	0.079	0.074
Isopropyl ether	0.089	0.083
Dimethyl sulfoxide	0.097	0.089
Carbon tetrachloride	0.104	0.079
Acetone	0.105	0.089
Methyl isobutyl ketone	0.122	0.144
1,1,1-Trichloroethane	0.132	0.105
Methylethylketone	0.150	0.127
2-Pentanone	0.150	0.150
Diethyl ether	0.157	0.151
N,*N*-Dimethylformamide	0.167	0.160
1,2-Dichloroethane	Large	0.155
Chloroform	Large	0.185
Methylene chloride	Large	Large
Ethyl acetate	Large	Large
Tetrahydrofuran	Large	Large
1,2-Dimethoxyethane	Large	Large
Benzene	Large	Large
Toluene	Large	Large
o-Xylene	Large	Large

[a] From Robinson *et al.* (*37*), with permission from Elsevier Publishing Co., Amsterdam.

[b] Solvents are ranked in order of increasing strength using the ϵ_0 values in the first column.

[c] An average calculated from the individual values obtained for benz(*a*)anthracene/XAD-2 and benzo(*a*)pyrene/XAD-2 systems.

[d] An average obtained from the four values calculated for the three absorbents SM-2, XAD-2, and XAD-4.

V. ION-PAIR CHROMATOGRAPHY

There has been much controversy surrounding the principles involved in what is referred to as ion-pair chromatography, especially with chemically bonded reversed phases. This type of chromatography is used to separate ionic species by adding a counter-ion to the mobile or stationary phase. This causes a change in the solute–solvent interactions, resulting in a change in retention behavior compared to a similar system which does not contain counter-ions. In reversed-phase chromatography the addition of counter-ions to the system usually increases retention of the ionic species. In normal-phase chromatography the ionic species are eluted faster. The controversy at the moment surrounds the basic processes involved in the retention mechanism when a counter-ion is added. Many factors such as pH, counter-ion strength, and types of mobile and stationary phases all influence the retention. The following summarizes some of the theories put forth to explain experimental results.

A. Ion-Pair Partition Chromatography

Ion-pair partition chromatography makes use of physically coated stationary phases, usually on silica gel type supports. It can be carried out in two ways: the first is by normal-phase chromatography, where the stationary phase consists of an aqueous mixture of counter-ion (e.g., 0.1 *M* $HClO_4$) with an immiscible mobile phase comprising solvent mixtures of, e.g., hexane, methylene chloride, 1-butanol, or similar solvents *(38–40)*. In the reversed-phase mode, the mobile phase is usually an aqueous solution of counter-ion, whereas the stationary phase consists, for example, of hydrophobic amines.

The theory of ion-pair partition between two immiscible liquid phases in LC can be derived from the principles of classical ion-pair extraction. In an ion-pair partition chromatography system where the stationary phase consists of an aqueous solution of a counter-ion and an organic mobile phase, an appropriately ionic solute will be distributed between the two. The separation of different solutes will depend on their respective distribution constants in the system. In developing a suitable theory, some assumptions are necessary. For example, the solute can only exist in the organic phase as an ion-pair with the counter-ion. Thus, if the mobile phase were organic, the solute would migrate down the column as the ion pair. Also, a further generalization is that the solute exists predominantly as the free ion in the aqueous stationary phase.

The strength (or stability) of the ion pair is directly related to its retention in the chromatographic system. For example, if the solute existed only as the ion pair it would be eluted unretained by the aqueous stationary phase. On the other hand, if the ionic solute did not form an ion pair with the counter-ion, it would remain as the free ion in the stationary phase and thus not be eluted from the column. These are the two extremes of ion-pair chromatography. Other factors such as temperature, pH, type and concentration of counter-ion, and nature of the organic solvent all affect the ion-pairing process and thus retention.

If we consider a partition chromatography system consisting of an organic and an aqueous phase containing a counter-ion, B^-, the equilibrium process set up upon the introduction of a solute ion, A^+, into the system may be defined as:

$$AB_{org} \rightleftharpoons A_{aq}{}^+ + B_{aq}{}^- + AB_{aq} \tag{13}$$

where org represents the organic phase and aq, the aqueous phase. If the ionization of AB_{aq} in the aqueous phase is such that it exists only as $A_{aq}{}^+$ and $B_{aq}{}^-$, then the amount of AB_{aq} will be zero and the extraction constant, E_{AB}, for the extraction of AB into the organic phase would be

$$E_{AB} = [AB]_{org}/[A^+]_{aq}[B^-]_{aq} \tag{14}$$

The distribution ratio, D_A, of the solute A between the two phases is defined as

$$D_A = [AB]_{org}/[A^+]_{aq} \tag{15}$$

This is related to the retention of A in the chromatographic system. If we substitute this term into Eq. (14) above, the following is obtained:

$$D_A = E_{AB}[B^-]_{aq} \tag{16}$$

This equation is valid if the stationary phase is organic. If it were aqueous, then the retention of A would be related to D'_A where

$$D'_A = [A^+]_{aq}/[AB]_{org} \tag{17}$$

which is the reciprocal of D_A. Thus retention of A by the aqueous stationary phase increases if D'_A increases. If we consider a system where the mobile phase is organic and the stationary phase aqueous, the capacity factor k'_A (discussed in Section 6.I) for solute A may be defined as

$$k'_A = D'_A\, V_s/V_m \qquad \text{or} \qquad k'_A = (E_{AB}\,[B^-]_{aq})^{-1}\, V_s/V_m \tag{18}$$

where V_s and V_m are the volumes of the stationary and the mobile phases, respectively. If the separation of two different solutes, say A1 and A2, is considered, then the retention of A2 (the second peak) relative to the first is described as

$$d_{A2} = K'_{A2}/k'_{A1} = E_{A1B}/E_{A2B} \tag{19}$$

where d_{A2} is the relative retention of A2. This equation, then, relates the separation of two solutes to the ratio of their extraction constants in the same system.

B. Ion-Pair Chromatography on Chemically Bonded Reversed Phases

Although ion-pair chromatography in liquid–liquid partition systems can be described in terms of classical ion-pair extraction, reversed-phase ion-pair chromatography on chemically bonded stationary phases appears to involve other processes more predominant than the formation of an ion-pair which passes from one phase to another. The major difference between liquid–liquid partition and partition with bonded phases is that the bonded phases are on the average only one molecule thick on the surface of the support, whereas physically coated stationary phases for liquid–liquid partition are many layers thick and thus contain "bulk" liquid. What this means is that the interactions with the bonded phases are most likely to be surface phenomena unlike the true partitioning possibil-

TABLE 6.6

Reversed-Phase Ion-Pair Chromatography Parameters[a]

Parameter	Effect on solute retention
Type of counter-ion	Strong ion-pairing ability causes increased retention.
Hydrophobicity of counter-ion	An increase in hydrophobicity increases quantity of counter-ion adsorbed onto surface, thus increasing solute retention.
Concentration of counter-ion	Higher concentrations cause increased quantity of counter-ion to be adsorbed (to a limit) onto the stationary phase, thus increasing retention.
pH	Retention increases as the concentration of the ionic form of the solute increases, an effect dependent upon the nature of the solute.
Type of organic modifier	Increasing lipophilicity decreases retention.
Concentration of modifier	Increasing concentration decreases retention.
Types of bonded phase	The more hydrophobic, the greater the coverage with counter-ion, which increases retention.
Temperature	Increased temperature decreases retention.

[a] From Gloor and Johnson (*46*), with permission from Preston Publications, Inc.

ity of a solute into the bulk liquid of a physically coated liquid stationary phase.

In recent reports (*19.41–45*) it has been shown that ion-pair chromatography on bonded phases is most likely a form of ion-exchange chromatography. Experimental results have shown that the counter-ion in an aqueous-based mobile phase is adsorbed onto the reversed phase and acts as an ion-exchange agent (*43*). Gloor and Johnson (*46*) have presented many practical aspects of employing reversed-phase ion-pair chromatography for separations and analyses. Table 6.6 lists some of the parameters involved and how they influence retention. The control of pH is very important for reversed-phase ion-pair chromatography. If the pH is such that the solute is ionized in the mobile phase, then retention will occur predominantly via an ion-exchange mechanism. If the solute is chromatographed at a pH where it is completely un-ionized, then retention will be based mainly on its hydrophobic character. Table 6.7 illustrates what effect pH changes will have on the mechanism of retention for a variety of acidic and basic solutes. Table 6.8 lists some of the types of counter-ions useful for reversed-phase ion-pair chromatography.

TABLE 6.7

pH Effects on Retention Mechanism in Reversed-Phase Ion-Pair Chromatography[a]

Solute	pH	Comments
1. Strong acids pKa < 2; e.g., sulfonic acids	2–8	Ionized throughout the pH range. Ion-exchange mechanism for retention.
2. Weak acids pKa > 2; e.g., carboxylic acids, amino acids, phenols	6–8	Solutes ionized. Retention is by ion-exchange mechanism.
	2–5	Ionization is suppressed. Retention is based on hydrophobicity, not on ion exchange.
3. Strong bases pKa > 8; e.g., quaternary amines	2–8	Solutes are ionized throughout the pH range. Retention is by ion-exchange mechanism.
4. Weak bases pKa < 8; e.g., catecholamines	6–8	Ionization is suppressed. Retention based on hydrophobicity, not ion exchange.
	2–5	Solutes are ionized. Retention is by ion-exchange mechanism.

[a] Adapted from Gloor and Johnson (*46*), with permission from Preston Publications, Inc.

TABLE 6.8

Types of Counter-Ions and Applications

Type	Application
Quaternary amines (e.g., tetraalkylammonium ions)	Strong and weak acids
Tertiary amines (e.g., trioctylamine)	Strong acids such as sulfonic acids
Alkyl or Arylsulfonic acids (e.g., heptane or octane sulfonic acids)	Strong and weak bases
Perchloric acid[a]	Strong and weak bases
Alkyl sulfates[a]	Strong and weak bases

[a] Can offer different selectivities from alkyl or aryl sulfonic acids.

VI. ION-EXCHANGE CHROMATOGRAPHY

The separation of solutes by an ion-exchange mechanism is dependent upon differences in electrostatic field strength around the solute ions, as well as the activity coefficient of the ion-exchange material. The most commonly employed materials are resin ion exchangers which for LC have found use as surface coatings on solid glass supports (i.e., pellicular beads $\simeq 40\ \mu m$ diameter). In-depth treatment of ion-exchange theory may be found in the literature (*47–49*). The following is a brief description of the ion-exchange process.

Consider a column packed with ion-exchange material which is presaturated with a weakly bound species, *W*. If an ionic solute *S* is placed on the column it will replace the weakly bound ions. At equilibrium the solute ions will be in a constant state of motion between the ion-exchange material and the mobile phase which contains excess *W* ions. The solute is ionically held by the ion-exchange resin. When an *S* ion leaves the resin to enter the mobile phase, a *W* ion takes its place. While in the mobile phase, *S* migrates down the column where it again comes in contact with the ion-exchange resin to replace another *W* ion. Thus *S* migrates down the column in this manner at a rate dependent upon the equilibrium distribution between mobile phase and the resin. The separation of different solutes is related to the differences in their distribution coefficients. Strongly bound solute ions will migrate more slowly down the column than weakly bound ones, and therefore will be retained longer. The selectivity of ion-exchange chromatography is determined by the type of resin,

TABLE 6.9

Applications of Ion-Exchange Resins

Type	Material	pH Range	Application
1. Strongly acidic cation exchange	Sulfonated polystyrene	1–14	Weak organic cations (e.g., amino acids, peptides)
2. Weakly acidic cation exchange	Carboxylic polymethacrylate	5–14	Weak and strong organic cations (e.g., organic bases, some antibiotics, amino acids)
3. Strongly basic anion exchange	Quaternary ammonium polystyrene	0–12	Weak organic anions, (e.g., alkaloids, fatty acids)
4. Weakly basic anion exchange	Polyamine polystyrene or phenol–formaldehyde	0–9	Strong and weak organic anions (e.g., vitamins, amino acids, sulfonic acids)

or by varying the ion-exchange sites as well as the type and concentration of the *W* ions in the mobile phase. Table 6.9 lists some common ion-exchange materials and their application.

Since acidic ion-exchange materials have a tendency to swell when they are converted to the ionic salt form, care must be taken when preparing and conditioning columns. The reason for the change in particle size is that in the unconditioned form H^+ ions take up a small area of the particle surface compared to, for example, sodium ion, which has a much larger ionic radius than H^+. Thus, when converting to the salt form, the resin is forced to spread to accommodate the layer of sodium ions. The reverse process can also cause problems related to shrinkage and thus voids in the column when returning to the H^+ state.

VII. ION CHROMATOGRAPHY

This type of chromatography is a form of ion-exchange chromatography devised especially for making conductometric measurements on separated species (*50–53*). In normal ion-exchange chromatography the eluted solute ions are always present in a solution containing a large excess of the eluting electrolyte, which is the mobile phase. Although such mobile phases may not interfere in UV absorption or fluorescence detection of the solutes, they make conductometric detection virtually impossible when trace quantities of solutes need to be determined.

Small *et al.* (*50*) devised a scheme to eliminate the background electrolyte from the mobile phase, permitting the conductometric detection of the solutes in distilled water. An example of this approach is shown in Fig. 6.12. Consider the separation of some cationic species such as Li^+, Na^+, and K^+. The mobile phase containing dilute HNO_3 is pumped through column 1 and 2 in series. Column 1 contains a cation-exchange resin whereas column 2 (the "stripper" column) is packed with a strong base resin in OH^- form. After the sample is injected onto the first column, the cations will be resolved from one another, eluting from column 1 at different times in a background of HNO_3. Upon entering the stripper column (2), two important reactions occur. First, the HNO_3 is removed from the mobile phase.

$$H^+NO_3^- + Resin^+OH^- \longrightarrow Resin^+NO_3^- + H_2O$$

Then the metal ions, as their nitrate salts, are converted to their respective bases.

$$M^+NO_3^- + Resin^+OH^- \longrightarrow Resin^+NO_3^- + M^+OH^-$$

The bases pass through the stripper column unretained. They elute from the stripper column in a background of deionized water, readily detected in the conductivity detector. An analogous system would be used for anions where the resolving column would be an anion-exchange column and the stripper column would be a strong acid cation-exchanger in the H^+ form to remove background electrolytes such as dilute NaOH.

One important aspect of this approach that differs from other chromato-

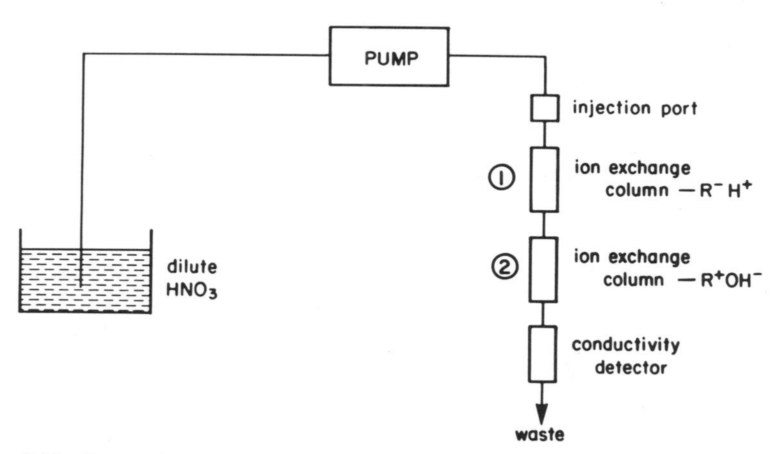

Fig. 6.12 Ion-exchange system set up for ion chromatography with conductometric detection.

graphic techniques is the necessity of regeneration of the stripper column. Factors affecting this are described in detail elsewhere (*50*) and include the size of the stripping bed and its specific capacity. These are limited by the fact that efficiency must be maintained in order to keep resolution optimized.

This technique is very new and has much potential as a trace analytical technique employing electrochemical means of detection. It has been applied to organic ammonium compounds and organic acids (*50*).

VIII. GEL PERMEATION CHROMATOGRAPHY

Gel permeation chromatography (GPC) is a type of LC that makes use of certain gels that provide separations of solutes primarily on the basis of molecular size. Although the chromatography principles involved are complex, a simple model may be described. The gel (polystyrene, cross-linked dextran, agarose, porous silica, etc.) is composed of a porous matrix whose pores are closely controlled in size. The separation mechanism involves the differences in ability of molecules to penetrate the pores. Those molecules that are too large cannot permeate into the gel pores and thus migrate down the column through the interstitial volume between particles (i.e., the void volume of the column). The sample in this case would be unretained. Molecules that are somewhat smaller may be able to penetrate into the gel pores to some degree; thus they will be retarded in their rate of migration through the column. Very small molecules that are able to penetrate completely into the pores of the gel will be retained the most, and therefore elute from the column last.

Gel-type packing materials may be considered to offer solutes a selection of path lengths based on molecular size. In other words, large molecules have short path lengths since they are restricted from entering the pathways available in the small pores. Small molecules that are able to pass into the pores therefore have a longer average path length and will be eluted later. Molecular shape is also a criterion for separation, spherical molecules being retained longer than rod-shaped solutes.

From this then, GPC separates solutes according to molecular size, with the largest-molecular-weight compounds eluting first. Many gels are available with pore sizes that permit separations of solutes with molecular weights ranging from 100 up to about 1,000,000. Most applications of GPC have been with aqueous-based mobile phases, although some work has been carried out with organic solvents (*54,55*). Other factors that may be involved in solute retention are adsorption effects and ion exchange. Also, some gels possess the ability to retain aromatic solutes more than aliphatic ones even though they may have similar molecular weights. Theoretical treatments of the factors involved in GPC, including various proposed models, have been put forth *(56–61)*.

Applications of GPC have been directed toward the separation of macromolecules, usually for preparative isolation or as a cleanup technique in trace analysis.

IX. LIGAND-EXCHANGE CHROMATOGRAPHY

Ligand-exchange chromatography may be considered as a special type of ion-exchange chromatography. It involves the separation of organic solutes by their ability to act as ligands toward certain metal ions. The metal ions are usually immobilized, either by impregnating the metal as a salt into a support such as silica gel or by binding to a strong ion-exchange resin such as Chelex 100, which has iminodiacetate groups as active sites. These sites strongly bind to transition metals while at the same time permitting them to form coordination complexes with other ligands. The useful characteristic of ligand-exchange chromatography is that selectivity of the system can be varied by appropriate choice of the support material, metal, and mobile phase.

Separations that are difficult to achieve by adsorption or partition chromatography may be easily achieved using ligand-exchange. The separation of polynuclear aromatic hydrocarbons using a silver-impregnated solid support is an example of this *(62,63)*. Figure 6.13 shows the separation of five similar polyaromatics. Four of the five compounds differ only in the position of the nitrogen, which directly affects their ability to form silver complexes. Applications of ligand-exchange chromatography have

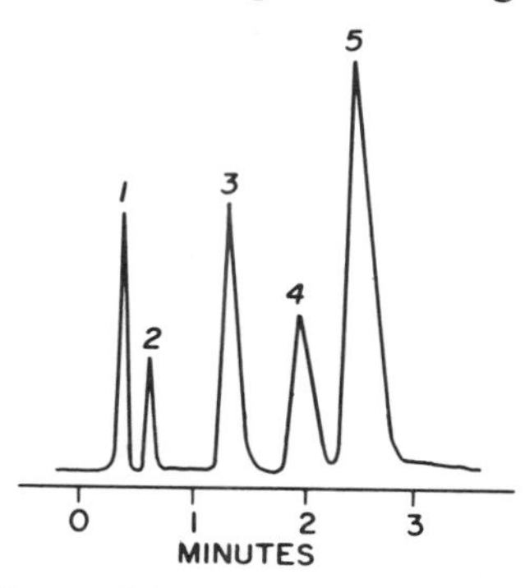

Fig. 6.13 Separation of benzo (h) quinoline (1), phenazine (2), acridine (3), benzo (c) quinoline (4), and benzo (f) quinoline (5). The column consisted of 0.3% silver oxide coated on Zipax pellicular packing. Mobile phase was 1.0% acetonitrile in *n*-hexane. From Vivilecchia *et al.* (62), with permission from Preston Publications, Inc.

been reviewed (*64*) and include compounds such as amines and polyamines, polyhydric alcohols, amino acids and peptides, nucleosides and nucleic acid bases, polyhydroxy compounds, and sugars.

The use of reversed-phase chromatography for the LC separation of a large number of compounds as silver complexes has been reported (*65–67*). The Ag^+ ion is added to the mobile phase and the neutral complex retained by the CH_2-bonded reversed-phase packing. Many different organics have been separated, including unsaturated aliphatics, sesquiterpenes, fatty acids, triglycerides, and many polynuclear aromatics. The addition of the Ag^+ tends to reduce k' values selectively by causing additional molecular interactions in the mobile phase.

X. CHARGE-TRANSFER CHROMATOGRAPHY

This type of chromatography involves complex formation between "electron-rich" donor molecules and "electron-deficient" acceptor molecules. Such complexation normally involves Π-bonding systems. Electronegative or electron-withdrawing substituents tend to produce strong acceptor molecules. The electron donors usually have sterically unhindered Π systems. When donor and acceptor molecules are appropriately placed together, a charge-transfer complex is formed. The strength and stability of the bond is the basis for chromatographic separation. This means that either the donor or acceptor must be immobilized on a support. For example, benzoquinone, an acceptor, has been used as a stationary phase for the thin-layer chromatographic (TLC) separation of some aromatic hydrocarbons (*68*). Although the technique has some potential for selective separations, only limited work has been reported to date by TLC (*69–71*), and none employing LC.

REFERENCES

1. L. R. Snyder, "Principles of Adsorption Chromatography." Dekker, New York, 1968.
2. L. R. Snyder and J. J. Kirkland, "Introduction to Modern Liquid Chromatography." Wiley (Interscience), New York, 1974.
3. R. P. W. Scott, "Contemporary Liquid Chromatography." Wiley, New York, 1976.
4. J. C. Giddings, "Dynamics of Chromatography, Part 1, Principles and Theory." Dekker, New York, 1965.
5. A. J. P. Martin and R. L. M. Synge, *Biochem. J.* **35,** 91 (1941).
6. J. H. Purnell, *J. Chem. Soc.* p. 1268 (1960).
7. R. P. W. Scott and P. Kucera, *J. Chromatogr. Sci.* **12,** 473 (1974).
8. E. Soczewinski, *Anal. Chem.* **41,** 179 (1969).
9. R. P. W. Scott and P. Kucera, *Anal. Chem.* **45,** 749 (1973).

10. R. P. W. Scott and P. Kucera, *J. Chromatogr.* **149,** 93 (1978).
11. L. R. Snyder, *Anal. Chem.* **46,** 1384 (1974).
12. E. Soczewinski, *J. Chromatogr.* **130,** 23 (1977).
13. D. L. Saunders, *Anal. Chem.* **46,** 470 (1974).
14. G. A. Howard and A. J. P. Martin, *Biochem. J.* **46,** 532 (1950).
15. I. Halasz and I. Sebestian, *Angew. Chem.* **8,** 453 (1969).
16. H. Colin and G. Guiochon, *J. Chromatogr.* **141,** 289 (1977).
17. C. Horvath and W. Melander, *J. Chromatogr. Sci.* **15,** 393 (1977).
18. C. Horvath and W. Melander, *Am. Lab.* **10,** 11 (1978).
19. R. P. W. Scott and P. Kucera, *J. Chromatogr.* **142,** 213 (1977).
20. J. H. Knox and A. Pryde, *J. Chromatogr.* **112,** 171 (1975).
21. O. Sinanogler and S. Abdulnur, *Fed. Proc., Fed. Am. Soc. Exp. Biol.* **24,** 12 (1965).
22. T. Haliciogler and O. Sinanogler, *Ann. N.Y. Acad. Sci.* **158,** 308 (1969).
23. H. Colin and G. Guiochon, *J. Chromatogr.* **137,** 19 (1977).
24. A. Nakae and G. Muto, *J. Chromatogr.* **120,** 47 (1976).
25. K. Karch, I. Sebestian, and I. Halasz, *J. Chromatogr.* **122,** 3 (1976).
26. D. C. Locke, J. T. Schmermund, and B. Banner, *Anal. Chem.* **44,** 90 (1972).
27. S. R. Bakalyar, R. McIlwrick, and E. Roggendorf, *J. Chromatogr.* **142,** 353 (1977).
28. L. R. Snyder, *J. Chromatogr.* **179,** 167 (1979).
29. W. A. Saner, J. R. Jadamec, and R. W. Sager, *Anal. Chem.* **50,** 749 (1978).
30. J. Chmielowiec and H. Sawatzky, *J. Chromatogr. Sci.* **17,** 245 (1979).
31. R. J. Perchalski and B. J. Wilder, *Anal. Chem.* **50,** 774 (1978).
32. R. G. Baum, R. Saetre, and F. F. Cantwell, *Anal. Chem.* **52,** 15 (1980).
33. F. F. Cantwell, *Anal. Chem.* **48,** 1854 (1976).
34. A. Y. Mohammed and F. F. Cantwell, *Anal. Chem.* **50,** 491 (1978).
35. D. J. Pietrzyk, E. P. Kroeff, and T. D. Rotsch, *Anal. Chem.* **50,** 497 (1978).
36. T. Uematsu and R. J. Sukadolnik, *J. Chromatogr.* **123,** 347 (1976).
37. J. L. Robinson, W. J. Robinson, M. A. Marshall, A. D. Barnes, K. J. Johnson, and D. S. Salas, *J. Chromatogr.* **189,** 145 (1980).
38. J. Crommen, B. Fransson, and G. Schill, *J. Chromatogr.* **142,** 283 (1977).
39. B. A. Persson and B. L. Karger, *J. Chromatogr. Sci.* **12,** 521 (1974).
40. B. A. Persson and P.-O. Lagerström, *J. Chromatogr.* **122,** 305 (1976).
41. J. C. Kraak and J. F. K. Huber, *J. Chromatogr.* **102,** 333 (1974).
42. C. P. Terweij-Groen and J. C. Kraak, *J. Chromatogr.* **138,** 245 (1977).
43. R. P. W. Scott and P. Kucera, *J. Chromatogr.* **175,** 51 (1979).
44. J. C. Kraak, K. M. Jonker and J. F. K. Huber, *J. Chromatogr.* **142,** 671 (1977).
45. P. T. Kissinger, *Anal. Chem.* **49,** 883 (1977).
46. R. Gloor and E. L. Johnson, *J. Chromatogr. Sci.* **15,** 413 (1977).
47. H. F. Walton, *Anal. Chem.* **40** 51R (1968).
48. C. D. Scott, *Science* **186,** 226 (1974).
49. J. Inczedy, *Analytical Applications of Ion Exchangers,* Permagon Press, New York (1971).
50. H. Small, T. S. Stevens and W. C. Bauman, *Anal. Chem.* **47,** 1801 (1975).
51. C. J. Hill and R. P. Lash, *Anal. Chem.* **52,** 24 (1980).
52. R. P. Lash and C. J. Hill, *Anal. Chim. Acta.* **108,** 405 (1979).
53. T. S. Stevens, V. T. Turkelson and W. R. Abe, *Anal. Chem.* **49,** 1176 (1977).
54. W. Halter, *Nature* (*London*) **206,** 693 (1965).
55. W. Setermann, G. Lueben, and T. Wieland, *Makromol. Chem.* **73,** 168 (1964).
56. D. L. Bly, *Science* **168,** 527 (1970).
57. G. K. Ackers, *Biochemistry* **3,** 723 (1964).

58. A. J. DeVries, M. LePage, R. Bean, and C. I. Guillemin, *Anal. Chem.* **39,** 935 (1967).
59. M. E. van Kreveld and N. van Der Hoed, *J. Chromatogr.* **83,** 111 (1973).
60. J. C. Giddings and K. L. Malik, *Anal. Chem.* **38,** 997 (1966).
61. J. C. Giddings, *Anal. Chem.* **40,** 2143 (1968).
62. R. Vivilecchia, M. Tiébaud, and R. W. Frei, *J. Chromatogr. Sci.* **10,** 411 (1972).
63. B. DeVries, *J. Am. Chem. Soc.* **40,** 184 (1963).
64. J. D. Navratil and H. F. Walton, *Am. Lab.* **8,** 69 (1976).
65. B. Vonach and G. Schomburg, *J. Chromatogr.* **149,** 417 (1978).
66. G. Schomburg and K. Zegarski, *J. Chromatogr.* **114,** 174 (1975).
67. R. J. Tscherne and G. Capitano, *J. Chromatogr.* **136,** 337 (1977).
68. G. H. Shenk, G. L. Sullivan, and P. A. Fryer, *J. Chromatogr.* **89,** 49 (1974).
69. A. Berg and J. Lam, *J. Chromatogr.* **16,** 157 (1964).
70. J. P. Sharma and S. Ahuja, *Fresenius' Z. Anal. Chem.* **367,** 368 (1973).
71. G. H. Shenk, ''Organic Functional Group Analysis.'' Pergamon, Oxford, 1968.

Chapter 7
Chemical Derivatization

I. INTRODUCTION

Most LC analyses at present are carried out by direct means; that is, the analyses are performed on the intact molecule without alteration. This is of course the simplest approach. However, in many cases direct analyses are difficult and often not possible at the desired level of sensitivity. It is here that recourse is often made to chemical modification of the analyte.

A. GC versus LC Derivatization

In GC, derivatization has long been used both to aid chromatography and to increase sensitivity (*1–5*). However, since most compounds can be successfully chromatographed by the various modes of LC (e.g., adsorption, partition, ion-exchange, ion-pair), derivatization for this approach is normally used to increase detectability. The LC detectors commonly employed (absorbance, fluorescence, electrochemical) dictate the types of derivatives that are useful. Most chemical derivatization techniques for LC have involved absorbance or fluorescence detection (*6–8*), although the potential of forming electrochemically active derivatives has been pointed out (*9*).

If derivatization is necessary, the analyst must decide whether to use a derivative suitable for GC or one for LC. This depends, of course, on the nature of the analyte, the reagents available, and the type of equipment used for the analysis. Since GC with EC detection is available in most residue analysis laboratories, a stable halogenated or other strongly electron-capturing derivative might be preferred. There are two reasons why this would be so. First, there are many reagents available that have been extensively used for the formation of GC–EC derivatives. Thus the methodology in most cases has been well worked out. Second, picogram quantities of derivative can be detected by GC–EC. This enables the analyst to inject 100- to 1000-fold less quantities of sample material into the chromatographic system compared to LC–UV, for example.

However, there can be instances where GC–EC is unsuitable for an analysis even when extremely sensitive derivatives are prepared. It is possible that with careful selection, a strongly UV-absorbing derivative can be effectively used which, because of the position of its absorbance maximum, enables the detection of low levels of analyte in some samples where GC–EC analysis would not.

Most reagents used for chemical derivatization in LC are derived from classic spectrophotometric methods. The application of this technique for LC up to now has been limited but continues to grow at a steady pace. Two books have appeared that deal with derivatization in LC (*2,10*).

Derivatization for UV or fluorescence detection involves the attachment of a strongly absorbing or fluorescent chromophore to the analyte by a suitable reaction. Table 7.1 lists some common chromophoric groups for both UV absorbance and fluorescence detection. Fluorescent reagents such as fluorescamine and *o*-phthaladehyde, which are weakly or nonfluorescent, change their structures upon reaction with primary amines to yield fluorescent products. The Table also includes some references of work done with the reagents.

Although sensitivity is of prime importance when employing chemical derivatization for LC, selectivity must also be considered. This can be achieved through the reaction itself. Some reactions do not offer much selectivity, such as the use of heat treatment (*42,43*) or ammonium bicarbonate (*44*) for the development of TLC plates. At the other extreme are

TABLE 7.1A

Some Chromophoric Substituents for UV Derivatization

Chromophore	λ_{max}	λ_{254}	Reference
2,4-Dinitrophenyl	—	$>10^4$	*11–13*
Benzyl	254	200	*14*
p-Nitrobenzyl	265	6200	*15,16*
3,5-dinitrobenzyl	—	$>10^4$	*11*
Benzoyl	230	low	*17*
p-Nitrobenzoyl	254	$>10^4$	*17*
Toluoyl	236	5400	*18*
p-Chlorobenzoyl	236	6300	*18*
Anisoyl	262	16000	*19*
Phenacyl	250	$\sim 10^4$	*20*
p-bromophenacyl	260	18000	*21,22*
2-Naphthacyl	248	12000	*23,24*

TABLE 7.1B

Some Chromophoric Substituents for Fluorescence Derivatization[a]

Reagent	λ_{max} ex	λ_{max} em	Reference
Dansyl chloride	365	500	*8,10,25–28*
Fluorescamine	390	475	*29–31*
Br-Mmc	328	375	*32–34*
o-Phthalaldehyde	340	>418	*35*
	229	>470	*36*
NBD	480	530	*37–39*
Dansyl hydrazine	365	500	*40,41*

[a] Excitation and emission wavelength will vary significantly, depending on the solvent used.

reagents such as fluorescamine or *o*-phthalaldehyde, which react specifically with primary amines. Selectivity is also achieved through the chemical, electrochemical, physical, or optical properties of the product. As the uniqueness of the product increases in the terms described above, detection of the product in complex samples becomes easier.

B. Chemical Confirmatory Tests

Confirmatory tests have great importance in all chromatographic techniques. Often, to confirm the identity of a peak, a sample is analyzed with a different chromatographic system. For example, one may change the column in GC, or the mobile phase (and possibly the column as well) in LC, to obtain a different retention value. However, to many analysts this may not be completely satisfactory as a confirmation. In GC much work has appeared on the use of chemical derivatives as a means of confirming pesticide residue identity (*45,46*). Such reactions are applicable to a host of other compounds that are not pesticides. If a response (peak) is assumed to be due to a certain compound, then it should provide a known product when carried through an appropriate chemical reaction. The product can then be chromatographed and compared to the product resulting from an authentic standard of the compound. In this case the confirmation is superior since the selectivity of the reaction and the chromatographic properties of the derivative can be used in making a judgment.

For confirmatory tests, there is no requirement for increased sensitivity of the derivative, as long as it is similar to the parent. Such tests are usually qualitative or semiquantitative and are considered positive when

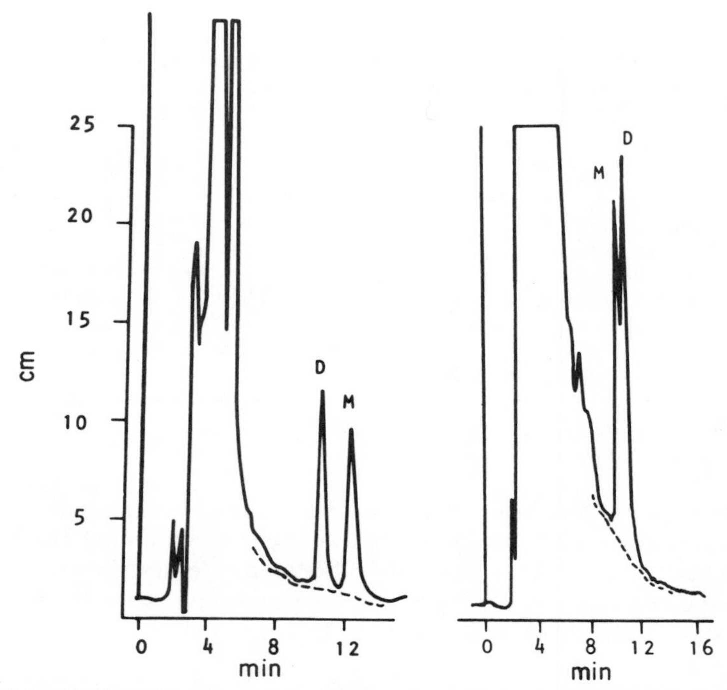

Fig. 7.1 Confirmation of monuron (M) and diuron (D) in corn at 0.1 ppm. Left chromatogram represents the direct LC analysis on a silica column with 20% 2-propanol/isooctane as the mobile phase. Right chromatogram is after methylation with methyl iodide/sodium hydride, obtained on the same column with 10% 2-propanol/isooctane as the mobile phase.

the product of the unknown compound elutes with the same retention value as that of a standard carried through the reaction.

Chemical derivatization for confirmation has not been widely used in LC. However, Fig. 7.1 illustrates an example of the confirmation of two urea herbicides via a methylation technique (*47*). The addition of a methyl group to these compounds serves only to make them less polar, causing a substantial shift in retention, which is required for the confirmation.

C. Pre- and Postchromatographic Reactions

In GC, chemical derivatization reactions are normally performed prior to chromatographic analysis. This is referred to as prechromatographic derivatization. With some specialized reagents, on-column reactions are possible. These reactions [such as methylation (*48*)] take place in the hot injection port or in the GC column during chromatography. The approach is very simple since all that is required is that a mixture of reagent and analyte be injected into the GC. These on-column reactions must use specialized reagents that cause few interferences in the detection. This is important since there is no means of removing excess reagent or side products once injection is made.

Chemical derivatization in LC can be carried out either before chromatography as in GC or after chromatographic separation through post-column reactions. The advantages of prechromatographic derivatization in LC are (1) there is no time limitation on the reaction; (2) vigorous reagents may be used; and (3) additional cleanup may be carried out on the derivative if necessary or desired. Because reaction times and conditions can be greatly varied, many reagents become useful for derivative formation, providing the analyst with a wide selection from which to choose.

This is not the case with postcolumn reactions where the derivatives are formed in a dynamic system. Normally the LC effluent is mixed with the reagents in a mixing (or reaction) coil, which may be heated to an elevated temperature. The reaction time is governed by the flow rate of the stream and reaction coil length. When the reaction is complete the stream enters the detector for quantitation of the product.

There are several factors that must be considered when developing a postcolumn reaction technique. First, the reagents themselves must not interfere in the detection of the product. The reaction time should be as short as possible and the reagents should offer some selectivity. Reaction time can be increased if bubble (*49*) or solvent segmentation (*50*) is used in the post-column system. Such segmented streams minimize band broadening. The solvent-segmentation technique may also be used to extract the product dynamically from the reaction mixture, permitting a further degree of selectivity. This also offers the possibility of using reagents that could not have been used previously because of interference with the detection of the product. Those reagents that can be eliminated by the extraction process may be used with such an approach. This technique is rather new and has been used for the postcolumn ion-pair extraction of some basic drugs and pesticides separated by reversed-phase (*50*) and normal-phase (*51*) chromatography. In these cases the excess counter-ion is removed through the partitioning process, permitting only the ion pair to be detected in the organic phase.

The most important difference between pre- and postchromatographic reactions is that in the former case chromatographic separations are based on the differences in the nature of the products, not of the parent compounds, as in postchromatographic derivatization. It should be noted that when derivatives are made of two similar compounds, the differences between them diminishes, especially when bulky chromophoric groups are attached. This may increase the difficulty in analyzing the two in the presence of one another. In addition, unlike prechromatographic derivatization, there is no need to form a single, stable, or well-defined product in a postchromatographic reaction. Techniques based on ion-pairing, complexation, or ligand-exchange reactions may be effectively used in such detection systems. The major criterion is that the postchromatographic system operate reproducibly. Discussions on postchromatographic detection systems for LC, including reaction coil geometry, theoretical treatment, and practical applications have appeared in the literature (*52,53*).

Before chemical derivatization is attempted, one should try to have a good understanding of the reaction mechanism and the factors affecting the ultimate yield. Many reactions that create suitable derivatives with pure standards may be unacceptable when applied to complex samples such as soil or food extracts. As mentioned earlier, this is where reaction selectivity becomes important. It is not necessary that the parent compound be converted 100% to the product. Normally, an excess of reagent is employed to help ensure high yields. However, caution must be exercised, since too much reagent might interfere with detection of the product. The main requirement is that the yield be obtained reproducibly in the presence of sample material. This is especially important in prechromatographic reactions where no prior chromatographic separation has been employed. The sample coextractives can dramatically influence the reaction by reacting with the reagent at a faster rate than the analyte or by altering the reaction conditions (e.g., pH, polarity). Often it is necessary to increase the molar excess of the reagent in a sample compared to a standard reaction in order to maintain reproducible yields of product.

Other considerations related to derivatization are the cost and stability of the reagents. In addition, the reactions should not require specialized equipment unless it is to be done on a routine basis. The corrosive nature and toxicological properties of the reagents, should be known, but when this is not so the analyst must respect the chemicals employed and handle them with caution. It should also be pointed out that even if the chemical and toxicological properties of the reagent are known, those of the product usually are not. In light of this, it is only good sense to treat the product with more respect than the reagent, especially if reactions are carried out on a milligram scale for initial characterization of the products.

The following sections present many reactions available for the formation of derivatives of many types of compounds. It is organized on the basis of reagents, since one reagent may be useful for a diverse range of compounds. Dansyl chloride, for example, has been used for the determinations of biogenic amines, barbiturates, and carbamate insecticides, to

TABLE 7.2

Prechromatographic Derivatization Reactions

Compound	Reagent	References
1. Amines		
Amines	NBD-chloride	*193*
	p-Nitrobenzoyl chloride	*63*
	4-Methoxybenzoyl chloride	*18*
	2,4-Dinitrofluorobenzene	*65–67*
	Acridine isothiocyanate	*227*
	Fluorescein isothiocyanate	*224–226*
	2-(4-Isocyanatophenyl)-6-methylbenzthiazole	*223*
	1,2-Naphthoquinone-4-sulfonate	*137*
	Bansyl chloride	*191*
	Mansyl chloride	*192*
Amino acids	NBD-chloride	*194,195*
	Dansyl chloride	*25,162–164,246, 247,267*
	Phenylisothiocyanate	*72–77,240,245*
	Methylisothiocyanate	*71*
	Fluorescamine	*31,205*
	Disyl chloride	*211*
	o-Phthalaldehyde	*35*
	Fluorescein isothiocyanate	*224–226*
	Pyridoxal	*230,231*
Amino acid enantiomers	*d*-10-Camphorsulfonylchloride/*p*-nitrobenzylbromide and others	*134,135*
Biogenic amines	Dansyl chloride	*26,161,166,265*
	o-Phthalaldehyde	*214*
	Acetic anhydride	*96,97*
	Quinoline-8-sulfonyl chloride	*100*
	p-Toluene-sulfonyl chloride	*101*
	NBD-chloride	*37,196*

(*continued*)

TABLE 7.2 (*Cont'd*)

Compound	Reagent	References
Polyamines	Dansyl chloride	*249–253*
	Fluorescamine	*29,203*
Polyfunctional amines	3,5-Dinitrobenzoyl chloride	*58*
	m-Toluoyl-chloride	*64*
Nitrosamines	NBD-chloride	*202*
	2,4-Dinitrofluorobenzene	*65–67*
	Dansyl chloride	*254*
Amino sugars	Disyl chloride	*212*
Cyclic AMP	Dansyl chloride	*255*
Amphetamines	*p*-Nitrobenzoyl chloride	*62*
	NBD-chloride	*38,197,198*
Catecholamines	*o*-Phthalaldehyde	*215*
	Fluorescamine	*30,204*
	Dansyl chloride	*248,256–258*
Histamines	*o*-Phthalaldehyde	*216*
Alkaloids	NBD-chloride	*200*
	Dansyl chloride	*28,170–173*
Vitamin B_6	Disyl chloride	*213*
Tocainide	Dansyl chloride	*259*
Imidazole	*p*-Nitrobenzoyl chloride	*57*
Amine-*N*-oxides	Dansyl chloride	*190*
Chlorozoxazone	Fluorescamine	*206*
Antithyroids	NBD-chloride	*39*
2. Carboxylic Acids		
Carboxylic acids	*o,p*-Nitrobenzyl-*N,N'*-diisopropyl-isourea	*15*
	2-Naphthacyl bromide	*24*
	Phenacyl bromide	*20,81*
	p-Nitrophenacyl bromide	*82,84*
	p-Bromophenacyl bromide	*85,88*

	p-Methoxyphenacyl bromide	*15*
	Br-Mmc	*32–34*
Fatty acids	Phenacyl bromide	*260*
	1-(*p*-Nitro)benzyl-3-*p*-tolyltriazene	*14,92,93*
	Triphenylphosphine/*p*-methoxyaniline	*95*
	Phenacyl bromide	*20*
	p-Nitrophenacyl bromide	*82*
	p-Methoxyphenacyl bromide	*15*
Prostaglandins	*p*-Nitrophenacyl bromide	*83,84*
	Diazomethane/pentafluorobenzylhydroxylamine hydrochloride	*114*
	Diazomethane/*p*-nitrobenzylhydroxylamine hydrochloride	*114*
Gibberellins	*o,p*-Nitrobenzyl-*N,N'*-diisopropyl-isourea	*15*
	p-Bromophenacyl bromide	*88*
Penicillins	*p*-Bromophenacyl bromide	*87*
Bile acids	Diazomethane	*145*
2-Naphthoxyacetic acid	Diazomethane	*261*
Dipropylacetic acid	Phenacyl bromide	*81*
3. Alcohols, phenols		
Carbohydrates	*p*-Nitrobenzoyl chloride	*56*
Hydroxybiphenyls	Dansyl chloride	*176*
Saccharides	Acetic anhydride	*98*
Chlorophenols	Dansyl chloride	*177*
Polyalcohols	3,5-Dinitrobenzoyl chloride	*58*
Digitalis glycosides	*p*-Nitrobenzoyl chloride	*55,56*
Glycosphingolipids	*p*-Nitrobenzoyl chloride	*60*
Alcohols	Phenylisothiocyanate	*79*
	EDTN	*229*
2-Naphthol	Diazomethane	*261*
Sapogenins	*p*-Nitrobenzoyl chloride	*61*
Hydroxy steroids	*p*-Nitrobenzoyl chloride	*19*
Metformin	*p*-Nitrobenzoyl chloride	*11*

(*continued*)

TABLE 7.2 (*Cont'd*)

Compound	Reagent	References
Hexachlorophene	*p*-Nitrobenzoyl chloride	*19*
Cannabinoids	Dansyl chloride	*175,262,263*
	Disyl chloride	*210*
4. Steroids		
Corticosteroids	EDTN	*229*
Estrogens	Dansyl chloride	*27,168,264,266*
	Azobenzene-4-sulfonyl chloride	*99*
Hydroxy steroids	*p*-Nitrobenzoyl chloride	*19*
Ketosteroids	2,4-Dinitrophenylhydrazine	*104,105,108*–111
	Dansyl hydrazine	*219*
Cortisol	Dansyl hydrazine	*335*
5. Aldehydes and ketones		
Aldehydes	Dansyl hydrazine	*40,41,219,220*
	2-Diphenacetyl-1,3-indandione-1-hydrazone	*221,222*
	p-Nitrobenzylhydroxylamine hydrochloride	*13,92,113*
	2,4-Dinitrophenylhydrazine	*16,104,108–111, 268*
Ketones	2,4-Dinitrophenylhydrazine	*104*
	p-Nitrobenzylhydroxylamine hydrochloride	*13,92*
	Dansyl hydrazine	*40,41,219,220*
	2-Diphenacetyl-1,3-indandione-1-hydrazone	*221,222*
Ketoacids	2,4-Dinitrophenylhydrazine	*107–109*
6. Others		
Metal chelates	Acetylacetonate complexes	*115–118*
	Dithizonates	*117–119*
	β-Ketoamine chelates	*121,122*
	Schiff-base chelates	*120*

	Thiosemicarbazones	*123,124*
	Dithiocarbamate complexes	*119,124,126–129*
	Salicylaldimine chelates	*122*
	Dialkylbis(thiobenzyl)hydrazone chelates	*124,125*
Ureas	Methyl iodide/sodium hydride	*140,141*
	Dansyl chloride	*182,183*
Ethyleneimine	1,2-Naphthoquinone-4-sulfonate	*137*
Serotonin	Dansyl chloride	*169*
Dianhydrogalactitol	Sodium diethyldithiocarbamate	*136*
Carbamates	Dansyl chloride	*178–181*
	NBD-chloride	*189,193*
Thromboxanes	*p*-Nitrobenzylhydroxylamine hydrochloride	*112*
Organophosphates	Dansyl chloride	*184*
Ecdysones	Phenanthrene boronic acid	*232*
Triazines	Dansyl chloride	*185*
Reserpine	Vanadium pentoxide/phosphoric acid	*234,235*
Thiamine	Potassium ferricyanide	*238*
Morphine	Potassium ferricyanide	*239*
Adenine, adenosine	Chloroacetaldehyde	*240*
Barbiturates	2-Naphthacyl bromide	*23*
	Dansyl chloride	*174*
Thiolcarbamates	NBD-chloride	*199*
Thiols	NBD-chloride	*201*
Aflatoxins	Trifluoroacetic acid	*241–243*

name but a few. However, for easy reference, Table 7.2 lists a large number of compounds with various reagents that have been employed for their analysis.

II. UV-ABSORBANCE DERIVATIZATION

A. Introduction

The most common LC detector in use for organic trace analysis is the UV-absorbance detector. Thus, most attempts at forming LC derivatives have been aimed at those that are strong absorbers of UV light. Table 7.1 lists a number of chromophoric substituents, most of which have high molar absorptivities and have found use in LC derivatization reactions.

It should be pointed out that although fluorescence detection inherently

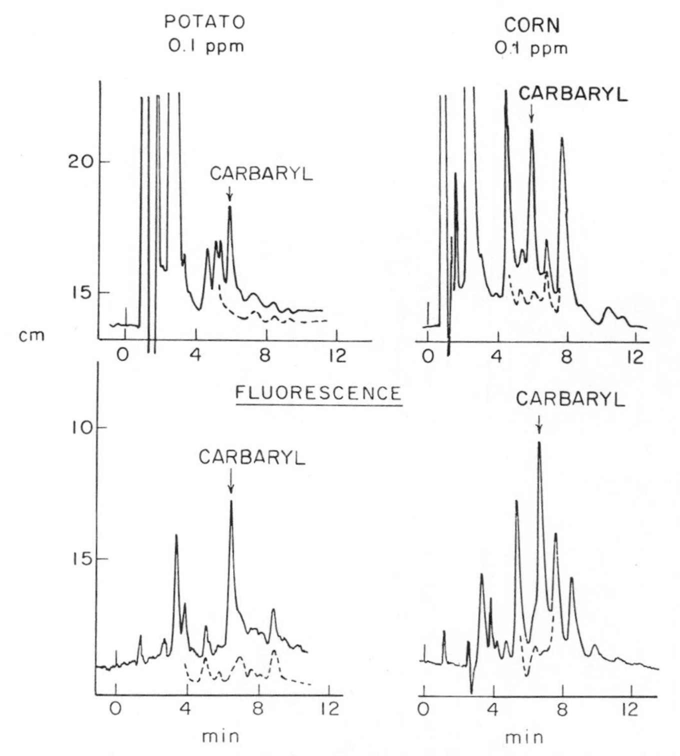

Fig. 7.2 Chromatograms of carbaryl in potato and corn after derivatization with dansyl chloride. From Lawrence and Leduc (*54*), with permission from the Association of Official Analytical Chemists.

is more selective than UV absorbance, in many instances, both may produce comparable results even in complex samples. Figure 7.2 illustrates this point by comparing chromatograms obtained from the same two food extracts containing the insecticide carbaryl (*54*). The derivatization involved hydrolysis of the insecticide, then coupling the phenolic portion with the highly fluorescent (and UV-absorbing) dansyl chloride (see Table 7.1). The resulting extracts were monitored with both UV and fluorescence detectors. Although it might be concluded from Fig. 7.2 that the fluorescence results are superior, the UV tracing is nevertheless acceptable.

Most UV derivatization reactions originate from classical methods of spectrophotometric or qualitative organic analysis. The literature is abundant with many variations of these reactions. This enables the analyst to obtain more information related to his needs in LC. New reagents continue to be developed that improve detection characteristics or offer new or different selectivities through the reaction process.

It should be noted that the fluorescent reagents mentioned in Tables 7.1 and 7.2 are also strong absorbers of UV light and thus may be used for UV-absorption derivatization as well.

The selection of the most appropriate UV-absorbing derivative depends upon several factors. If only a single-wavelength detector is employed, for example at 254 nm, then a derivative that absorbs strongly at that wavelength is required. (Table 7.1 lists the absorbances of several chromophoric groups at 254 nm.) If a variable wavelength detector is used, then a derivative should be employed that will absorb UV light in a region where the sample appears relatively clean. This is difficult to know beforehand, but usually selectivity increases as the absorption maxima increase. Thus, in general it would be preferable to have derivatives absorb near or into the visible region.

Care must be taken to ensure that excess reagent or reagent impurities do not interfere in the detection. Usually, the products have physical or chemical properties that vary significantly from those of the reagent, which enables some cleanup before chromatographic analysis. This is necessary because the reagent normally absorbs UV light as strongly as the product, and it is usually present in the reaction mixture in a far greater concentration. If reagent impurities interfere in the final analysis, then the reagent must be purified. This becomes especially important when ultratrace (low ppb levels, for example) quantities are to be determined where the sample must be concentrated after derivatization.

Even if the reagent is pure, reaction side products or products derived from the sample coextractives may interfere. These must be dealt with on an individual basis. Usually, chromatographic systems are developed so that such interferences are minimal. If this is not possible then further cleanup will be required.

The following describes reactions that have been used for the formation of UV-absorbing derivatives for LC. Also included are reagents that may not have been used for LC determinations but have potential in this area.

B. Benzoylation

Several reagents including benzoyl chloride (*55*), *p*-nitrobenzoyl chloride (*19*,*56*,*57*), 3,5-dinitrobenzoyl chloride (*58*), and *p*-methoxybenzoyl chloride (*59*) have been used for the formation of UV-absorbing derivatives of amines, phenols, and alcohols. Figure 7.3 illustrates a generalized reaction scheme for the benzoylation of an alcohol. The stable organic products are easily analyzed by adsorption or reversed-phase chromatography. The derivatives are usually purified by hydrolyzing the excess reagent to the water-soluble acids, then extracting the product with organic solvent. All of the reagents mentioned above react in a manner similar to benzoyl chloride. Stable derivatives are produced in each case, with UV-absorption maxima depending upon the substitution on the benzene ring.

Table 7.1 compares the absorption maxima of several of these chromophoric groups. From this it can be seen that the nitro-substituted benzoates absorb much more strongly than the unsubstituted benzoyl chloride. It has been shown, for example, that the *p*-nitrobenzoate derivatives of some steroids are about 10 times more sensitive than the corresponding benzoates (*56*). For compounds containing several hydroxyl groups, the absorbtivity of the derivative is proportional to the number of benzoate groups attached. The absorbtivity for the *p*-nitrobenzoate product of desacetyllanatoside *C* (which has 8 *p*-nitrobenzoate groups attached) is about 118,400 (*55*), based on the molecular weight of the parent compound. This shows that the absorbance of each chromophoric substituent is approximately additive, offering good potential for the sensitive detection of polyhydroxy compounds.

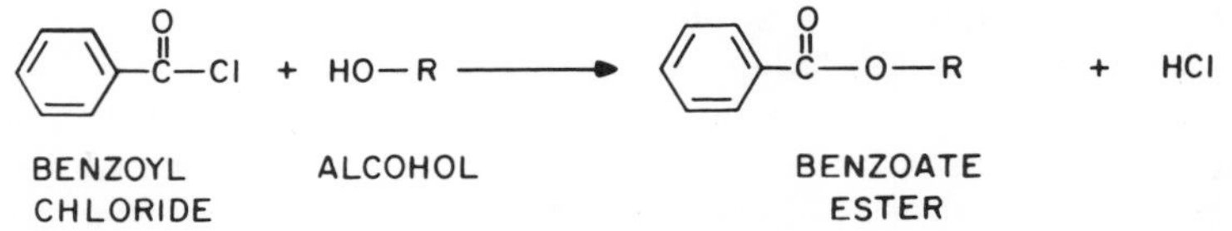

$$C_6H_5-C(=O)-Cl + HO-R \longrightarrow C_6H_5-C(=O)-O-R + HCl$$

BENZOYL CHLORIDE ALCOHOL BENZOATE ESTER

Fig. 7.3 Formation of a benzoate ester from an alcohol using benzoyl chloride.

Other compounds analyzed as benzoate derivatives include hydroxysteroids (*19*), digitalis glycosides (*55,56*), carbohydrates (*56*), hexachlorophene (*19*), glycosphingolipids (*60*), sapogenins (*61*), amphetamines (*62*), imidazolone (*57*), and other amines (*63*). Metformin, a biquanide compound, has been shown to react with *p*-nitrobenzoyl chloride to yield an *s*-triazine ring system suitable for UV detection by LC (*11*). The reagent 3,5-dinitrobenzoyl chloride has been used to derivatize polyalcohols and polyfunctional amines (*58*). Polyfunctional amines also have been successfully determined in industrial samples after conversion to their respective 2-methylbenzoyl derivatives (*64*). For the determination of some amines at 254 nm, 4-methoxybenzoyl chloride has been employed (*18*).

C. 2,4-Dinitrofluorobenzene (DNFB)

DNFB has been often used for the determination of amines and amino acids (*65–67*). It reacts with most primary and secondary amines as well as with phenols to form derivatives that absorb strongly at 254 nm. They also have strong electron-capturing properties and thus have also been employed as derivatives for GC (*67,68*). Unlike the benzoyl chloride type of reagents, DNFB reacts poorly with alcohols. This fact may be useful if phenols or amines need to be determined in the presence of carbohydrates or other aliphatic polyhydroxy compounds. A typical reaction scheme for the formation of the DNFB derivative of an amine is presented in Fig. 7.4. The sensitivity of these products is similar to the 2,4-dinitrobenzoyl derivatives since the major chromophoric groups are similar. DNFB has been used for the LC determination of some nitrosamines after reduction to their amines, as well as carbamate (*67*) and triazine (*68*) pesticides by GC. The latter two should also be suitable for LC analysis.

D. Phenyl and Methylisothiocyanate Reactions

Phenyl- or methylisothiocyanate reacts with amino acids to form thiohydantoins. The reaction has been found very useful for amino acid se-

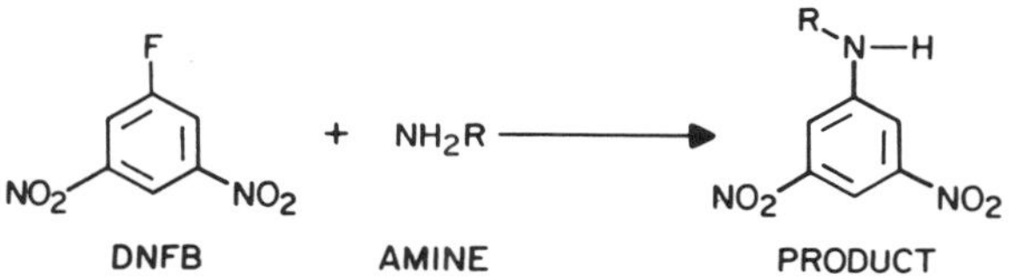

DNFB + NH_2R (AMINE) → PRODUCT

Fig. 7.4 Reaction of DNFB (2,4-dinitrofluorobenzene) with an amine.

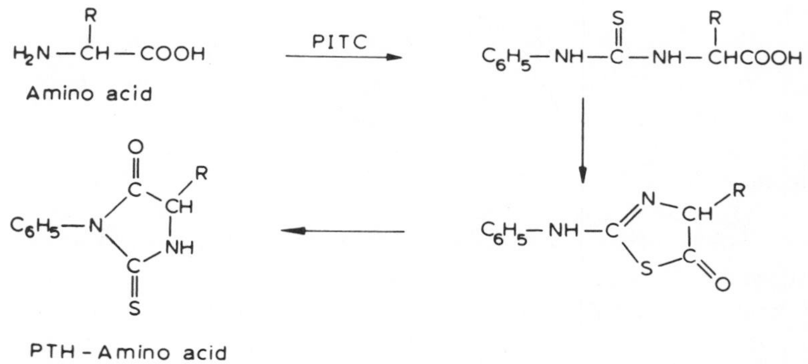

Fig. 7.5 Formation of the phenylthiohydantoin (PTH) derivative of an amino acid. PITC, phenylisothiocyanate.

quencing of peptides. Figure 7.5 shows a typical reaction scheme. After initial attachment of the isocyanate, a ring closure occurs with the carboxylic acid moiety producing a thiohydantoin. Because the derivatives are stable under acid conditions, *N*-terminal amino acids can be determined in acid hydrolysates of proteins (*69,70*). However, for UV-absorption detection in LC, the phenylthiohydantoins would be preferred over the methylanalogs, since the former are about 200-fold more sensitive and can be detected in the 10–100 pmole range at 260 nm (*71*). Phenylthiohydantoin derivatives of amino acids have been separated by LC with adsorption (*71,72*) and reversed-phase (*72–77*) chromatography.

Methylisocyanate has been employed in the analysis of some hydroxyarylamines (*78*). Figure 7.6 illustrates the reaction. Detection was at 254 nm and a reversed-phase system was employed for separation.

Phenylisocyanate has also been used for the analysis of some alcohols after conversion to the corresponding phenylurethanes (carbamates). Figure 7.7 illustrates a typical reaction. The carbamates are then separated by LC with absorption detection at 230 nm (*79*).

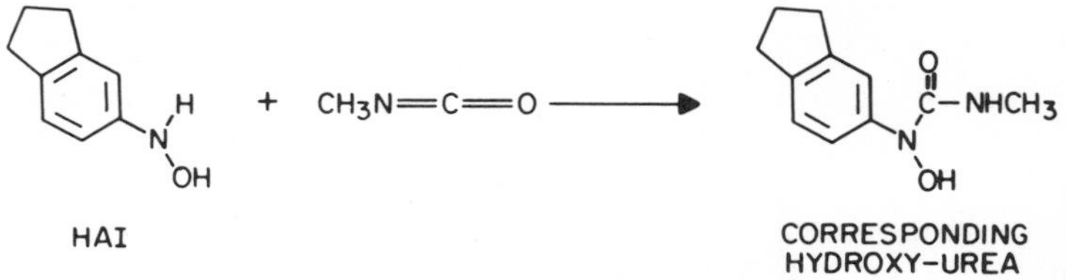

Fig. 7.6 Reaction of methylisocyanate with a hydroxyarylamine (5-hydroxyaminoindan, HAI).

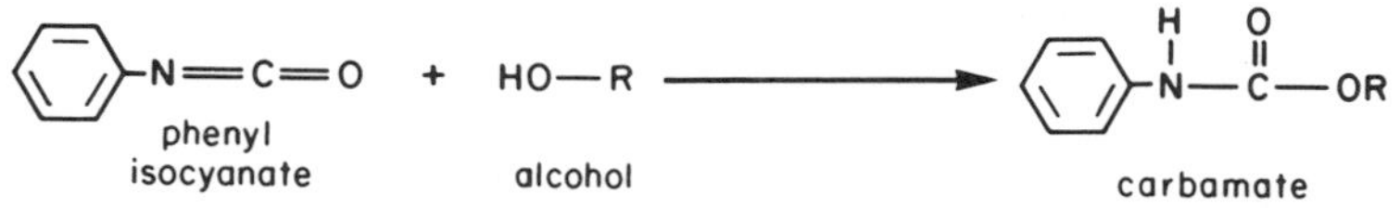

Fig. 7.7 Conversion of an alcohol to a carbamate by treatment with phenylisocyanate.

E. Esterification of Carboxylic Acids

Acyl bromides have found much use in the formation of esters of carboxylic acids. For example, 2-naphthlacyl bromide has been used for fatty acids and barbiturates to provide products for LC separation (*23,24*). Since the naphthyl moiety is a strong absorber of UV light, low nanogram quantities of the derivatives may be determined. Figure 7.8 illustrates a reaction of 2-naphthacyl bromide with a carboxylic acid. Other reagents that provide similar derivatives include phenacyl bromide (*20,80,81*), *p*-nitrophenacyl bromide (*82–84*), *p*-bromophenacyl bromide (*21,85–87*), and *p*-methoxyphenacyl bromide (*88*). Table 7.1 provides data on the spectrophotometric properties of some of these compounds.

These esterification reactions are carried out in polar organic solvents such as acetonitrile, tetrahydrofuran, or acetone in the presence of catalysts like potassium ion (solubilized by crown ethers), triethylamine, or *N,N*-diisopropylethylamine. It has been shown that triethylamine and potassium ion–crown ether catalysts produce similar results in acetonitrile for the esterification of some penicillins (*86*). Final yields were the same, as were the overall reaction rates, although the triethylamine-catalyzed reaction appeared to have an initially faster reaction rate.

Applications of acyl bromides to derivatization and LC of fatty acids (*80,82,88–90*), prostaglandins (*83,84*), gibberellins (*87*), penicillins (*86*), and dipropylacetic acid (*81*) have been reported.

The formation of benzyl or *p*-nitrobenzyl esters of carboxylic acids has been accomplished using reagents such as *o,p*-nitrobenzyl-*N,N'*-diisopropyl urea (*15,91*), 1-(*p*-nitro)benzyl-3-*p*-tolyltriazene (*14,92,93*), or their non-nitro analogs. The products are the corresponding benzyl esters. However, the urea reagent is preferable to the triazene since the latter has been shown to be carcinogenic (*94*) and nitrogen gas is a by-

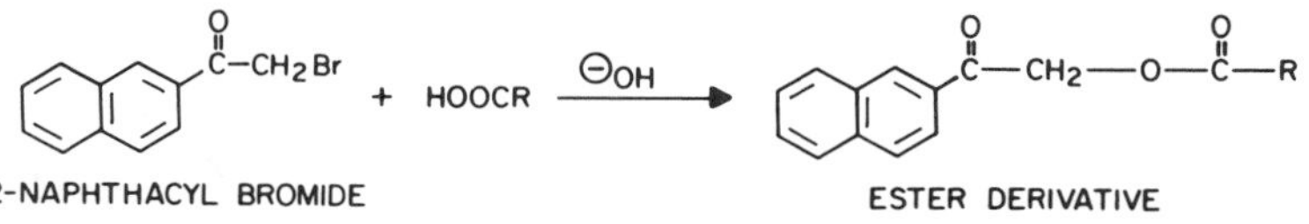

Fig. 7.8 Reaction of 2-naphthacyl bromide with a carboxylic acid to yield the corresponding ester.

product, which might prevent the use of sealed reaction tubes. The nitrobenzyl esters can be detected in the low nanogram range.

p-Methoxyanilides of fatty acids have been prepared for LC analysis by converting the acids to their corresponding acid chlorides by treatment with triphenylphosphine (*95*) followed by derivatization with *p*-methoxyaniline to yield the derivatives, which have an absorption maximum near 254 nm. Minimum detectable levels were in the low nanogram range. However, this two-step reaction may be less desirable than the acyl bromide reaction, which is a single-step reaction under somewhat milder conditions.

Naphthyldiazoalkanes have been used for derivatization of C_{10} to C_{18} fatty acids (*96*) and some polar and nonpolar bile acids (*97*). The reaction proceeds with 1-naphthyldiazomethane or 1-(2-naphthyl)-diazoethane, and yields derivatives with strong absorbance in the 250–280 nm range. This is particularly useful since it enables the use of the less expensive fixed-wavelength UV-absorbance detectors. Figure 7.9 illustrates a generalized reaction scheme for derivative formation. The derivatives have been separated by adsorption chromatography on silica gel with a mobile phase of hexane–tetrahydrofuran–methanol (300:120:8).

F. Acetylation

Acetic anhydride has been used for the formation of acetamide derivatives of biogenic amines and peracetylated derivatives of mono- and disaccharides (*98,99*). In these cases the derivatization serves to aid chromatography rather than increase molar absorptivity. For the disaccharides, for example, UV detection was required at 220 nm. Also, the detection limits were worse than those found for the perbenzoylated products mentioned in an earlier section.

In spite of this, the above-mentioned work has shown that anhydride-type reactions can be accomplished successfully for LC. The use of other anhydrides that could produce strongly UV-absorbing derivatives with amines or alcohols has much potential for trace organic analysis by LC. As has been found in GC, anhydrides are often superior to acid halides for derivatization. This may be also true for reagents more suited to LC.

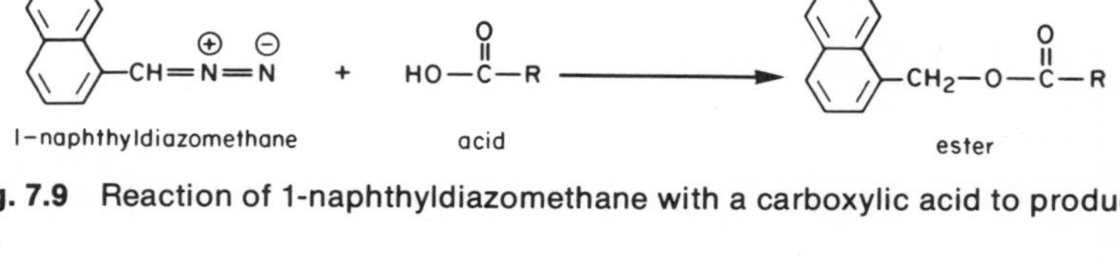

Fig. 7.9 Reaction of 1-naphthyldiazomethane with a carboxylic acid to produce an ester.

It should also be pointed out that for separation purposes only, small derivatives such as acetates might be preferable to, for example, perbenzoylated products (*100*). The reason for this is that large bulky groups tend to mask small differences in molecular structure, therefore making chromatographic separation of two similar compounds more difficult.

G. Arylsulfonyl Chloride Reactions

Azobenzene-4-sulfonyl chloride reacts with amines and phenols under mild base catalysis to form strongly UV-absorbing derivatives. It has been used for the analysis of estrogens in biological extracts (*101*). The azo group is responsible for increasing the wavelength of the absorption maxima to 313 nm, thereby offering some selectivity over other derivatives that absorb at lower wavelengths (for example, 220–280 nm).

A similar reagent, quinalone-8-sulfonyl chloride, provides highly absorbing derivatives of biogenic amines (*102*). Detection limits at 230 nm for compounds such as spermidine, cadaverine, or cystamine have been reported to be about 20 ng per injection.

Biogenic amines have also been determined as their *p*-toluenesulfonyl derivatives with UV-absorption detection (*103*).

The above reactions are normally carried out in aqueous acetone or acetonitrile in the presence of a mild base.

H. Hydrazones and Oximes of Carbonyl Compounds

Ketones and aldehydes react with 2,4-dinitrophenylhydrazine (DNPH) to form the corresponding hydrazones (*104*). A reaction scheme is shown in Fig. 7.10. Hydrazones have been prepared for both TLC (*105*) and LC (*12,106*) analyses of compounds such as ketosteroids, ketoacids (*107–109*), and other carbonyl compounds (*110–112*). Detection limits are usually in the low nanogram range when monitored at 254 nm or 336 nm.

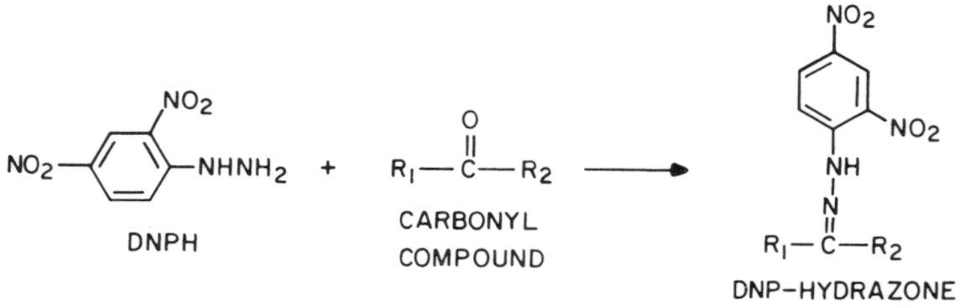

DNPH + CARBONYL COMPOUND (R_1—C(=O)—R_2) → DNP-HYDRAZONE

Fig. 7.10 Reaction of 2,4-dinitrophenylhydrazine with an aldehyde to yield the hydrazone.

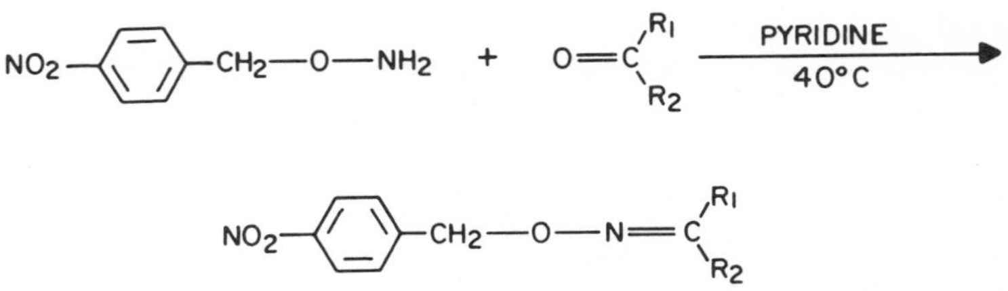

Fig. 7.11 Reaction of *p*-nitrobenzylhydroxylamine with a ketone.

Detection limits by TLC have been reported to be about 20–40 ng/spot at 367 nm when quantitatively measured *in situ* on the plate.

p-Nitrobenzylhydroxylamine hydrochloride (PNBA) (*13,92,113*) reacts with aldehydes and ketones to produce UV-absorbing oxime derivatives. The reaction is base-catalyzed, unlike that above for 2,4-dinitrophenylhydrazine, which is carried out under weakly acidic conditions. These two reagents might complement one another in this regard. For example, compounds that may not be stable under acidic conditions could be derivatized with PNBA, whereas DNPH could be used for compounds unstable in base. PNBA has been applied to the analysis of prostaglandins and thromboxanes by LC (*114*). The compounds are first converted to methyl esters by treatment with diazomethane. After this the carbonyl moieties are derivatized with PNBA to form the highly absorbing (at 254 nm) *p*-nitrobenzyloximes. Figure 7.11 illustrates a typical reaction for a ketone. Pentafluorobenzyl hydroxylamine has also been studied for the LC derivatization of prostaglandins (*114*). However, since the derivatives contain halogens they would be more suited to GC analysis with electron-capture detection. The UV response of these derivatives is about four- to fivefold less sensitive to UV detection at 254 nm than the *p*-nitrobenzyl analogs.

I. Metal Chelates

In recent years, much work has been done on the use of metal chelation reactions for the separation and detection of metals by LC using UV-absorbance detection. Results have been reported on metal separations as their acetylacetonate complexes (*115–118*), dithizonates (*119–121*), Schiff base chelates (*122*), β-ketoamines (*123*), fluorinated and nonfluorinated β-ketoamine and salicyladimine chelates (*124*), 1-(pyridylazo)-2-naphthol chelates (*125*), glyoxalbis(2,2,3,3-tetramethylbutyl)thiosemicarbazones, and diacetylbis(cyclohexyl)thiosemicarbazones (*126,127*) dithiocarbamate complexes (*121,126,128–131*), and pyridine-2-carbaldehyde-2-quinolylhydrazone chelates (*132*).

Most of these chelates are strong absorbers in the UV-visible region and thus are well suited to LC with absorbance detection. The chelates are prepared in a classical manner and extracted into organic solvents such as chloroform. The chelate is then injected into the LC directly or after exchanging the solvent for mobile phase. Both normal-phase adsorp-

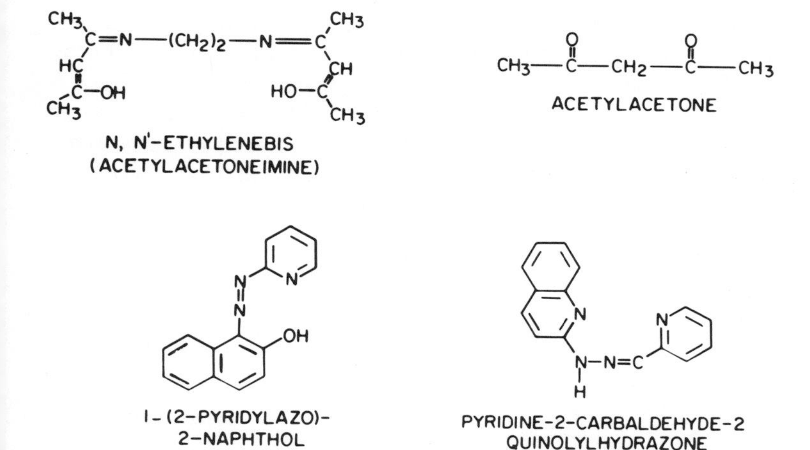

Fig. 7.12 Structures of ligands used for metal chelation.

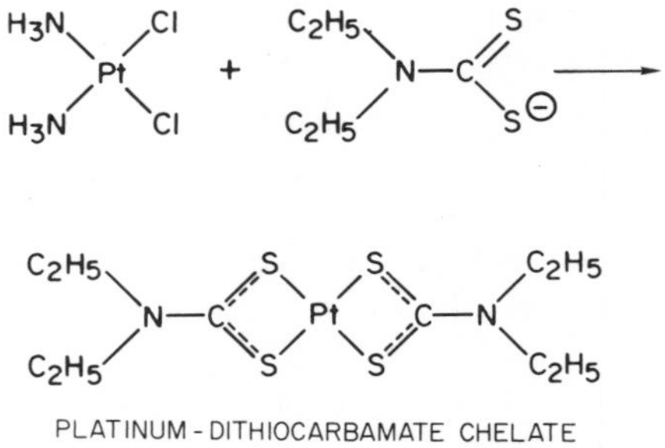

PLATINUM - DITHIOCARBAMATE CHELATE

Fig. 7.13 Ligand exchange reaction between sodium ethylenebisdithiocarbamate and *cis*-dichlorodiamine platinum (II).

tion and partition chromatography have been employed for the separations. Figure 7.12 shows the structures of several of the ligands used for metal complexation and chromatography.

Ligand-exchange reactions have been employed for the analysis of *cis*-dichlorodiamine platinum (II) in urine using sodium ethylenebisdithiocarbamates (*133*). Figure 7.13 illustrates the reaction. The resulting dithiocarbamate chelate is easily extracted from the urine. Detection limits using 254 nm absorption were about 25 ng/ml urine.

J. Separation of Enantiomers

Enantiomeric amino acids have been separated by LC after formation of distereomeric derivatives (*134–137*). The reaction first involves the introduction of a second asymmetric center into the amino acid. This is usually accomplished by reaction at the amino group with reagents such as *d*-10-camphorsulfonyl chloride, (+)-neomenthyl isothiocyanate, (−)-1,7-dimethyl-7-norbornyl isothiocyanate, or others. A second derivatization is required to block the carboxylic acid moiety. Reagents such as

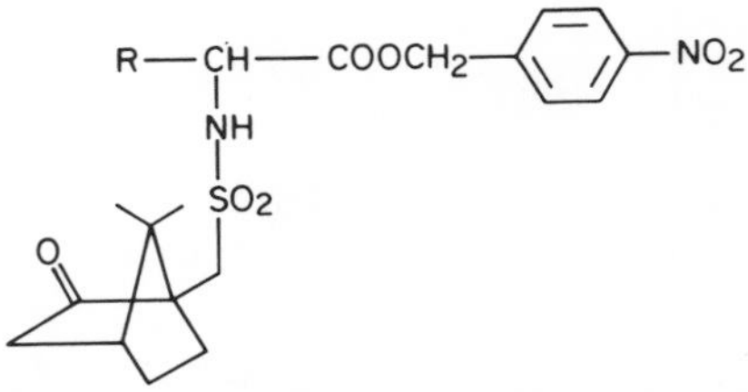

Fig. 7.14 Structure of an amino acid derivatized with *d*-10-camphorsulfonyl chloride followed by esterification of the carboxylic acid group with *p*-nitrobenzyl bromide.

p-nitrobenzyl bromide or silylating agents have been employed for this (*134,135*). Figure 7.14 shows the structure of an amino acid derivatized by this technique.

K. Alkylation

The addition of alkyl groups such as methyl or ethyl to molecules usually serves to block labile hydrogen atoms such as found in N—H or O—H functional groups (*138,139*). The technique is often used in GC (for example, for esterification of carboxylic acids using diazomethane or alkylation of ureas, triazines, and other pesticides (*140–143*).

Diazomethane has been used for esterification of some bile acids (*144*) and prostaglandins (*106*) (as part of a double derivatization technique) for analysis by LC. Methyl or ethyl iodide has been employed for the analysis of some urea herbicides by both LC and GC (*140,141*).

Alkylation does not aid detection of the compounds in LC. It merely changes the retention time. In this sense the method is best used for confirmatory tests. Figure 7.1 illustrates the confirmation of two urea herbicides (using methyl iodide in the presence of sodium hydride for the derivation). Figure 7.15 shows a typical reaction with a urea herbicide. The products always are less polar than the parent compounds, thus elute with a smaller k' in an adsorption chromatography system. The opposite is true if a reversed-phase column is employed. One feature of the alkylation reaction is that since the derivatives are usually thermally stable, further confirmation of the LC results can be carried out by GC (*141*).

L. Esterification of Dianhydrogalactitol

Dianhydrogalactitol has been determined in plasma (*145*) after conversion to the corresponding bis (dithiocarbamoyl) ester using the reagent sodium diethyldithiocarbamate. The derivative strongly absorbs at 254 nm, which permits detection at levels below 50 ng/ml in plasma. Fig-

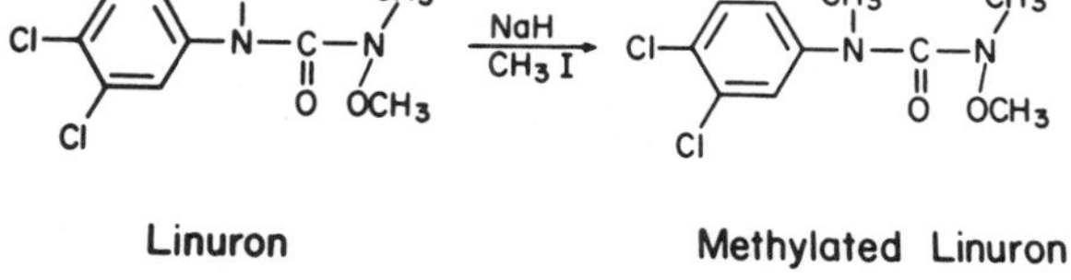

Cl, Cl, H, N, C, O, CH_3, N, OCH_3 — NaH / CH_3I → Cl, Cl, CH_3, N, C, O, CH_3, N, OCH_3

Linuron — Methylated Linuron

Fig. 7.15 Methylation of a urea herbicide with methyl iodide in the presence of sodium hydride and dimethylsulfoxide.

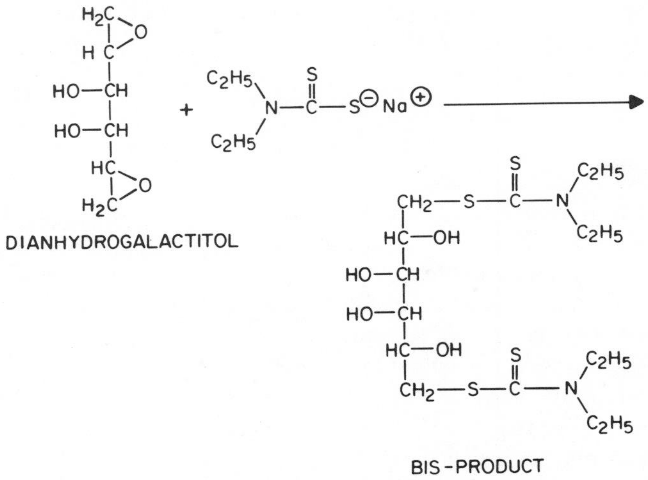

DIANHYDROGALACTITOL

BIS-PRODUCT

Fig. 7.16 Reaction of dianhydrogalactitol with sodium diethyldithiocarbamate to form the bis-product.

ure 7.16 shows the reaction. Chromatographic separation was achieved on a CN-bonded column using heptane–chloroform–acetic acid as the mobile phase.

M. Ethyleneimine and Other Amines—Derivatization With Folin's Reagent

Ethyleneimine can be converted to a naphthoquinone using Folin's reagent (1,2-naphthoquinone-4-sulfonate) (*146*). The product, 4-(1-aziridinyl)-1,2-naphthoquinone, is detected at 420 nm after LC separation. This method has been applied to other amines for separation by TLC (*146*).

N. Ion-Pair Formation

Several classes of compounds have been analyzed by ion-pair chromatography. These include thyroid hormones (*147*), sulfa drugs (*147*), alkaloids (*148*), and many other organic acids and bases (*149–153*). The principles of ion-pair chromatography have been discussed in detail in another chapter of this book and elsewhere (*154–158*). Research in this area is continuing at a rapid pace. Ion pairing also has much potential as a selective detection technique and is discussed in Section V,C,6 of this chapter. More work is required on investigations of ion-pairing systems, stationary phases, and distribution studies in general.

III. FLUORESCENCE DERIVATIZATION

A. Introduction

Fluorescence derivatization was initially employed for TLC analysis and found use in the amino acid sequencing of proteins (*159*). The fluorescent derivatives (dansyl sulfonamides) were sensitive enough that minute quantities of sample material could be used. Detection was normally made visually under a UV light at an appropriate wavelength. The separated spots usually appeared as bright shades of blue, green, or yellow on a dark background. Quantitation was carried out by removing the substance from the TLC layer and then measuring it in solution in a classical manner. In the early 1970s, TLC scanners became available from a number of manufacturers. These permitted *in situ* quantitation of the fluorescent products directly on the TLC plate. However, with the great popularity of LC, TLC scanning is declining in importance for quantitation.

One of the earliest applications of fluorescence derivatization applied to LC was the dansylation of some carbamate insecticides (*160*). Since then the use of fluorescence derivatives for LC analysis has continued at a steady pace.

Unlike UV derivatization, there are only a few fluorescent reagents that have been examined for application to trace analysis. Fluorogenic reagents have been used for both pre- and post-column derivatization reactions. The latter are discussed in a later section. Prechromatographic derivatization is best carried out with reagents which themselves are not fluorescent but which produce highly fluorescent products with the compounds of interest. This can be advantageous, since excess reagent would not interfere in the detection of the derivatives. However, since this is not always possible, a preliminary cleanup may be required before chromatography. Also, when doing derivatization reactions at ultratrace levels, reagent impurities that are often fluorescent may pose problems. In this case, it is wise to purify the reagent first before any reactions are carried out.

The following discusses a number of reagents and reactions suitable for prechromatographic derivatization for LC analysis of organic compounds at trace levels.

B. Dansyl Chloride Reaction With Amines and Phenols

Dansyl chloride (DNS-chloride, DNS-Cl; 5-dimethylaminonaphthalene-1-sulfonyl chloride) is a reagent that reacts with phenols and primary or secondary amines under slightly basic conditions to form sulfonate esters or sulfonamides. The derivatives are highly fluorescent, whereas the reagent itself is not. However, the hydrolysis product, dansyl-OH (dansylic acid), is strongly fluorescent and can cause interferences with water-soluble derivatives. But, even then, dansyl-OH can be separated during the chromatographic process. For nonpolar derivatives such as dansyl-phenols, the hydrolyzed reagent can be removed by partition.

Figure 7.17 shows a general reaction scheme for the formation of dansyl derivatives of an amine and a phenol. The reaction is usually carried out with a 5- to 10-fold molar excess of dansyl chloride in a mixture of acetone and aqueous sodium bicarbonate (or carbonate) at various concentrations. The use of potassium fluoride, solubilized with 18-crown-6 (1,4,7,10,13,16-hexaoxacyclooctadecane), has also been investigated for catalyzing the dansylation of some biogenic amines (*161*). The derivatives are formed by leaving the reaction mixture for several hours at room tem-

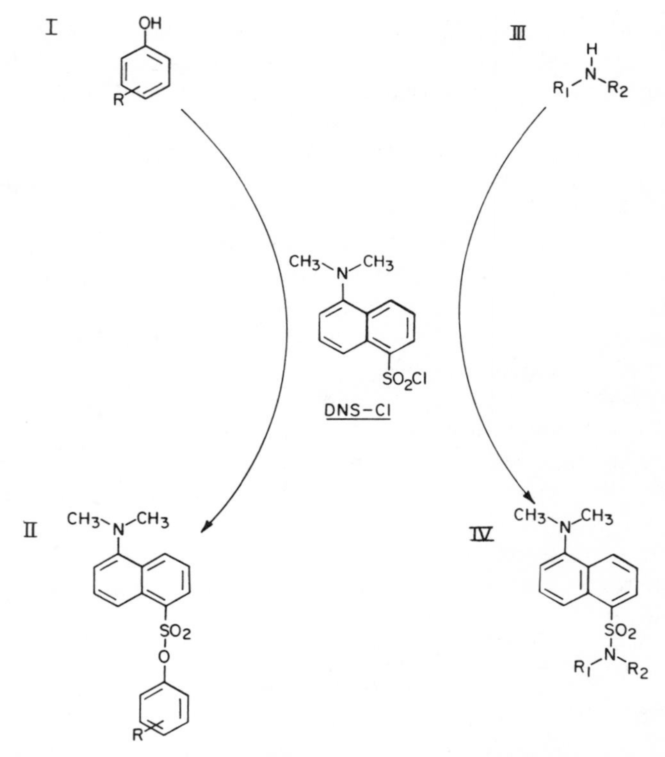

Fig. 7.17 Dansylation of a phenol(I) and an amine(III) to produce the dansyl sulfonate(II) and dansyl sulfonamide(IV).

perature or for shorter periods at elevated temperatures (30°–50°C). The organic soluble products are removed from the reaction mixture by direct extraction of the aqueous mixture with solvents such as hexane or benzene. The organic phase is then used for chromatographic analysis.

Variations of this reaction scheme have been used to detect amino acids (*162–164*), polyamines (*165*), biogenic amines (*166,167*), optical isomers (*168*), estrogens (*169*), serotonin (*170*), alkaloids (*171–173*), barbiturates (*174*), cannabinoids (*175*), hydroxybiphenyls (*176*), chlorophenols (*177*), and pesticides including carbamates (*178–181*), ureas (*182,183*), organophosphates (*184*), and triazines (*185*). Additional applications are listed in Table 7.2.

The detection limits are often in the low nanogram range for these compounds. Figure 1.7 shows the added sensitivity gained by LC for the car-

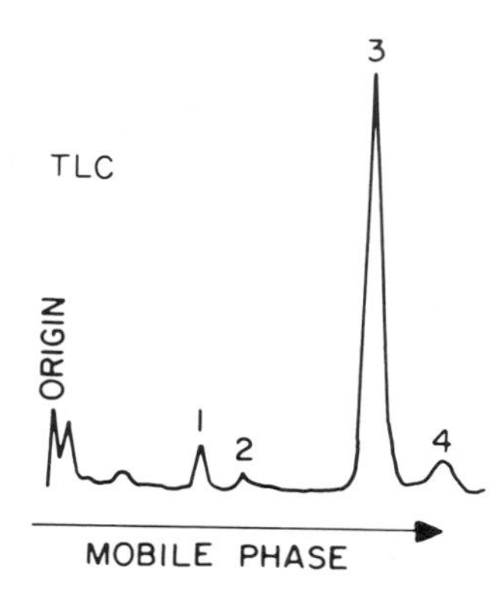

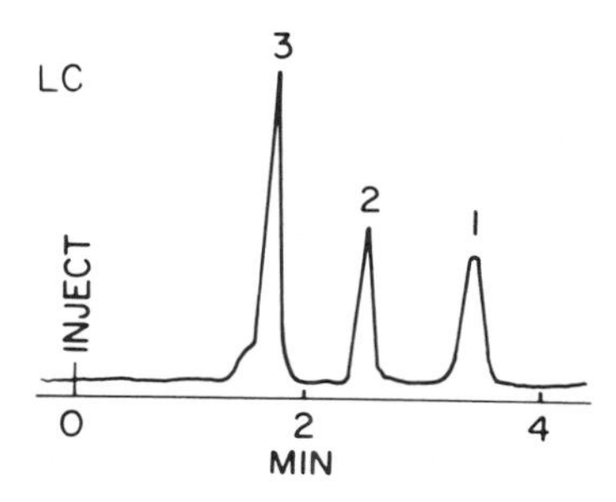

Fig. 7.18 Comparison of typical TLC and LC chromatograms of a dansylated cough syrup extract. 1, dansyl emetine; 2, dansyl cephalin; 3, dansyl ephedrine; 4, dansyl chloride. Adapted from Frei *et al.* (*172*), with permission from Elsevier Scientific Publ. Co.

bamate carbofuran, in a potato extract at 0.1 ppm. The dansylation results are about tenfold better than the direct UV analysis (*180*). Figure 7.18 compares chromatographic results by TLC and LC for the separation of some dansylated alkaloids. It can be seen that both modes can successfully separate the compounds (*172*).

The excitation (ex) and emission (em) maxima can vary (350–370 nm ex; 490–540 nm em) for the derivatives, depending upon the compound derivatized as well as the solvent used to measure the fluorescence. For TLC analysis, the plates are usually sprayed with triethanolamine, which greatly enchances the fluorescence compared to a dry plate. However, it has been reported that reproducibility is better when the spots are left unsprayed before quantitation (*183*). Excitation and emission spectra obtained *in situ* are usually within the range mentioned above.

Since dansyl chloride reacts with amines and phenols, any coextracted materials or impurities containing these groups might produce interferences in chromatography. Ammonia from the air and in solvents will readily react with dansyl chloride to form dansyl sulfonamide. Although such products may pose no problems if separated by chromatography or partition, they still consume the reagent and thus might affect the yield of the desired product. Dimethylsulfoxide cannot be used as a solvent because of reaction with dansyl chloride under basic conditions (*186*). Some amino acids (especially basic amino acids) tend to form a small percentage of secondary products during dansylation (*187*). Such side reactions and effects should be known before attempts are made to analyze samples.

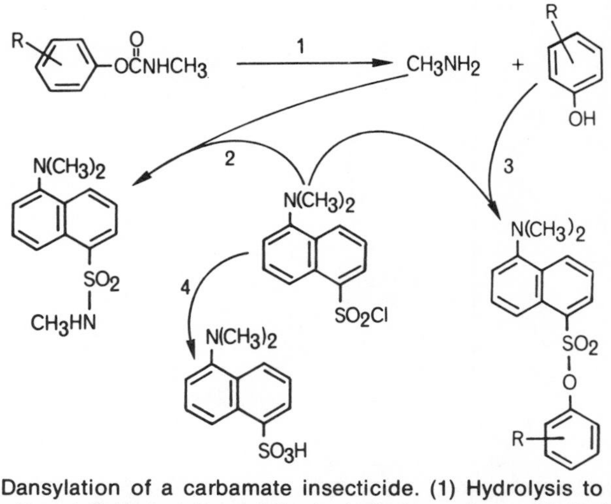

R–C₆H₄–OC(O)NHCH₃ —1→ CH₃NH₂ + R–C₆H₄–OH

$N(CH_3)_2$ / SO_2 / CH_3HN ; $N(CH_3)_2$ / SO_2Cl ; $N(CH_3)_2$ / SO_2 / O / R ; $N(CH_3)_2$ / SO_3H ; 2 ; 3 ; 4

Fig. 7.19 Dansylation of a carbamate insecticide. (1) Hydrolysis to methylamine and a phenol. (2 and 3) Dansylation of the hydrolysis products. (4) Hydrolysis of the reagent during the reaction.

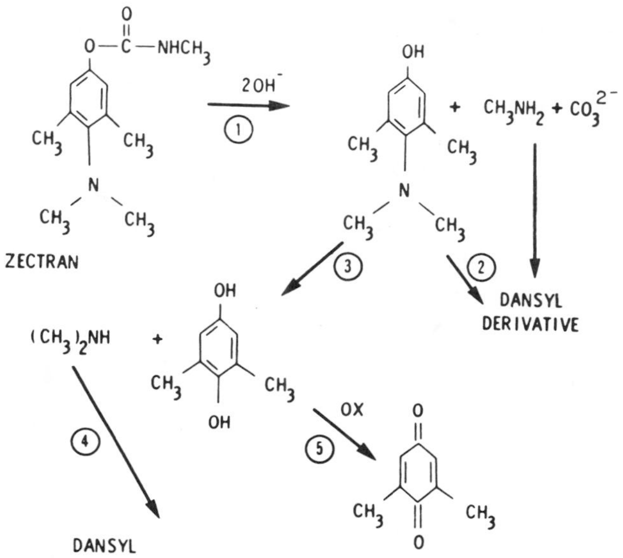

Fig. 7.20 Reaction scheme for the derivatization of Zectran (carbamate). (1) Hydrolsis). (2) Dansylation of phenol and amine moieties. (3) Competing reaction that liberates dimethylamine. (4) Dansylation of dimethylamine. (5) Conversion of the hydroquinone to the corresponding benzoquinone, which does not react with dansyl chloride.

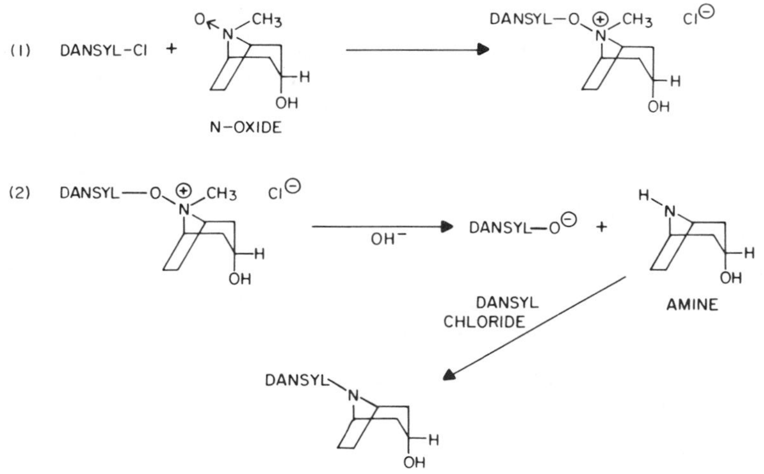

Fig. 7.21 Partial reaction scheme for dansyl chloride and tropine *N*-oxide to yield *N*-dansylnortropine. (1) Initial reaction. (2) Cleavage of dansyl-$O^{(-)}$ to ultimately yield the free amine, which reacts with a second molecule of dansyl chloride to form the product.

Multifunctional alkaloids containing both N—H and phenolic O—H groups produce multiple dansyl derivatives (*170*,*171*) as determined by mass spectrometry. Also, if preliminary hydrolysis is required before dansylation, as for carbamate and urea pesticides, more than one hydrolysis product may be formed that could produce fluorescent products (*179*,*188*). This is illustrated in Figs. 7.19 and 7.20. The first depicts a typical reaction scheme for *N*-methylcarbamate insecticides. The reaction involves hydrolysis of the compounds to methylamine and a phenol. When dansyl chloride is added to the reaction mixture, both hydrolysis products yield derivatives that are separated and detected by LC (*160*). However, when this same method was used for the carbamates Zectran or Matacil, a third product was always present (*189*). Subsequent investigations showed that a secondary reaction was occurring during the hydrolysis step, which liberated the dimethylamine that yielded the third dansyl derivative.

Dansyl chloride has been shown to react with the *N*-oxides of some aliphatic tertiary amines to ultimately yield the corresponding *N*-demethylated, *N*-dansyl derivatives (*190*). Figure 7.21 illustrates an overall reaction scheme. The initial attachment of the dansyl moiety is at the oxygen position. However, in the presence of base, O—N and C—N cleavages occur, yielding the free amine which reacts with additional dansyl chloride to produce the dansylated secondary amine. This reaction has been applied to the determination of tropine *N*-oxide (*190*).

Similar reagents, bansyl chloride (5-di-*N*-butylamino-naphthalene-1-sulfonyl chloride) and mansyl chloride [6-(*N*-methylanilino)naphthalene-2-sulfonyl chloride] also react with amines and provide comparable derivatives (*191*,*192*). However, since they do not offer many real advantages, little work has been carried out on these.

C. NBD-Chloride Reaction With Aliphatic Amines

NBD-chloride (4-chloro-7-nitrobenz-2,1,3-oxadiazole) reacts with primary and secondary aliphatic amines to form highly fluorescent derivatives. Other compounds such as anilines, phenols, and thiols yield weakly or nonfluorescent derivatives. Thus, the method is selective toward aliphatic amines. Figure 7.22 shows a reaction scheme for the formation of the NBD derivative of an aliphatic amine. The reaction can be carried out in the same manner as described for dansyl chloride since both require basic conditions for the reaction. A two-phase reaction has also been evaluated for forming NBD derivatives of aliphatic amines (*193*). The products are extractable from aqueous mixtures into solvents such as benzene or ethyl acetate and exhibit most intense fluorescence in polar

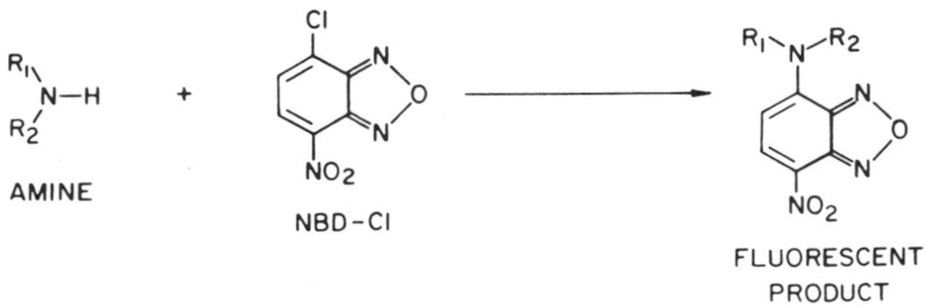

AMINE + NBD-Cl → FLUORESCENT PRODUCT

Fig. 7.22 NBD-chloride reaction with an aliphatic amine. R_1 is aliphatic or hydrogen; R_2 is aliphatic.

organic solvents. Since the fluorescence is greatly reduced in aqueous solution, moisture should be avoided. Detection limits are in the low picomole range for many amines, either as spots on a TLC plate or by LC.

The reagent has advantages over dansyl chloride since the reagent and its hydrolysis product, NBD-OH, are nonfluorescent. Thus, they cause no problems in the fluorescence measurement of the derivatives. Figure 7.23 demonstrates the selectivity by comparing a TLC separation observed in daylight with a fluorescence scan from a soil sample analysis. The TLC plate has much colored material whereas the fluorescence scan is relatively clean (*193*). For *in situ* TLC analysis the plates do not have to be sprayed to optimize fluorescence intensity. The excitation maxima are

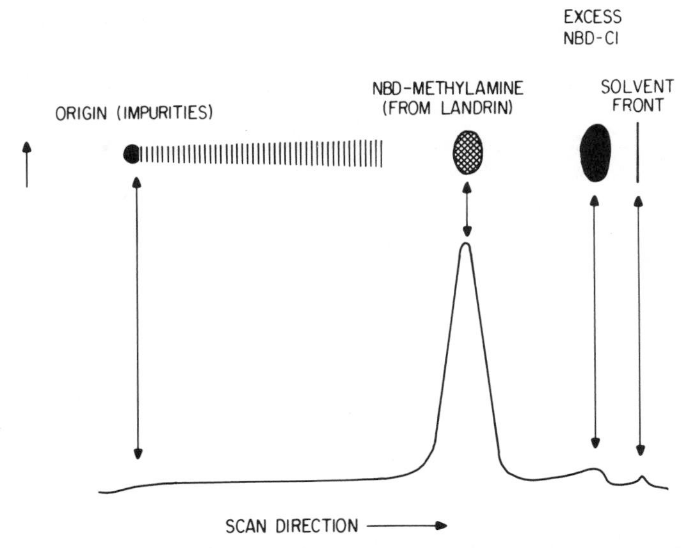

Fig. 7.23 Chromatogram and corresponding fluorescence scan of a water-extract analysis of landrin (10 μg/kg) after hydrolysis and reaction with NBD-chloride. From Lawrence and Frei (*193*), with permission from the American Chemical Society.

much higher (~480 nm) than for the dansyl derivatives, although the emission maxima are similar (~530 nm). NBD-chloride has been used for the analysis of amino acids (*194,195*), biogenic amines (*37,196*), amphetamines (*38,197,198*), methylcarbamate insecticides (*189,193*), thiolcarbamates (*199*), antithyroids (*39*), alkaloids (*200*), thiols (*201*), and nitrosamines (*202*).

D. Fluorescamine

Fluorescamine, 4-phenylspiro[furan-2-(3*H*),1′-phthalan]-3,3′-dione, is a selective reagent for primary amines. It reacts almost instantly with these substrates while the excess is hydrolyzed to a nonfluorescent product. The reagent itself is also nonfluorescent. Figure 7.24 illustrates the reaction with a primary amine. The conversion is carried out in aqueous acetone mixtures at a pH near 8–9, while the test tube is shaken on a vortex mixer. The derivatives may then be directly chromatographed on a reversed-phase system (*203*). Since this reaction proceeds so rapidly, it is ideally suited to postchromatographic reactions after the components have been separated, either by TLC or LC. The excitation of the products usually occurs at 390 nm, with fluorescence emission at 475 nm.

Fluorescamine has been used for the prechromatographic derivatization of polyamines (*29,203*), catecholamines (*30,204*), and amino acids (*31,205*). In a comparison study of the use of dansyl chloride and fluorescamine for analysis of chlorzoxazone (*206*), it was found that fluorescamine provided the more sensitive derivative by about one order of magnitude. The reactions, however, were not applied to chromatographic analyses but were carried out in organic solvents after extracting the products from the aqueous reaction mixture. NBD-chloride failed to react with the semicarbazide of chlorzoxazone to produce a fluorescent product.

Although the fluorescamine reaction is more selective and by far much faster than either the dansyl chloride or NBD-chloride reactions, it has several undesirable features. The cost of fluorescamine is much higher

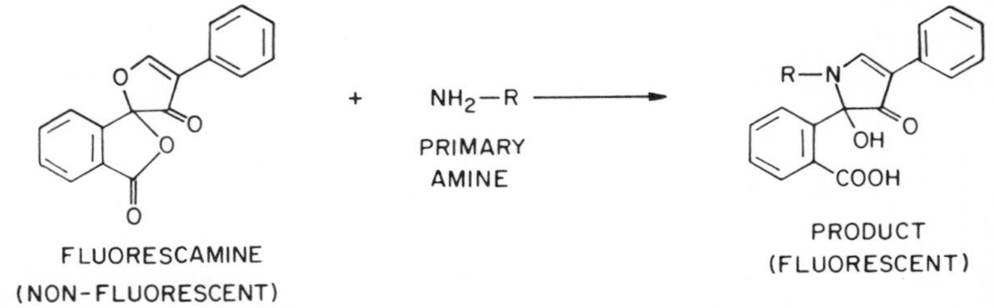

O O O O + NH_2—R → R—N O OH COOH

FLUORESCAMINE (NON-FLUORESCENT) PRIMARY AMINE PRODUCT (FLUORESCENT)

Fig. 7.24 Fluorescamine reaction with a primary amine. R, aliphatic or aryl.

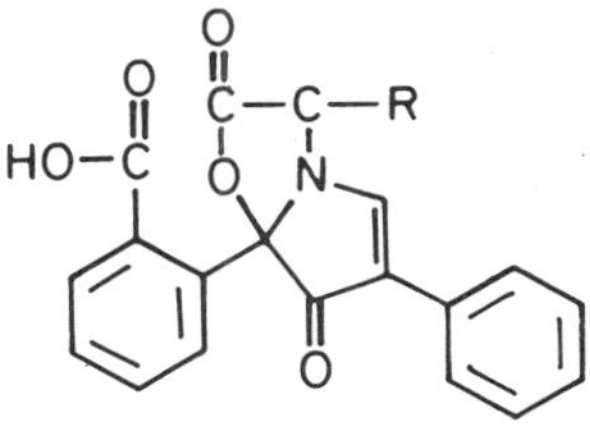

Fig. 7.25 Lactone product resulting from the reaction of fluorescamine with some amino acids.

than the other two. Both reaction rate and fluorescence intensity is significantly influenced by pH. For manual analyses, reproducibility for replicate samples is often worse than for the other reagents described above. Also, the stability of the fluorescamine products is not nearly as good as the others; thus, storage of the products for longer than a few hours is often impossible. McHugh *et al.* (*205*) found that fluorescamine produces two fluorescent derivatives with several amino acids, the proportions of which were dependent upon the amino acid. They concluded that separation of a complex mixture of amino acids would be difficult after fluorescamine derivatization. The two products appear to be the normal fluorescamine derivative, as shown in Fig. 7.24, and a lactone product, as indicated in Fig. 7.25. This, coupled with its fast and selective reactivity, makes it more suited to postcolumn techniques. In fact, most work with this reagent has been done in this manner (*6,52*). Further discussion of postcolumn reactions is included in a later section.

E. 4-Bromomethyl-7-methoxycoumarin Reaction with Carboxylic Acids

4-Bromomethyl-7-methoxycoumarin (Br-Mmc) reacts with monocarboxylic acids in acetone in the presence of solid K_2CO_3 to form highly fluorescent derivatives which may be separated by TLC (*32,33*) and by LC (*34*). Figure 7.26 shows a scheme for the reaction with propionic acid. The derivatives are excited near 328 nm and fluorescence is observed near 375 nm (*33*). Although very useful for most monocarboxylic acids,

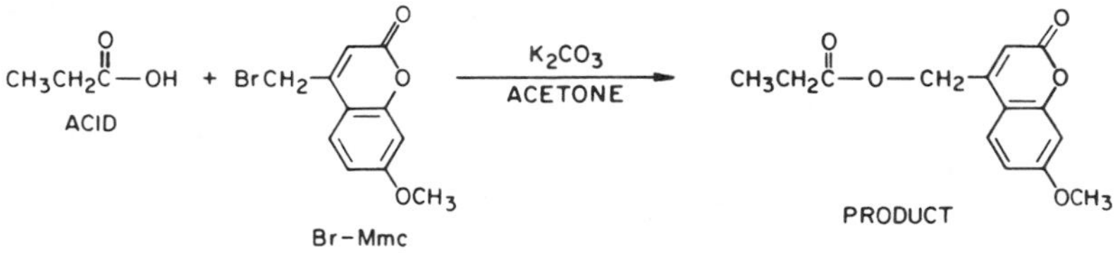

Fig. 7.26 Br-Mmc reaction with a carboxylic acid to form the fluorescent ester.

the reaction under the above-described conditions does not work for dicarboxylic acids. However, dicarboxylic acids were successfully reacted with Br-Mmc with the aid of crown ethers as solubilizing agents (*207*). Reaction of Br-Mmc with phenols and amines in aqueous media was unsuccessful (*208,209*), which indicates that the reaction has some degree of selectivity. More work needs to be carried out to fully assess the utility of this technique. Detection limits approach 50 pmole per spot by TLC. The method has been examined with fatty acids (*32*) and some acidic herbicides (*33*).

F. Diphenylindenonesulfonyl Chloride (Disyl Chloride)

Disyl chloride reacts in an analogous fashion to dansyl chloride with amines and phenols. A reaction scheme is shown in Fig. 7.27. The amine or phenol is treated with a molar excess of the reagent in acetonitrile or acetone in the presence of Na_2CO_3 or $NaHCO_3$. The products are then extracted from the reaction mixture for TLC. However, unlike the dansyl derivatives, the disyl products require treatment with sodium ethoxide to induce fluorescence (*109,110*). By TLC this would mean the use of a spray, whereas by LC it would need a postcolumn addition of the ethoxide, thus requiring extra equipment. The fluorescence of the derivatives (green emission when excited at 365 nm) is only stable for a few hours after spraying, which makes it difficult to store the plates for later analysis if needed. The sensitivity, however, appears to be better than the previous reagents discussed, since 0.1 ng of disyl tetrahydrocannabinol can be visually detected on a TLC plate (*210*). The reagent has been used for the determination of amino acids (*211*), amino sugars (*212*), vitamin B_6 (*213*), and tetrahydrocannabinol (*210*).

G. *o*-Phthalaldehyde

o-Phthalaldehyde (OPA), although more suited to postcolumn derivatization, has been used for the prechromatographic reaction and analysis of

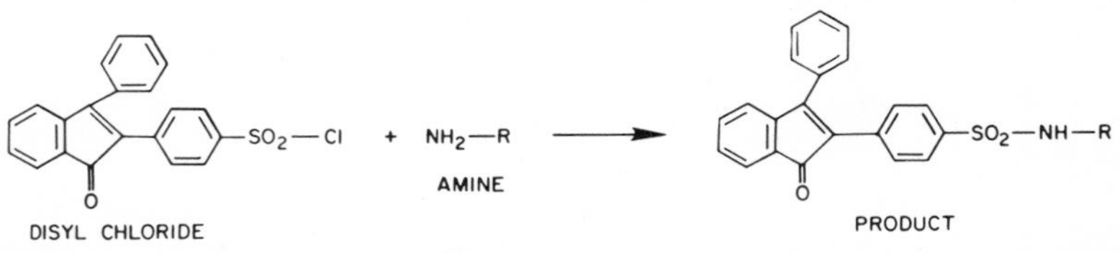

Fig. 7.27 Disyl chloride reaction with an amine to form a fluorescent sulfonamide.

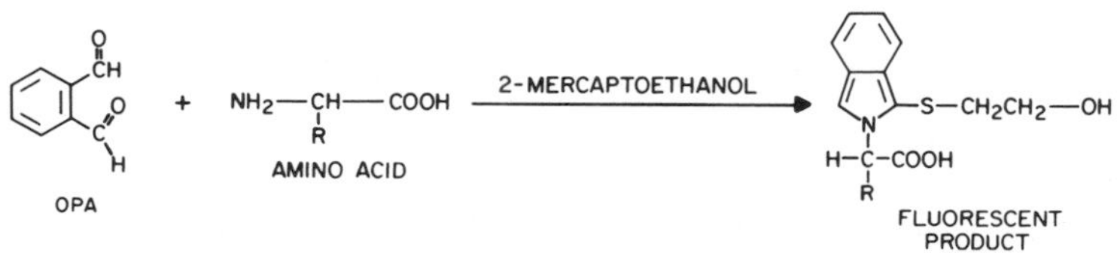

OPA + $NH_2—CH(R)—COOH$ (AMINO ACID) —2-MERCAPTOETHANOL→ FLUORESCENT PRODUCT ($—S—CH_2CH_2—OH$; $H—C(R)—COOH$)

Fig. 7.28 Reaction of OPA (*o*-phthalaldehyde) with amino acids, in the presence of 2-mercaptoethanol to form a fluorescent derivative.

several biogenic amines (*214*), amino acids (*35*), catecholamines (*215*), and histamines (*216*). Detection limits are in the order of 0.1 ng of compounds such as serotonin, dopamine, and norepinephrine (*214*), and subpicomole quantities of amino acids (*35*). The derivatives are stable for about 24 hr and thus are not suitable for storage for longer periods, as are the dansyl or NBD derivatives mentioned earlier. Figure 7.28 illustrates a generalized reaction scheme for the formation of OPA amino acids. Simons and Johnson (*217,218*) found that the reaction product was a thioalkyl-substituted isoindole. They postulated that the instability of this product was due to a spontaneous molecular rearrangement (*218*). Lindroth and Mopper (*35*) carried out an in-depth study on reaction and chromatography conditions for application to amino acids. Under optimum conditions they could separate 25 amino acids in less than 25 min. Figure 7.29 shows a typical isocratic separation by reversed-phase chromatography.

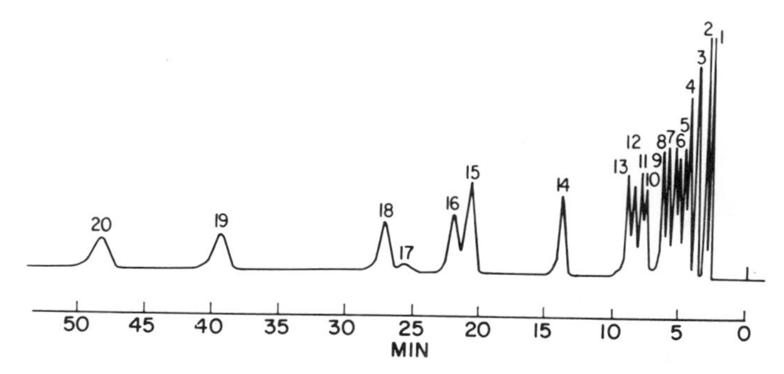

Fig. 7.29 Isocratic separation of 20 amino acid *o*-phthalaldehyde derivatives on a 200 × 7.6 mm (i.d.) Nucleosil RP-18 (5 μm) column with a mobile phase consisting of methanol/phosphate buffer, 0.1 *M*, pH 6.8 (43/57). Peaks: 1, Asp; 2, Glu; 3, Asn; 4, Ser; 5, Gln; 6, His; 7, methionine sulfone; 8, Thr; 9, Gly; 10, Arg; 11, β-Ala; 12, Tyr; 13, Ala; 14, α-aminobutyric acid; 15, Trp; 16, Val; 17, NH_4^+; 18, PHe; 19, Ile; and 20, Leu. From Lindroth and Mopper (*35*), with permission from the American Chemical Society.

H. Fluorescent Hydrazones

Dansyl hydrazine (5-dimethylaminonaphthalene-1-sulfonyl hydrazine) has been used to form fluorescent hydrazones of aldehydes and ketones (*40,41,219,220*), including ketosteroids, glycoproteins and reducing sugars. The reaction proceeds in ethanolic solution (or DMSO) with a trace of HCl. Figure 7.30 shows a reaction scheme for the formation of the fluorescent products. The hydrazones are extracted from the basified reaction mixture with diethyl ether for chromatography. The fluorescence maxima are 340 nm ex and 525 nm emission. The method has been applied to the TLC analysis of keto steroids (*219*) with a detection limit of 1–2 nmoles. Unlike most of the other reagents described above, dansyl hydrazine itself is very fluorescent and must be removed from the derivatives. Normally the excess reagent is converted to a water-soluble product with substances like sodium pyruvate, after completion of the reaction.

2-Diphenylacetyl-1,3-indandione-1-hydrazone reacts with aldehydes and ketones (*221*) in a similar manner to dansyl hydrazine and has been used for paper-chromatographic analyses (*222*). Detection limits are about 2 ng/spot by this technique. TLC or LC methods would probably improve sensitivity significantly, although no work in this area has yet been carried out. The derivatives are excited at 400–415 nm, with emission at about 525 nm.

I. Fluorescent Isothiocyanates and Isocyanates

The reagent 2-(4′-isocyanathophenyl)-6-methylbenzthiazole is an isocyanate capable of reacting with amines, amino acids, phenols, and alcohols to produce fluorescent substituted carbamates or ureas (*223*). Figure 7.31 illustrates a reaction with a primary amine to produce the corresponding urea derivative. Upon TLC, as little as 0.1–0.01 μmoles of

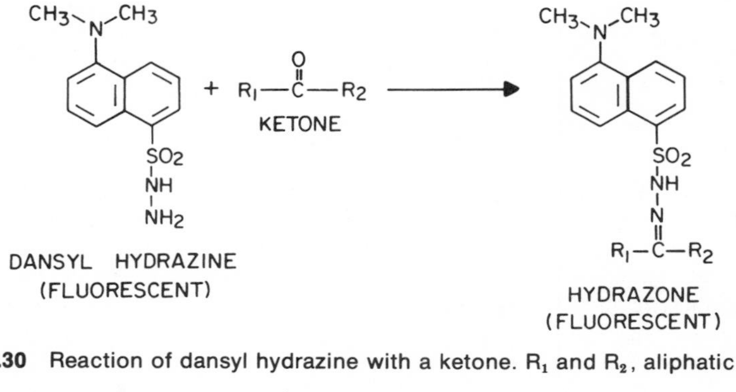

DANSYL HYDRAZINE (FLUORESCENT) + KETONE → HYDRAZONE (FLUORESCENT)

Fig. 7.30 Reaction of dansyl hydrazine with a ketone. R_1 and R_2, aliphatic or aryl.

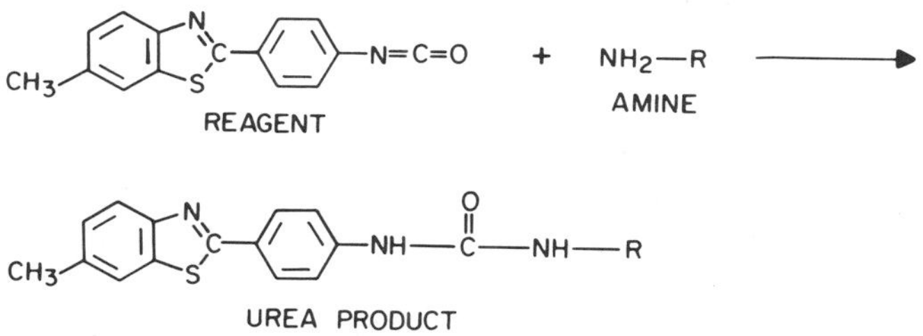

Fig. 7.31 Reaction of 2-(4′-isocyanatophenyl)-6-methylbenzthiazole with an amine to form the fluorescent urea derivative.

aliphatic amines can be detected when irradiated at 350 nm. Emission of the derivatives is around 383 nm, whereas the reagent itself emits at 374 nm.

Fluorescein isothiocyanate has also been used as a reagent for the formation of fluorescent derivatives of amines and amino acids for chromatographic purposes (*224–226*). Figure 7.32 shows the structure of fluorescein isothiocyanate.

Acridine isothiocyanate (9-isothiocyanatoacridine) reacts with primary and secondary amines to form highly fluorescent products (*227*). Figure 7.33 illustrates the reaction. The mechanism is somewhat different than that shown in Fig. 7.31, since the derivatization involves a ring closure after photo-oxidation. This was found necessary since without the ring closure, the derivative is significantly less fluorescent than the reagent itself (*228*).

J. EDTN Reaction With Alcohols and Phenols

EDTN [1-ethoxy-4-(dichloro-5-triazinyl)naphthalene] has been used for the formation of fluorescent derivatives of primary alcohols and phenols, but not secondary aliphatic alcohols. The reaction takes place in aqueous acetone with a small quantity of sodium carbonate as a catalyst. The

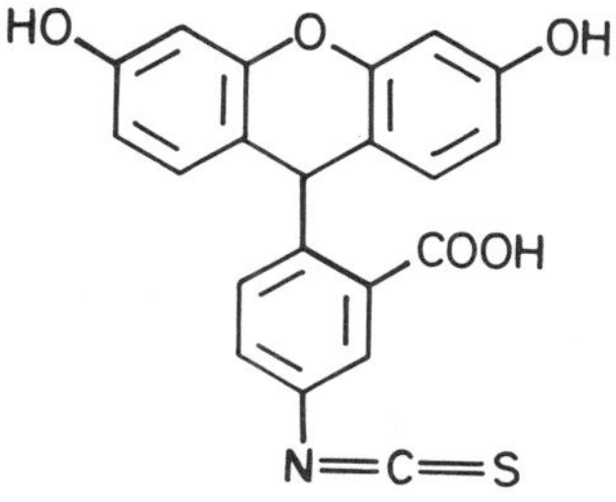

Fig. 7.32 Fluorescein isothiocyanate.

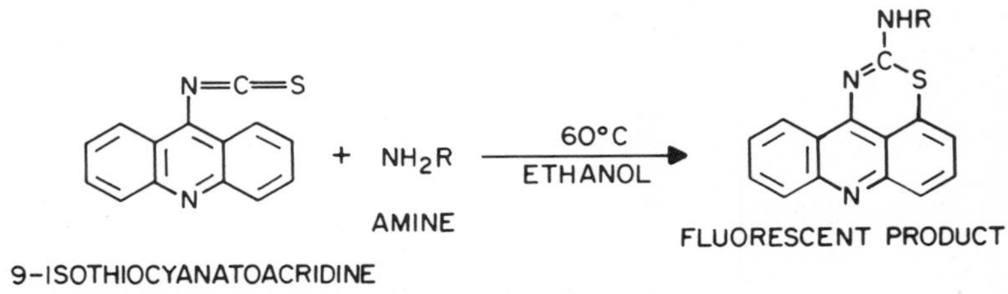

N=C=S
9-ISOTHIOCYANATOACRIDINE + NH_2R AMINE —60°C, ETHANOL→ FLUORESCENT PRODUCT

Fig. 7.33 Reaction of 9-isothiocyanatoacridine with an amine.

derivatives are extracted from the mixture with methylene chloride for chromatographic analysis. Excitation occurs at 352 nm in alcohol solution, and emission is at 379 nm. The sensitivity is significantly less than the other reagents described, being about 100 ng for corticosteroids (*229*).

K. Pyridoxal Derivatives of Amines and Amino Acids

Pyridoxal forms fluorescent derivatives with amines and amino acids by first forming a Schiff base in a weakly alkaline medium; this is then reduced with sodium tetrahydroborate to form a stable derivative. Fluorescence excitation and emission maxima appear at 332 nm and 400 nm, respectively. As little as 0.5 nmoles of amino acids may be determined by LC (*230,231*), which is about tenfold less sensitive than several of the other reagents discussed for amino acid analysis.

L. Arylboronic Acid Reactions With Bifunctional Compounds

Phenanthrene boronic acid and other polynuclear aromatic boronic acids were studied for application to the fluorescence labeling of some 1,2-, 1,3-, or 1,4-bifunctional compounds (*232*). The reaction is analogous to that of forming GC-stable alkyl boronates. Preparation of the reagents have been described (*233*). This approach has been used to determine two ecdysones in whole insects by TLC (*232*).

M. Vanadium Pentoxide Oxidation of Reserpine

The oxidation of reserpine with vanadium pentoxide in phosphoric acid to produce a fluorescent product has been utilized for its determination in plasma by LC. Ion-pair chromatography was employed with heptane sulfonate as counter-ion (*234*). The fluorophor is sensitive enough to be detected at 100 pg/ml plasma. This same reaction was used for the detection of reserpine after separation by TLC (*235*). The procedure is highly specific, since it has been shown that the 2,3,4,9-tetrahydro-7-methoxy-1*H*-pyrido[3,4,6]indole moiety is a requirement for oxidative conversion to the fluorescent product (*236*). Thus, related alkaloids that do not contain this group are not detected (*237*).

N. Ferricyanide Oxidation of Thiamine and Morphine

Oxidation with potassium ferricyanide has been used for the fluorescence determination of thiamine in urine samples (*238*). The product, thiochrome, is extracted from the reaction mixture and then subjected to LC for quantitative analysis. Minimum detectable levels are in the area of 30 ng/ml of urine. An on-column oxidation technique using potassium ferricyanide for the conversion of morphine to the fluorescent "pseudomorphine" has been described (*239*). Levels as low as 0.01 μg/ml in urine could be quantitatively detected. Figure 7.34 illustrates the overall reaction scheme. The on-column reaction was found most suitable, since the derivatives are not stable. However, the relatively fast reaction would most likely be more suitable for postcolumn detection, since the reagents themselves are not fluorescent. This would alleviate injection of the reagents onto the column.

O. Chloroacetaldehyde Reaction with Adenine

The reaction of chloroacetaldehyde with adenine and adenosine (and its nucleotides) to produce fluorescent 1,N^6-etheno derivatives has been used for a selective LC method capable of determining low picomole levels of these compounds (*240*). The derivatives were separated on a porous polystyrene resin using 0.1 *M* phosphate buffer as the mobile phase.

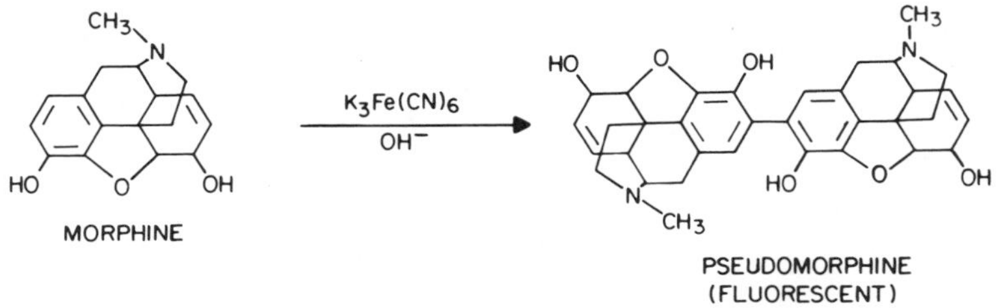

Fig. 7.34 Oxidation of morphine to the fluorescent "pseudomorphine" using potassium ferricyanide.

P. Aflatoxins

Aflatoxins B_1 and G_1 have been determined by LC after conversion to their highly fluorescent hemiacetals, B_{2a} and G_{2a} (*241–243*). The reaction, partially shown in Fig. 7.35, takes place in the presence of trifluoroacetic acid. The products are extracted into mobile phase (acetonitrile to water, 9:1) and analyzed by reversed-phase chromatography on Spherisorb ODS. Detection limits have been reported to be in the low picogram range, with fluorescence detection at 365 nm (ex) and 450 (em).

Q. Conclusion

Although the above describes reagents that have been used to form fluorescent products prior to chromatography, derivatization is by no means limited to those reagents. Many chemicals exist that may be used to form suitable fluorescent products. One can make use of organic chemical principles for reactions and substitute a reagent that would produce a fluorescent product rather than one that would not. However, it is necessary to have a basic understanding of fluorescence theory in order to be able to make a useful selection (see Chapter 5, Fluorescence Detectors). Books such as *Spot Tests in Organic Analysis* (*244*), *Handbook of Tables for Organic Compound Identification* (*245*) and others (*2,10*) provide many types of reactions for the formation of chemical derivatives.

IV. DERIVATIZATION FOR OTHER DETECTION MODES

Most prechromatographic derivatization procedures for LC are employed either for fluorescence of UV-visible absorption detection. However, several other detection modes are in use in LC, and derivatization aimed at these systems may provide distinct advantages over fluorescence or absorption.

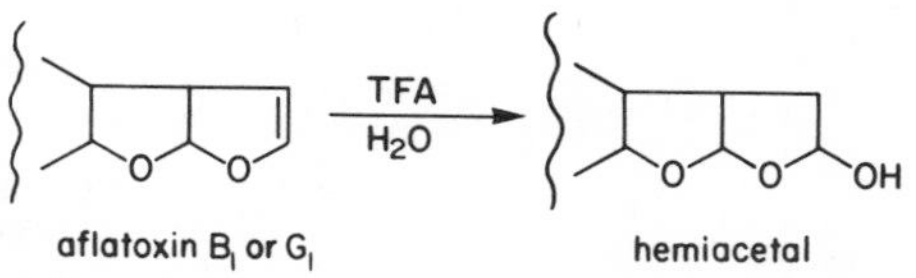

TFA
H_2O
aflatoxin B_1 or G_1
hemiacetal

Fig. 7.35 Conversion of aflatoxin B_1 or G_1 to their respective hemiacetals by catalysis with trifluoroacetic acid (TFA).

A. Derivatization for Electrochemical Detection

The potential of pre- and postcolumn derivatization for LC with electrochemical detection has been reviewed (*9*). There are a number of reagents available that can produce electrochemically sensitive derivatives of hydroxyl (*15*), amino and amino acid (*269,270*), carboxylic acid (*156,271*), and carbonyl (*92,108*) compounds. All of these contain one or two reducible nitro groups and thus may be also used for sensitive UV detection as well. However, electrochemical detection can be more sensitive and much more selective than UV absorption, providing the analyst with a valuable approach to analysis at trace concentrations in complex substrates such as biological fluids, foods, or soils. The nitro-containing reagents listed in Table 7.1 are suitable for electrochemical detection. In addition, related reagents producing 3,5-dinitrobenzoyl (*58*) and 2,4-dinitrobenzenesulfonyl (*270*) derivatives should be useful.

B. LC–Atomic Absorption

Derivatization for LC with atomic absorption (AA) detection can be carried out for two purposes, both involving metal chelation. First, for the chromatographic separation of metals, chelates can be formed with appropriate ligands such as those mentioned in Fig. 7.12. Since the AA detection is based on atomic absorption of the metal, the chelation only serves to aid chromatography. Second, various ligands may be determined by complexing with metals; the metal in this case provides a mechanism for selective AA detection. Applications of metal chelates in LC with AA detection have been reported for the analysis of amino carboxylic acid ligands such as EDTA, NTA, EGTA, and CDTA as copper chelates (*272*). Other applications for metal chelates (*273–276*) and organometallic compounds (*277–279*) have also been reported. In addition to these, AA detection is also suitable for the metal analyses mentioned earlier for UV detection (Section II,I).

C. Radiochemical Derivatization

The application of radiochemical detection in LC has been hindered by the fact that long counting times are required for low levels of ^{14}C or ^{3}H. Although research is continuing in this area, it will be some time before derivatization with a radioactive tag will find much practical use in LC. TLC has been used for many years to separate radioactive species, and radioscanning of TLC plates is also in much use. Counting times can be very long because the chromatographically separated compounds are held stationary by the TLC layer. Discussion of the practice and potential of radiochemical analysis associated with chromatography has appeared (*10,280,281*).

V. POSTCOLUMN REACTIONS

A. Introduction

Postcolumn reactions have become more popular in the last few years as a means of selective and sensitive detection of compounds eluting from an LC apparatus. The purpose is to form a UV–visible-absorbing, fluorescent, or electrochemically active derivative after separation, but before the mobile phase enters the detector. This approach has been used routinely for many years in the commercial amino acid analyzers.

One advantage of postcolumn detection is that the reaction need not go to completion or even produce well-defined products, as long as the reaction is reproducible. Any side reactions or artifacts formed because of reaction with the solute molecules pose no problem in a postcolumn reaction system since all enter the detector at the same time, contributing to the total response. It is also possible to combine detectors for dual detection. For example, UV detection can be carried out immediately after the column; then the postcolumn reaction is performed afterward, before the products enter a second detector for detection of the reaction products.

Chromatographic separations are most likely to be superior when postcolumn reactions are carried out. The reason is, of course, that chromatography is based on the separation of the compounds themselves and not the derivatives. As mentioned earlier, derivatization with reagents that produce bulky products tends to mask small physical or chemical differences between similar molecules, making separation more difficult.

Postcolumn reactions are, in reality, automated chemical derivatizations carried out continuously in a moving stream. Such automation offers the distinct advantage of saving time. Rather than carry out manual chemical reactions involving usually an extraction step or some other preliminary cleanup, an automated postcolumn technique may be preferred.

Some restrictions on the use of postcolumn reactions are as follows. The most important consideration is that the reagents themselves do not interfere in the detection. This is critical, since excess reagent cannot usually be removed. Recent work on the use of postcolumn reactions (ion-pair formation) with dynamic extraction for removal of excess reagent has been shown to offer much potential in this regard (*50,51,282,283*). However, more work is required to apply the technique on a broader scale.

The reaction kinetics are also important since it is preferable to have rapid reactions (not necessarily to 100% completion) so that the time in the postcolumn system is minimal. This will keep band spreading as low as possible, thus maintaining the efficiency of the total system as high as possible.

Another restriction on postcolumn reactions is the reaction medium. This always must include the mobile phase, which is not necessarily (is seldom, in fact) the best solvent mixture for the reactions. Also, gradient elution may be difficult to carry out because it results in a change of mobile phase that can influence the reaction. The following discusses certain types of postcolumn reaction systems as well as applications to organic trace analysis.

B. Types of Postcolumn Reactors

There are several types of postcolumn reactors available that have been applied to various analytical problems. All have particular advantages and disadvantages with regard to mixing of streams, band broadening, geometry, and applications.

1. Air-Segmented Reaction Systems

This type of system is used in most commercially available autoanalyzers. The column effluent is segmented into small portions by means of air bubbles introduced into the stream at regular intervals (*284*). The purpose of this is to reduce longitudinal diffusion, thus minimizing band broadening. The principle is very efficient and particularly useful with postcolumn reactions of 5 min or greater.

Theoretical treatment of mixing and dispersion in air-segmented streams has been reported (*49,285–288*). Axial dispersion is related to the thickness of the liquid film between the air bubble and tubing wall, the length of the tube, the length of the liquid segment, and the flow rate. An optimized system would include small liquid segments, a high frequency of air bubbles, short reaction tubes, high flow rates, and small tube diameter. However, other factors—including dead volume of connecting tees and the debubbler unit—can make a large contribution to band broadening.

2. Nonsegmented Continuous Flow Tubular Reactors

Nonsegmented flow systems normally make use of capillary tubing for the reaction unit. This is preferable for fast reactions (e.g., 30 sec or less) where little time is required for the formation of the products. Theoretical treatment of band broadening in this type of system has appeared (*289*,*290*). It has been shown that a coiled reactor design results in reduced band broadening because of better radial mixing.

3. Packed-Bed Reactors

Packed-bed reactors usually consist of a straight piece of tubing packed with glass beads or other materials that do not retain the solute molecules or reagents. Such a system can be considered analogous to a chromatography column under nonretention conditions. It has been suggested by a number of authors that well-designed bed reactors should be preferred when reactor times between 30 sec and several minutes are required (*290–293*).

C. Applications of Postcolumn Reactors in LC

The following sections describe applications of postcolumn reactors to LC analyses. The reactions, reagents, geometries, etc., used indicate the versatile nature of postcolumn reaction detection. Also, the methods described show applications to compounds over a diverse range, indicating the general potential of the approach for LC analysis.

1. Fluorescamine

Fluorescamine (Fig. 7.25) is particularly suited to post-column reactions. It is selective for primary amines and reacts in a few seconds with compounds such as amino acids. The products are strongly fluorescent, whereas the reagent itself or its hydrolysis products are not. Because of the short reaction times, tubular reactors are almost always employed. Primary and secondary amino acids have been determined by several authors (*294–298*). The secondary amino acids were first converted to primary amines through oxidative decarboxylation with *N*-chlorosuccinimide (*294*). The amino acids are normally separated by reversed-phase or ion-exchange chromatography. Detection limits range down to about 100 pmoles of amino acid at fluorescence wavelengths of 390 nm (ex) and 475 nm (em).

When reactions are fast, reagent mixing becomes important. It has been shown that mixing-tee design can greatly affect the overall reaction because of differences in mixing rate, especially if the densities of the individual streams are significantly different (*299*,*300*).

Other applications include indoles (*301*), hydrazines (*302*), aminophosphoric acids (*303*), and cefatrizine (an orally active cephalosporin) (*304*).

Fluorescamine has also been used for precolumn derivatization (see Section III,D) of a wide variety of primary amino compounds.

2. Orthophthalaldehyde

Orthophthalaldehyde (OPA) reacts with primary amines in the presence of mercaptoethanol to form highly fluorescent products. Like fluorescamine, the reagents themselves are nonfluorescent, making removal of excess reagent unnecessary. Segmented flow (*305*), continuous flow (*306,307*), and bed-reactor (*289*) systems have been employed for detection of amino acids and other amines via this reaction. Carbamate pesticides have been detected after a postcolumn hydrolysis step in which methylamine is liberated (*308,309*). The resulting amine reacts with OPA, and the product is detected fluorometrically at 340 nm (ex) and 455 (em). Other applications include peptides (*310*), polyamines (*311*) and 5-hydroxytryptophan (*312*), and other amino compounds (*313*). OPA has also been used for precolumn derivation as mentioned in Section III,G.

3. NBD-Cl Reaction with Secondary Amines

NBD-Cl (4-chloro-7-nitrobenz-2,1,3-oxadiazole) has been evaluated as a reagent for the postcolumn detection of secondary amino acids (*314*) which cannot be detected with either fluorescamine or OPA. This technique was suitable for proline and proline-containing peptides. Some properties of NBD amines are given in Section III,C, where precolumn derivatization reactions employed the same reagent.

4. Enzyme Reactions

Enzyme reactions are often extremely selective and thus can provide a very useful means of detecting substances eluting from an LC. Carbamate pesticides, for example, were detected using an enzymatic postcolumn reaction with cholinesterase (*315*). The reaction involves conversion of a substrate by the enzyme to a fluorescent product. When an inhibiting insecticide appears, the reaction stops and a decrease in fluorescence is measured. Detection limits ranged from 0.2–2 ng per injection. A similar enzyme technique was used to monitor for a number of insecticides (*316*). In this case, a colorimetric reaction was used between enzyme and substrate. As low as 10 pgs could be detected, depending on the inhibition strength of the insecticide.

The use of enzyme–substrate reactions have been employed for the postcolumn detection of various enzymes (*292,317*) of some importance clinically. A bed reactor was used in this case. As low as 6 ng of enzyme could be detected with this system.

5. Oxidation–Reduction Reactions

Various oxidation–reduction reactions have been employed in postcolumn reactions. Cerate oxidation has been useful in the detection of phenolics, polythionates, and organic acids (*318–320*). These compounds convert Ce^{4+} to the fluorescent Ce^{3+}, which can be monitored and related to substrate concentration. The reaction takes about 4 min in a segmented system. A similar reaction was applied to tetra- and triiodothyronine hormones (*321*). In this case, the hormones catalyzed a redox reaction between Ce^{4+} and As^{3+}. Detection was based on the disappearance of the Ce^{4+} absorbance at 365 nm.

Reducing sugars have been detected using a redox system employing ferricyanide. The reducing sugar converts $Fe(CN)_6^{3-}$ to $Fe(CN)_6^{4-}$ which can be detected electrochemically (*292,322*). Redox reactions have also been used to convert compounds such as dopa, dopamine, adrenaline, metanephrine, and others to fluorescent products (*323,324*).

6. Ion-Pair Formation

Postcolumn ion-pair formation or complexation has been employed in several cases for selective detection of compounds separated by LC. *N,N*-dimethylbenzylamine was determined by means of an iodine charge-transfer complexation reaction (*325*). "True" serum creatinine has been detected by means of a post-column complexation reaction with picric acid in the presence of NaOH (*326*). Ion-pair detection of some pharmaceuticals and pesticide metabolites has been carried out using solvent segmentation to remove excess counter-ion (*282–284,327*). The latter have been shown to be especially useful for tertiary amines since they are difficult to derivatize. The technique involves the formation of an ion-pair as the basic compound elutes from LC column. The counter-ion is usually strongly fluorescent (*282–284*). The resulting ion-pair is continuously extracted into an organic phase, and the fluorescence of the ion-pair (based on the fluorescence of the counter-ion) is measured and related to the quantity of base present. Excess counter-ion remains in the aqueous phase and is directed to waste. This same approach may be used to detect organic acids, employing a basic counter-ion that is highly fluorescent or UV–visible-absorbing.

7. Acid Treatment of the LC Effluent

Many organic compounds are converted to UV–visible or fluorescent chromophores upon treatment with strong acid. This has been exploited for postcolumn reactions in LC. Carbohydrates (*328*) have been determined after postcolumn treatment with sulfuric acid at 110°C. Absorbance was measured at about 297 nm. The method has been applied to a large number of sugars.

Cardiac glycosides were treated postcolumn with concentrated HCl to yield a fluorescent product (*329*). Ascorbic acid and hydrogen peroxide were added to catalyze the reaction.

Condensed phosphates were hydrolyzed with 5 *N* sulfuric acid to yield orthophosphate, which was then colorimetrically determined at 420 nm via the phosphovanadomolybdate reaction (*289*).

8. Photochemical Reactions

Chemical reactions initiated by light have been examined by a few authors for application to LC analyses. For example, *N*-nitroso compounds have been detected after photochemical conversion to nitrite, which was then determined colorimetrically with the Griess reaction (*330*).

Cannabinoids were converted to highly fluorescent photoproducts after only a few seconds of irradiation (*331*).

Photochemical reactions have been used to confirm the presence of some ergot alkaloids (*332*). The fluorescence signal of the compounds disappears after about 20 sec of UV irradiation. This phenomenon was used to confirm the presence of ergots in urine.

9. Other Reactions

Other applications of postcolumn reactions employed for select purposes follow. The determination of cyclohexanone in cyclohexanone oxime was carried out after postcolumn derivation with 2,4-dinitrophenylhydrazine (*285*). The hydrazine was measured at 430 nm. A colorimetric reaction has been employed for the pesticide parathion and its oxygen analog paraoxon (*333*).

Postcolumn ionization can be very useful as a selective detection technique. Barbiturates (*334*) have been detected by UV after increasing the pH of the column effluent to ~ 10. The absorption of the barbiturates increases by a factor of 20 under these conditions.

REFERENCES

1. J. Drozd, *J. Chromatogr.* **113,** 303 (1975).
2. K. Blau and G. S. King, "Handbook of Derivatives for Chromatography." Heyden, London, 1977.
3. S. Ahuja, *J. Pharm. Sci.* **65,** 163 (1976).
4. J. D. Nicholson, *Analyst* (*London*) **103,** 193 (1978).
5. J. D. Nicholson, *Analyst* (*London*) **103,** 1 (1978).
6. J. F. Lawrence, *J. Chromatogr. Sci.* **17,** 147 (1979).

7. T. Jupille, *J. Chromatogr. Sci.* **17,** 160 (1979).
8. J. F. Lawrence, *in* "Chemical Derivatization and Modification Techniques in Analytical Chemistry" (R. W. Frei and J. F. Lawrence, eds.), in press. Plenum, New York, 1981.
9. P. T. Kissinger, K. Bratin, G. C. Davis, and L. A. Pachla, *J. Chromatogr. Sci.* **17,** 137 (1979).
10. J. F. Lawrence and R. W. Frei, "Chemical Derivatization in Liquid Chromatography." Elsevier, Amsterdam, 1976.
11. M. S. F. Ross, *J. Chromatogr.* **133,** 408 (1977).
12. R. A. Henry, J. S. Schmidt, and J. F. Dieckman, *J. Chromatogr. Sci.* **9,** 513 (1971).
13. L. J. Papa and L. P. Turner, *J. Chromatogr. Sci.* **10,** 747 (1972).
14. I. R. Politzer, G. W. Griffith, B. J. Dowty, and J. L. Laseter, *Anal. Lett.* **6,** 539 (1973).
15. D. R. Knapp and S. Krueger, *Anal. Lett.* **8,** 603 (1975).
16. Regis Lab. Notes, No. 17. Regis Chem. Co., Morton Grove, Illinois, 1975.
17. F. A. Fitzpatrick and S. Siggia, *Anal. Chem.* **45,** 2130 (1973).
18. C. R. Clark and M. M. Wells, *J. Chromatogr. Sci.* **16,** 332 (1978).
19. P. J. Porcaro and P. Shubiak, *Anal. Chem.* **44,** 1865 (1972).
20. R. F. Borch, *Anal. Chem.* **47,** 2437 (1975).
21. H. D. Durst, M. Milano, E. J. Kitka, Jr., S. A. Connelly, and E. Grushka, *Anal. Chem.* **47,** 1797 (1975).
22. E. Grushka, H. D. Durst, and E. J. Kitka, Jr., *J. Chromatogr.* **112,** 673 (1975).
23. A. Hulshoff, H. Roseboom, and J. Renema, *J. Chromatogr.* **186,** 535 (1979).
24. M. J. Cooper and M. W. Anders, *Anal. Chem.* **46,** 1849 (1974).
25. E. Bayer, E. Grom, B. Kaltenegger, and R. Uhmann, *Anal. Chem.* **48,** 1106 (1976).
26. N. Seiler and M. Wiechman, *in* "Progress in Thin-Layer Chromatography and Related Methods" (A. Niederwieser and G. Pataki, eds.), Vol. 1, p. 95. Ann Arbor Sci. Publ., Ann Arbor, Michigan, 1970.
27. L. P. Penzes and G. W. Oertel, *J. Chromatogr.* **51,** 325 (1970).
28. F. Nachtmann, H. Spitzy, and R. W. Frei, *Anal. Chim. Acta* **76,** 57 (1975).
29. K. Samejima, M. Kawase, S. Sakamoto, M. Okada, and Y. Endo, *Anal. Biochem.* **76,** 392 (1976).
30. G. Schwedt, *J. Chromatogr.* **118,** 429 (1975).
31. F. Reiterer, F. Nachtmann, G. Knapp, and H. Spitzy, *Mikrochim. Acta* p. 115 (1978).
32. W. Dunges, *Anal. Chem.* **49,** 443 (1977).
33. W. Dunges, *Chromatographia* **9,** 624 (1976).
34. W. Dunges and N. Seiler, *J. Chromatogr.* **145,** 483 (1978).
35. P. Lindroth and K. Mopper, *Anal. Chem.* **51,** 1667 (1979).
36. D. W. Hill, F. H. Walters, T. D. Wilson, and J. D. Stuart, *Anal. Chem.* **51,** 1338 (1979).
37. G. L. Dadisch and P. Wolschann, *Fresenius' Z. Anal. Chem.* **292,** 219 (1978).
38. T. J. Hoper, R. C. Briner, H. G. Sadler, and R. L. Smith, *J. Forensic Sci.* **21,** 842 (1976).
39. H. F. DeBrabander and R. Verbeke, *J. Chromatogr.* **108,** 141 (1975).
40. G. Avigad, *J. Chromatogr.* **139,** 343 (1977).
41. A. E. Eckhardt, C. E. Hayes, and I. J. Goldstein, *Anal. Biochem.* **73,** 192 (1976).
42. G. L. Brun and V. Mallet, *Int. J. Environ. Anal. Chem.* **3,** 73 (1973).
43. H. Shanfield, F. Hsu, and A. J. P. Martin, *J. Chromatogr.* **126,** 457 (1976).
44. R. Segura and A. M. Grotto, Jr., *J. Chromatogr.* **99,** 643 (1974).
45. W. P. Cochrane, *J. Chromatogr. Sci.* **17,** 124 (1979).
46. W. P. Cochrane, *in* "Chemical Derivatization and Modification Techniques in Analytical Chemistry" (R. W. Frei and J. F. Lawrence, eds.), in press. Plenum, New York, 1981.
47. J. F. Lawrence, *J. Assoc. Off. Anal. Chem.* **59,** 1066 (1976).

48. W. C. Kossa, J. MacGee, S. Ramachandran, and A. J. Webber, *J. Chromatogr. Sci.* **17,** 177 (1979).
49. L. R. Snyder and H. J. Adler, *Anal. Chem.* **48,** 1022 (1976).
50. J. F. Lawrence, U. A. T. Brinkman, and R. W. Frei, *J. Chromatogr.* **171,** 73 (1979).
51. J. F. Lawrence, U. A. T. Brinkman, and R. W. Frei, *J. Chromatogr.* **185,** 473 (1979).
52. R. W. Frei and A. H. M. T. Scholten, *J. Chromatogr. Sci.* **17,** 152 (1979).
53. R. W. Frei, *in* "Chemical Derivatization and Modification Techniques in Analytical Chemistry" (R. W. Frei and J. F. Lawrence, eds.), in press. Plenum, New York, 1981.
54. J. F. Lawrence and R. Leduc, *J. Assoc. Off. Anal. Chem.* **61,** 872 (1978).
55. F. Nachtmann, H. Spitzy, and R. W. Frei, *J. Chromatogr.* **122,** 293 (1976).
56. F. Nachtmann, *Z. Anal. Chem.* **282,** 209 (1976).
57. W. H. Newsome and L. Panopio, *J. Agric. Food Chem.* **26,** 638 (1978).
58. M. A. Carey and H. E. Persinger, *J. Chromatogr.* **10,** 537 (1972).
59. F. Nachtmann and K. W. Budna, *J. Chromatogr.* **136,** 279 (1977).
60. M. D. Ullman and R. H. McCluer, *J. Lipid Res.* **18,** 371 (1977).
61. J. W. Higgins, *J. Chromatogr.* **121,** 329 (1976).
62. C. R. Clark, J. D. Teague, M. M. Wells, and J. H. Ellis, *Anal. Chem.* **49,** 912 (1977).
63. D. H. Niederhiser, R. K. Fuller, L. J. Hejduk, and H. P. Roth, *J. Chromatogr.* **117,** 187 (1976).
64. S. L. Wellons and M. A. Carey, *J. Chromatogr.* **154,** 219 (1978).
65. R. Schwyzer and H. Kappler, *Helv. Chim. Acta* **46,** 1550 (1973).
66. G. B. Cox, *J. Chromatogr.* **83,** 471 (1973).
67. E. R. Holden, *J. Assoc. Off. Anal. Chem.* **56,** 713 (1973).
68. J. F. Lawrence, *J. Agric. Food Chem.* **22,** 936 (1974).
69. J. Elion, M. Downing, and K. Mann, *J. Chromatogr.* **155,** 436 (1978).
70. V. M. Stepanov, S. P. Katrukha, L. A. Baratova, L. P. Belyanova, and V. P. Kouzhenko, *Anal. Biochem.* **43,** 209 (1971).
71. P. Frankhauser, P. Fried, P. Stahala, and M. Brenner, *Helv. Chim. Acta* **57,** 271 (1974).
72. M. Downing and K. Mann, *Anal. Biochem.* **74,** 298 (1976).
73. G. Frank and W. Strubert, *Chromatographia* **6,** 522 (1973).
74. C. L. Zimmerman, E. Appela, and J. J. Pisno, *Anal. Biochem.* **75,** 569 (1977).
75. W. F. Moo-penn, M. H. Johnson, K. C. Betchel, and D. L. Jue, *J. Chromatogr.* **172,** 476 (1979).
76. C. L. Zimmerman, E. Appela, and J. J. Pisano, *Anal. Biochem.* **75,** 77 (1976).
77. J. X. De Vries, R. Frank, and C. Birr, *FEBS Lett.* **55,** 65 (1975).
78. L. A. Sternson, W. J. De Witte, and J. G. Stevens, *J. Chromatogr.* **153,** 481 (1978).
79. F. A. Fitzpatrick and S. Siggia, *Anal. Chem.* **45,** 2310 (1973).
80. M. J. Cooper and M. W. Anders, *J. Chromatogr. Sci.* **13,** 407 (1975).
81. G. J. Schmidt and W. Slavin, *Chromatogr. Newsl.* **6,** 22 (1978).
82. F. A. Fitzpatrick, *Anal. Chem.* **48,** 499 (1976).
83. M. W. Merrit and G. E. Bronson, *Anal. Chem.* **48,** 1851 (1976).
84. W. Morozowich and S. L. Douglas, *Prostaglandins* **10,** 19 (1975).
85. P. T. S. Pei, W. C. Kossa, S. Ramachandran, and R. S. Henly, *Lipids* **11,** 814 (1976).
86. S. Lam and E. Grushka, *J. Liq. Chromatogr.* **1,** 33 (1978).
87. R. O. Morris and J. B. Zaerr, *Anal. Lett.* **A11,** 73 (1978).
88. R. A. Miller, N. E. Bussell, and C. Picketts, *J. Liq. Chromatogr.* **1,** 291 (1978).
89. E. O. Umeh, *J. Chromatogr.* **56,** 29 (1971).
90. H. C. Jordi, *J. Liq. Chromatogr.* **1,** 215 (1978).
91. E. Heftman and G. Saunders, *J. Liq. Chromatogr.* **1,** 333 (1978).
92. E. H. White, A. A. Baum, and D. E. Eitel, *Org. Syn.* **48,** 102 (1968).

93. Regis Lab. Notes, No. 16. Regis Chem. Co., Morton Grove, Illinois, 1974.
94. F. A. Schmidt and D. J. Hutchinson, *Cancer Res.* **34,** 1671 (1974).
95. N. E. Hoffman and J. C. Liao, *Anal. Chem.* **48,** 1104 (1976).
96. D. P. Mathees and W. C. Purdy, *Anal. Chim. Acta* **109,** 61 (1979).
97. D. P. Mathees and W. C. Purdy, *Anal. Chim. Acta* **109,** 161 (1979).
98. E. Roder and J. Merzhauser, *Fresenius' Z. Anal. Chem.* **272,** 34 (1974).
99. J. Merzhauser, E. Roder, and C. Hesse, *Klin. Wochenschr.* **51,** 883 (1973).
100. J. Thiem, J. Schwentner, H. Karl, A. Sievers, and J. Reimer, *J. Chromatogr.* **155,** 107 (1978).
101. L. P. Penzes and G. W. Oertel, *J. Chromatogr.* **51,** 322 (1970).
102. E. Roder, I. Pigulla, and J. Troschütz, *Fresenius' Z. Anal. Chem.* **288,** 56 (1977).
103. T. Sugiura, T. Hayashi, S. Kawai, and T. Ohno, *J. Chromatogr.* **110,** 385 (1975).
104. S. Selim, *J. Chromatogr.* **136,** 271 (1977).
105. L. Treiber, P. Knapstein, and J. C. Touchstone, *J. Chromatogr.* **37,** 83 (1968).
106. F. A. Fitzpatrick, S. Siggia, and J. Dingman, Sr., *Anal. Chem.* **44,** 2211 (1972).
107. H. Katsui, T. Yoshida, C. Tanegashima, and S. Tanaka, *Anal. Biochem.* **34,** 349 (1971).
108. N. Ariga, *Anal. Biochem.* **49,** 436 (1972).
109. H. Katsuki, T. Yoshida, C. Tanegashima, and S. Tanaka, *Anal. Biochem.* **24,** 119 (1968).
110. B. Sieferte and M. Kolbe, *Fresenius' Z. Anal. Chem.* **271,** 337 (1974).
111. H. Terada, T. Hayashi, S. Kawai, and T. Ohno, *J. Chromatogr.* **130,** 281 (1977).
112. S. Honda and K. Kakehi, *J. Chromatogr.* **152,** 405 (1978).
113. T. H. Jupille, *Am. Lab.* **8,** 85 (1976).
114. F. A. Fitzpatrick, M. A. Wynalda, and D. G. Kaiser, *Anal. Chem.* **49,** 1032 (1977).
115. J. F. K. Huber, J. C. Kraak, and H. Veening, *Anal. Chem.* **44,** 1554 (1972).
116. I. Jonas and B. Norden, *Nature (London)* **258,** 597 (1975).
117. K. Saitoh and N. Suzuki, *J. Chromatogr.* **109,** 333 (1975).
118. D. R. Jones and S. E. Manahan, *Anal. Lett.* **8,** 569 (1975).
119. K. Lorben, K. Mueller, and H. Spitzy, *Mikrochim. Acta* p. 602 (1975).
120. M. Lochmuller, P. Heizmann, and K. Ballschmiter, *J. Chromatogr.* **137,** 156 (1977).
121. J. W. O'Laughlin and T. P. O'Brien, *Anal. Lett.* **A11,** 829 (1978).
122. P. Uden and F. H. Walters, *Anal. Chim. Acta* **79,** 175 (1975).
123. E. Gaetani, C. F. Laureri, A. Mangia, and G. Parolari, *Anal. Chem.* **48,** 1725 (1976).
124. P. C. Uden, E. M. Parees, and F. H. Walters, *Anal. Lett.* **8,** 795 (1975).
125. A. Galik, *Anal. Chim. Acta* **57,** 339 (1971).
126. P. Heizmann and K. Ballschmiter, *J. Chromatogr.* **137,** 153 (1977).
127. P. Heizmann and K. Ballschmiter, *Fresenius' Z. Anal. Chem.* **266,** 206 (1973).
128. M. Moriyaser and Y. Hashimoto, *Anal. Lett.* **A11,** 593 (1978).
129. O. Liska, J. Lehotay, E. Brandsteterova, G. Guiochon, and H. Colin, *J. Chromatogr.* **172,** 384 (1979).
130. G. Schwedt, *Chromatographia* **12,** 289 (1979).
131. J. Lehotay, O. Liska, E. Brandsteterova, and G. Guiochon, *J. Chromatogr.* **172,** 379 (1974).
132. R. W. Frei, D. E. Ryan, and C. A. Stockton, *Anal. Chim. Acta* **42,** 59 (1968).
133. S. L. Bannister, L. A. Sternson, and A. J. Repta, *J. Chromatogr.* **173,** 333 (1979).
134. H. Furukawa, Y. Mori, Y. Takeuchi, and K. Ito, *J. Chromatogr.* **136,** 428 (1977).
135. T. Nambara, S. Ikegawa, M. Hasegawa, and J. Goto, *Anal. Chim. Acta* **101,** 111 (1978).
136. J. Goto, M. Hasegawa, S. Nakamura, K. Shimada, and T. Nambara, *J. Chromatogr.* **152,** 413 (1978).

137. G. Helmchen, H. Volter, and W. Schuhle, *Tetrahedron Lett.* p. 1417 (1977).
138. S. U. Khan, *Residue Rev.* **59,** 21 (1975).
139. R. Greenhalgh and J. Kovacicova, *J. Agric. Food Chem.* **23,** 325 (1975).
140. J. F. Lawrence and G. W. Laver, *J. Agric. Food Chem.* **23,** 1106 (1975).
141. J. F. Lawrence, C. van Buuren, U. A. T. Brinkman, and R. W. Frei, *J. Agric. Food Chem.* **28,** 630 (1980).
142. J. F. Lawrence, *J. Agric. Food Chem.* **22,** 936 (1974).
143. J. F. Lawrence, *J. Agric. Food Chem.* **24,** 1236 (1976).
144. W. E. Jefferson, Jr. and F. C. Chang, *Anal. Lett.* **9,** 429 (1976).
145. D. Munger, L. A. Sternson, A. J. Repta, and T. Higuchi, *J. Chromatogr.* **143,** 375 (1977).
146. D. J. Evans, R. J. Mayfield, and I. M. Russell, *J. Chromatogr.* **115,** 391 (1975).
147. B. L. Karger, S. C. Su, S. Marchese, and B. A. Persson, *J. Chromatogr. Sci.* **12,** 678 (1974).
148. W. Santi, J. M. Huen, and R. W. Frei, *J. Chromatogr.* **115,** 423 (1975).
149. J. C. Kraak and J. F. K. Huber, *J. Chromatogr.* **102,** 333 (1974).
150. S. Eksborg and G. Schill, *Anal. Chem.* **45,** 2092 (1973).
151. S. Eksborg, P. O. Lagerstrom, R. Modin, and G. Schill, *J. Chromatogr.* **83,** 99 (1973).
152. B. A. Person and P. O. Lagerstrom, *J. Chromatogr.* **122,** 305 (1976).
153. B. Fransson, K. G. Wahlund, I. M. Johansson, and G. Schill, *J. Chromatogr.* **125,** 327 (1976).
154. G. Schill, *in* "Assay of Drugs and Other Trace Compounds in Biological Fluids" (E. Reid, ed.), p. 87. North-Holland Publ., Amsterdam, 1976.
155. P. T. Kissinger, *Anal. Chem.* **49,** 883 (1977).
156. R. P. W. Scott and P. Kucera, *J. Chromatogr.* **142,** 213 (1977).
157. C. Horvath, W. Meander, I. Molnar, and P. Molnar, *Anal. Chem.* **49,** 2295 (1977).
158. J. C. Kraak, K. M. Jonker, and J. F. K. Huber, *J. Chromatogr.* **142,** 671 (1977).
159. B. S. Hartley and V. Massey, *Biochim. Biophys. Acta* **21,** 58 (1956).
160. R. W. Frei, J. F. Lawrence, J. Hope, and R. M. Cassidy, *J. Chromatogr. Sci.* **12,** 40 (1974).
161. B. A. Davis, *J. Chromatogr.* **151,** 252 (1978).
162. T. Kinoshita, F. Iinuma, and K. Atsumi, *Chem. Pharm. Bull.* **23,** 1166 (1975).
163. K. T. Hsu and B. L. Currie, *J. Chromatogr.* **166,** 555 (1978).
164. T. Seki and H. Wada, *J. Chromatogr.* **102,** 251 (1974).
165. N. D. Brown, R. B. Sweet, J. A. Kintzios, D. Cox, and B. P. Doctor, *J. Chromatogr.* **164,** 35 (1979).
166. R. Kitani, K. Imai, and Z. Iamura, *Chem. Pharm. Bull.* **18,** 1495 (1970).
167. Y. Saeki, N. Uehara, and S. Shirakawa, *J. Chromatogr.* **145,** 221 (1978).
168. J. N. Lepage, W. Lindner, G. Davies, D. E. Seitz, and B. L. Karger, *Anal. Chem.* **51,** 433 (1979).
169. R. Dvir and R. Chayen, *J. Chromatogr.* **45,** 76 (1969).
170. N. Seiler and K. Bruder, *J. Chromatogr.* **106,** 159 (1975).
171. H. Wachsmuth and R. Denusen, *J. Pharm. Belg.* **21,** 290 (1966).
172. R. W. Frei, W. Santi, and M. Thomas, *J. Chromatogr.* **116,** 365 (1976).
173. J. M. Neal and J. L. McLaughlin, *J. Chromatogr.* **73,** 277 (1972).
174. W. Dunges, G. Naundorf, and N. Seiler, *J. Chromatogr. Sci.* **12,** 655 (1974).
175. I. S. Forrest, D. E. Green, S. D. Rose, G. C. Skinner, and D. M. Torres, *Res. Commun. Chem. Pathol. Pharmacol.* **2,** 787 (1971).
176. R. M. Cassidy, D. S. LeGay, and R. W. Frei, *J. Chromatogr. Sci.* **12,** 85 (1974).
177. M. Frei-Hausler, R. W. Frei, and O. Hutzinger, *J. Chromatogr.* **84,** 214 (1973).
178. J. F. Lawrence and R. W. Frei, *J. Chromatogr.* **66,** 73 (1972).

179. R. W. Frei and J. F. Lawrence, *J. Chromatogr.* **67,** 87 (1972).
180. J. F. Lawrence and R. Leduc, *J. Chromatogr.* **152,** 507 (1978).
181. J. F. Lawrence and R. Leduc, *J. Assoc. Off. Anal. Chem.* **61,** 872 (1978).
182. J. F. Lawrence and G. W. Laver, *J. Assoc. Off. Anal. Chem.* **57,** 1022 (1974).
183. A. H. M. T. Scholten, C. van Buuren, J. F. Lawrence, U. A. T. Brinkmann and R. W. Frei, *J. Liq. Chromatogr.* **2,** 607 (1979).
184. J. F. Lawrence, C. Renault, and R. W. Frei, *J. Chromatogr.* **125,** 343 (1976).
185. J. F. Lawrence and G. W. Laver, *J. Chromatogr.* **100,** 175 (1974).
186. R. E. Boyle, *J. Org. Chem.* **31,** 3880 (1966).
187. C. R. Crevling, K. Kondo, and J. W. Daly, *Clin. Chem.* **14,** 302 (1968).
188. J. Pribyl and F. Herzel, *Fresenius' Z. Anal. Chem.* **286,** 95 (1977).
189. R. W. Frei and J. F. Lawrence, *J. Assoc. Off. Anal. Chem.* **55,** 1259 (1972).
190. M. Wiechmann, *Hoppe-Seyler's Z. Physiol. Chem.* **358,** 981 (1977).
191. W. O. Lehmann, H. O. Beckey, and H. R. Schulten, *Anal. Chem.* **48,** 1572 (1976).
192. N. N. Osborne, W. L. Stahl, and V. Neuhoff, *J. Chromatogr.* **123,** 212 (1976).
193. J. F. Lawrence and R. W. Frei, *Anal. Chem.* **44,** 2046 (1972).
194. P. B. Ghosh and M. W. Whitehouse, *Biochem. J.* **108,** 155 (1968).
195. A. A. Boulton, P. B. Ghosh, and A. R. Katritzky, *Tetrahedron Lett.* p. 2887 (1966).
196. H. J. Klimisch and L. Stadler, *J. Chromatogr.* **90,** 141 (1974).
197. J. Monforte, R. J. Bath, and I. Sunshine, *Clin. Chem.* **18,** 329 (1972).
198. F. van Hoof and A. Heyndrickx, *Anal. Chem.* **46,** 268 (1974).
199. F. van Hoof and A. Heyndrickx, *Meded. Fac. Landbouwwet., Rijksuniv. Gent* **38,** 911 (1973).
200. J. Reisch, H. J. Kommert, and D. Clasing, *Pharm. Ztg.* **115,** 754 (1970).
201. D. J. Biekett, N. C. Prince, G. K. Radda, and A. G. Salmon, *FEBS Lett.* **6,** 346 (1970).
202. H. J. Klimisch and D. Ambrosius, *J. Chromatogr.* **121,** 93 (1976).
203. K. Samejima, *J. Chromatogr.* **96,** 250 (1974).
204. K. Imai, *J. Chromatogr.* **105,** 135 (1975).
205. W. McHugh, R. A. Sandmann, W. G. Haney, S. P. Soed, and D. P. Wittmer, *J. Chromatogr.* **124,** 376 (1976).
206. J. T. Stewart and C. W. Chan, *Anal. Lett.* **B11,** 667 (1978).
207. E. Grushka, S. Lam, and J. Chassin, *Anal. Chem.* **50,** 1398 (1978).
208. B. B. Dey and K. Radhabai, *J. Indian Chem. Soc.* **11,** 635 (1934).
209. J. A. Secrist, J. R. Barrio, and N. J. Leonard, *Biochem. Biophys. Res. Commun.* **45,** 1262 (1971).
210. J. S. Vinson, D. D. Patel, and A. H. Patel, *Anal. Chem.* **49,** 163 (1977).
211. C. P. Ivanov and Y. Vladovska-Yukhnovska, *J. Chromatogr.* **71,** 111 (1972).
212. Y. Vladovska-Yukhnovska, C. P. Ivanov, and M. Malgrand, *J. Chromatogr.* **90,** 181 (1974).
213. I. Durko, Y. Vladovska-Yukhnovska, and C. P. Ivanov, *Clin. Chim. Acta* **49,** 407 (1973).
214. T. P. Davis, C. W. Gehrke, Jr., C. W. Gehrke, T.-D. Cunningham, K. C. Kuo, K. O. Gerhardt, H. D. Johnson, and C. H. Williams, *Clin. Chem.* **24,** 1317 (1978).
215. L. D. Mell, Jr., A. R. Dasler, and A. B. Gustafson, *J. Liq. Chromatogr.* **1,** 261 (1978).
216. R. E. Subden, R. G. Brown, and A. C. Noble, *J. Chromatogr.* **116,** 310 (1978).
217. S. S. Simons, Jr. and D. F. Johnson, *Anal. Biochem.* **82,** 250 (1977).
218. S. S. Simons, Jr. and D. F. Johnson, *J. Am. Chem. Soc.* **98,** 7098 (1976).
219. R. Chayen, R. Dvir, S. Gould, and A. Harell, *Anal. Biochem.* **42,** 283 (1971).
220. C. Apter, R. Chayen, S. Gould, and A. Harell, *Clin. Chim. Acta* **42,** 115 (1972).
221. R. Braun and W. Mosher, *J. Am. Chem. Soc.* **80,** 3048 (1958).

222. R. Brandt and N. D. Cheronis, *Mikrochim. Acta* **3,** 467 (1963).
223. D. Tocksteinova, J. Churacek, J. Slosar, and L. Skalik, *Mikrochim. Acta* p. 507 (1978).
224. H. Maeda and H. Kawauchi, *Biochem. Biophys. Res. Commun.* **31,** 188 (1968).
225. H. Kawachi, K. Kadooka, M. Tahaka, and K. Tazinura, *Agric. Biol. Chem.* **35,** 1720 (1971).
226. K. Muramoto, H. Kawauchi, Y. Yamamoto, and K. Tuzimura, *Agric. Biol. Chem.* **40,** 815 (1976).
227. A. Deheenheer, J. E. Sinsheimer, and J. H. Bruckhalter, *J. Pharm. Sci.* **62,** 1370 (1972).
228. J. E. Sinheimer, V. Jagodic, L. J. Polak, D. D. Hong, and J. H. Burchalter, *J. Pharm. Sci.* **64,** 925 (1975).
229. R. Chayen, S. Gould, A. Harell, and C. V. Stead, *Anal. Biochem.* **39,** 533 (1971).
230. N. Lustenberger, H. W. Lange, and K. Hempel, *Angew. Chem., Int. Ed. Engl.* **11,** 227 (1972).
231. H. W. Lange, N. Lustenberger, and K. Hemple, *Fresenius' Z. Anal. Chem.* **261,** 337 (1972).
232. C. F. Poole, S. Singhawangcha, A. Zlatkis, and E. D. Morgan, *J. High Res. Chromatogr.* **1,** 96 (1978).
233. C. F. Poole, S. Singhawangcha, and A. Zlatkis, *J. Chromatogr.* **158,** 33 (1978).
234. R. A. Sams, *Anal. Lett.* **B11,** 697 (1978).
235. R. A. Sams and R. Huffman, *J. Chromatogr.* **161,** 410 (1978).
236. R. P. Haycock and W. J. Mader, *J. Am. Pharm. Assoc., Sci. Ed.* **46,** 744 (1957).
237. C. R. Szalkowski and W. J. Mafer, *J. Am. Pharm. Assoc., Sci. Ed.* **45,** 613 (1956).
238. R. L. Roser, A. H. Andrist, W. H. Harrington, H. K. Naito, and D. Lonsdale, *J. Chromatogr.* **146,** 43 (1978).
239. I. Jane and J. F. Taylor, *J. Chromatogr.* **109,** 37 (1975).
240. M. Yoshioka and Z. Tamura, *J. Chromatogr.* **123,** 220 (1976).
241. R. M. Beebe, *J. Assoc. Off. Anal. Chem.* **61,** 1347 (1978).
242. D. M. Takahashi, *J. Chromatogr.* **131,** 147 (1977).
243. D. M. Takahashi, *J. Assoc. Off. Anal. Chem.* **60,** 799 (1977).
244. F. Feigl, "Spot Tests in Organic Analysis." Elsevier, Amsterdam, 1966.
245. Z. Rappoport, "Handbook of Tables for Organic Compound Identification." Chem. Rubber Publ. Co., Cleveland, Ohio, 1967.
246. G. Schwedt and H. H. Bussemas, *Fresenius' Z. Anal. Chem.* **285,** 381 (1977).
247. T. Yamabe, N. Takai, and H. Nakamura, *J. Chromatogr.* **104,** 359 (1975).
248. G. Schwedt and H. H. Bussemas, *Chromatographia* **9,** 17 (1976).
249. K. Heyns, H. Roper, and K. Stolzenburg, *Chem., Mikrobiol., Technol. Lebensm.* **5,** 155 (1978).
250. S. Fishman, *J. Pharm. Sci.* **64,** 674 (1975).
251. F. L. Vandemark, G. J. Schmidt, and W. Salvin, *J. Chromatogr. Sci.* **16,** 465 (1978).
252. M. M. Abdel-Monem, K. Ohno, N. E. Newton, and C. E. Weeks, *Adv. Polyamine Res.* **2,** 37 (1978).
253. A. S. Dion and E. J. Herbst, *Ann. N.Y. Acad. Sci.* **171,** 723 (1970).
254. J. Lammens and M. Verzele, *Chromatographia* **11,** 376 (1878).
255. K. O. Loeffler, D. E. Green, F. C. Chao, and I. S. Forrest, *Proc. West. Pharmacol. Soc.* **18,** 363 (1975).
256. J. H. Fleisher and D. H. Russell, *J. Chromatogr.* **110,** 335 (1975).
257. K. Igarashi, I. Izumi, K. Hara, and S. Hirose, *Chem. Pharm. Bull.* **22,** 451 (1974).
258. S. R. Abbott, A. Abu-Shumays, K. O. Loeffler, and I. S. Forrest, *Res. Commun. Chem. Path. Pharmacol.* **10,** 9 (1975).

259. N. N. Osborne and V. Neuhoff, *J. Chromatogr.* **134,** 489 (1976).
260. H. Khalaf and M. Rimpler, *Hoppe-Seyler's Z. Physiol. Chem.* **358,** 505 (1977).
261. T. E. Archer and J. D. Stokes, *J. Agric. Food Chem.* **26,** 452 (1978).
262. G. Schwedt and H. H. Bussemas, *Fresenius' Z. Anal. Chem.* **283,** 23 (1977).
263. N. Seiler and B. Knodgen, *J. Chromatogr.* **131,** 109 (1977).
264. P. J. Meffin, S. R. Harapat, and D. C. Harrison, *J. Pharm. Sci.* **66,** 583 (1977).
265. G. C. Boffey and G. M. Martin, *J. Chromatogr.* **90,** 178 (1974).
266. L. D. Mell, Jr., S. W. Joseph, and N. E. Bussell, *J. Liq. Chromatogr.* **2,** 407 (1979).
267. M. van Bezeij and M. W. Bosch, *Z. Klin. Chem. Klin. Biochem.* **13,** 381 (1975).
268. K. Kuwata, M. Uebori, and Y. Yamasaki, *J. Chromatogr. Sci.* **17,** 264 (1979).
269. C. Peraino and A. E. Harper, *Anal. Chem.* **33,** 1863 (1961).
270. A. D. Smith and J. B. Jepson, *Anal. Biochem.* **18,** 36 (1976).
271. E. Grushka, H. D. Durst, and E. J. Kikta, Jr., *J. Chromatogr.* **112,** 673 (1975).
272. D. R. Jones, IV and S. E. Manahan, *Anal. Chem.* **48,** 502 (1976).
273. D. R. Jones, IV and S. E. Manahan, *Anal. Lett.* **8,** 421 (1975).
274. D. R. Jones, IV and S. E. Manahan, *Anal. Chem.* **48,** 1897 (1976).
275. F. E. Brinckman, W. R. Blair, K. L. Jewett, and W. P. Iverson, *J. Chromatogr. Sci.* **15,** 493 (1977).
276. D. R. Jones, IV, H. C. Tung, and S. E. Manahan, *Anal. Chem.* **48,** 7 (1976).
277. C. Botre, F. Cacace, and R. Cozzani, *Anal. Lett.* **9,** 825 (1976).
278. R. M. Cassidy, M. T. Hurteau, J. P. Mislan, and R. W. Ashley, *J. Chromatogr. Sci.* **14,** 444 (1976).
279. F. J. Fernandez, *Chromatogr. Newsl.* **5,** 17 (1977).
280. T. R. Roberts, "Radiochromatography—The Chromatography and Electrophoresis of Radiolabelled Compounds." Elsevier, Amsterdam, 1978.
281. G. Pataki, "Techniques in Amino Acid and Peptide Chemistry." Ann Arbor Sci. Publ., Ann Arbor, Michigan, 1968.
282. R. W. Frei, J. F. Lawrence, U. A. T. Brinkman, and I. Honigberg, *J. High Res. Chromatogr.* **2,** 11 (1979).
283. C. van Buuren, J. F. Lawrence, U. A. T. Brinkman, I. L. Honigberg, and R. W. Frei, *Anal. Chem.* **52,** 700 (1980).
284. L. W. Skeggs, *Am. J. Clin. Pathol.* **28,** 311 (1957).
285. R. W. Deelder and P. J. H. Hendricks, *J. Chromatogr.* **83,** 343 (1973).
286. G. Ertinghaus, H. J. Adler, and A. S. Reichler, *J. Chromatogr.* **42,** 355 (1969).
287. L. R. Snyder and H. J. Adler, *Anal. Chem.* **48,** 1017 (1976).
288. L. R. Snyder, *J. Chromatogr.* **125,** 287 (1976).
289. R. W. Deelder, M. G. F. Kroll, A. J. B. Beeren, and J. H. M. van den Berg, *J. Chromatogr.* **149,** 699 (1978).
290. J. Ruzicka and E. H. Hausen, *Anal. Chim. Acta* **99,** 37 (1978).
291. R. S. Deelder, M. G. F. Kroll, and J. H. M. van den Berg, *J. Chromatogr.* **125,** 307 (1976).
292. T. D. Schlabach, S. H. Chang, K. M. Goodring, and F. E. Regnier, *J. Chromatogr.* **134,** 91 (1977).
293. K. M. Jonker, H. Poppe, and J. F. K. Huber, *Chromatographia* **11,** 123 (1978).
294. A. M. Felix and G. Terkelsen, *Anal. Biochem.* **60,** 78 (1974).
295. W. Voelter and K. Zech, *J. Chromatogr.* **112,** 743 (1975).
296. W. Voelter and K. Zech, *Chromatographia* **8,** 350 (1975).
297. A. M. Felix and G. Terkelsen, *Arch. Biochem. Biophys.* **157,** 177 (1973).
298. A. M. Felix and G. Terkelsen, *Anal. Biochem.* **56,** 610 (1973).
299. R. W. Frei, L. Michel, and W. Santi, *J. Chromatogr.* **126,** 665 (1976).

300. R. W. Frei, L. Michel, and W. Santi, *J. Chromatogr.* **142,** 261 (1977).
301. H. Nakamura and J. Pisano, *Arch. Biochem. Biophys.* **158,** 605 (1973).
302. R. W. Wecks, S. K. Yasuda, and B. J. Dean, *Anal. Chem.* **48,** 159 (1976).
303. J. Foucre, H. Jensen, and E. Neuzil, *Anal. Chem.* **48,** 155 (1976).
304. E. Crombez, G. van der Weken, W. van den Bossche, and P. de Moerloose, *J. Chromatogr.* **177,** 323 (1979).
305. G. Schwedt, *Anal. Chim. Acta* **92,** 337 (1977).
306. M. Roth, *J. Clin. Chem. Clin. Biochem.* **14,** 361 (1976).
307. M. Roth, *Anal. Chem.* **43,** 880 (1971).
308. H. A. Moye, S. J. Scherer, and P. A. St. John, *Anal. Lett.* **10,** 1049 (1977).
309. R. T. Krause, *J. Chromatogr. Sci.* **16,** 281 (1978).
310. E. H. Creaser and G. J. Hughes, *J. Chromatogr.* **114,** 69 (1977).
311. L. J. Marton and P. L. Y. Lee, *Clin. Chem.* **21,** 1721 (1975).
312. F. Engback and I. Magnussen, *Clin. Chem.* **24,** 376 (1978).
313. A. Kojina-Sudo, *Ind. Health* **15,** 109 (1977).
314. M. Roth, *Clin. Chim. Acta* **83,** 273 (1978).
315. H. A. Moye and T. W. Wade, *Anal. Lett.* **9,** 801 (1976).
316. K. A. Ramsteiner and W. D. Hörmann, *J. Chromatogr.* **104,** 438 (1975).
317. S. H. Chang, K. M. Goodring, and F. E. Regnier, *J. Chromatogr.* **125,** 103 (1976).
318. A. W. Wolkoff and R. H. Larose, *J. Chromatogr.* **99,** 731 (1977).
319. A. W. Wolkoff and R. H. Larose, *Anal. Chem.* **47,** 1003 (1975).
320. S. Katz and W. W. Pitt, Jr., *Anal. Lett.* **5,** 177 (1972).
321. F. Nachtmann, G. Knapp, and H. Spitzy, *J. Chromatogr.* **149,** 613 (1978).
322. Y. Takata and G. Muto, *Anal. Chem.* **45,** 1864 (1973).
323. G. Schwedt, *Fresenius' Z. Anal. Chem.* **287,** 152 (1977).
324. G. Schwedt, *J. Chromatogr.* **143,** 463 (1977).
325. C. R. Clark, C. H. Darling, J. Chan, and A. C. Nichols, *Anal. Chem.* **49,** 2080 (1977).
326. N. D. Brown, N. D. Sing, W. E. Neeley, and S. E. Koetitz, *Clin. Chem.* **23,** 1281 (1977).
327. J. C. Gfeller, G. Frey, J. M. Huen, and J. P. Thevenin, *J. Chromatogr.* **172,** 141 (1979).
328. S. Katz and L. H. Thacker, *J. Chromatogr.* **64,** 247 (1972).
329. J. C. Gfeller, G. Frey, and R. W. Frei, *J. Chromatogr.* **142,** 271 (1977).
330. W. Iwaoka and S. R. Tannenbaum, *J. Chromatogr.* **124,** 105 (1976).
331. P. J. Twitchett, P. L. Williams, and A. C. Moffat, *J. Chromatogr.* **149,** 683 (1978).
332. A. H. M. T. Scholten and R. W. Frei, *J. Chromatogr.* **176,** 349 (1979).
333. D. E. Ott, *Bull. Environ. Contam. Toxicol.* **17,** 261 (1977).
334. C. R. Clark and J. Chan, *Anal. Chem.* **50,** 635 (1978).
335. T. Kawasaki, M. Maeda, and A. Tsuji, *J. Chromatogr.* **163,** 143 (1979).

Chapter 8
Sample Extraction and Cleanup

I. CONSIDERATIONS

This chapter is intended to show the analyst what types of sample extraction and cleanup procedures have been studied and applied to organic trace analysis. It illustrates a number of techniques that may be used to isolate the analyte from a sample matrix. The methods may vary greatly in nature, a fact that is important in trace analysis since if one approach fails, a significantly different one may meet with success. There need be no explanation or discussion of good analytical technique. Accuracy and an orderly logical progression through a method is always required. When attempting any of the methods described herein, it is assumed that the reader is aware of the needs for standardization and reproducibility. Impurities in solvents, reagents, and chromatographic materials can interfere in final determinations; therefore, the analyst must make reagent blank runs before attempting samples so that necessary adjustments or changes may be made in order to optimize the extraction or cleanup.

A. Sampling Procedure

The selection of samples for a determination can be as important as the analysis itself. It is essential that samples be chosen so they represent as accurately as possible the population from which they came. In order to do this, strict statistical guidelines must be followed.

The actual sampling techniques vary greatly, depending upon the type of sample. Air samples are normally collected by passing large volumes through filters, adsorption columns, or liquid solutions in order to trap the desired organics. Water samples may be collected in suitable containers and at various depths. In moving streams, total flow rate and the flow profile (i.e., the variation in flow in different parts of the stream) should be determined. Even seasonal variations should be known, depending upon what correlations are to be made. Soil samples may be selected at various depths and over a random area. Food samples are usually selected by individual items either from a grocery store, warehouse, or farmer's field. In all cases, the selected sample should be as representative as possible.

Sampling techniques for biological samples, e.g., blood or urine, are somewhat different than for environmental samples. The quantities collected are usually much less, and only one sample per human or experimental animal is taken for a given analysis. In all cases, "control" samples are taken from a number of other donors to be used as a base for comparison to the test samples.

When a sample arrives in the laboratory, it is normally homogenized (if necessary), and a representative subsample is taken for analysis. The remainder of the sample is frozen and stored. The analyst carries out the extraction and cleanup on the subsample.

B. Nature of the Trace Organics

The characteristics of the trace substance should be considered before method development is attempted. The physical and chemical properties can be of valuable assistance in isolating the analyte from unwanted sample components. For example, if the analyte has some acidic or basic character, then many coextractives can be eliminated by an aqueous–organic partitioning at an appropriate pH. The reactivity of the trace compound toward solvents and other chemicals which may be used in the analysis should be known. Since the sample extract is usually concentrated to some degree before LC analysis, the volatility of the compound should be examined by observing any losses before and after a concentration step.

In general, polar compounds are more difficult to analyze than nonpolar substances. They tend to adsorb more strongly to solids, even to the point of chemical binding or conjugation in plant or animal systems. Even during an analysis, such compounds may be lost through adsorption to glassware and other surfaces.

Those properties exploitable for detection pruposes should also be known since they could be advantageous in aiding the selectivity of the detection. Such characteristics as wavelengths of absorption or fluorescence should be determined. Also, the electrochemical behavior of the analyte should be examined. This may be particularly useful since electrochemical detectors made expressly for LC are now commercially available at reasonable prices. They can offer good selectivity with sensitivity comparable to UV detection.

C. Degree of Sample Preparation

The need for sample preparation arises from the fact that most detection methods at present cannot directly distinguish between the analyte and the sample matrix at trace concentrations. Some exceptions may include air pollution monitors and in some cases, GC–mass spectrometry (MS) in instances where the sample may be directly analyzed (e.g., vapor samples such as the atmosphere or in head-space analysis [e.g., industrial contaminents (*1*)] and perhaps some water samples, etc.).

The extent of sample cleanup prior to LC analysis depends upon two factors. The extract must be sufficiently "clean" so that the analyte may be observed in the resulting chromatogram, and the extract should not cause irreversible column deterioration. The two are not necessarily related. A detector may be selective enough to detect the analyte in crude sample extracts, yielding good chromatograms for quantitation. However, much of the matter injected that is not detected may contaminate the column.

The degree of cleanup is also related to the type of sample matrix and to the concentration of the analyte in the sample. A method developed for one type of sample may not be suitable for another type or even the same sample when the analyte is at a much lower concentration, because of interferences. This inevitably leads to more cleanup.

Figure 8.1 shows a schematic of the steps involved in sample treatment prior to determination by LC. In the first step the analyte is removed from the sample by some means of extraction. This can be achieved by using

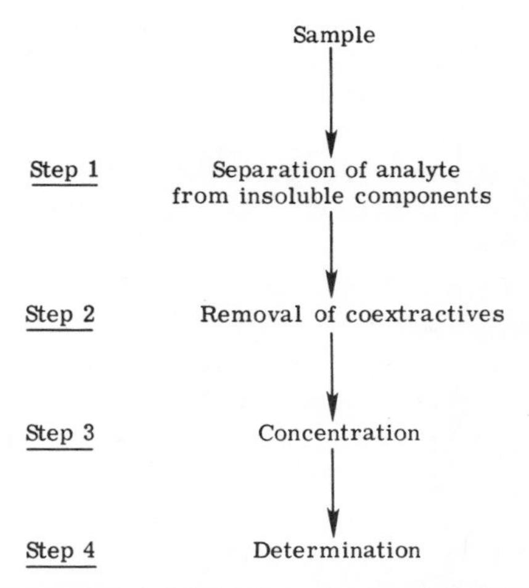

Fig. 8.1 Schematic illustration of the steps involved in sample treatment prior to chromatographic analysis.

appropriate solvents, trace enrichment techniques, or by means of heat, including sweep-codistillation or steam distillation if the analyte is reasonably volatile. These treatments remove 99.9% or more of the sample from the analyte. The remainder of the cleanup is concerned with removal of the final 0.1% or less of the sample components that were coextracted with the analyte (step 2 of Fig. 8.1). This part of the procedure can become the most involved and time-consuming. It is here that the chemical and physical nature of the analyte must be exploited for further purification. Solvent–solvent partitions and open-column chromatographic techniques are normally employed at this stage.

The final purified extract will most likely have to be concentrated to a

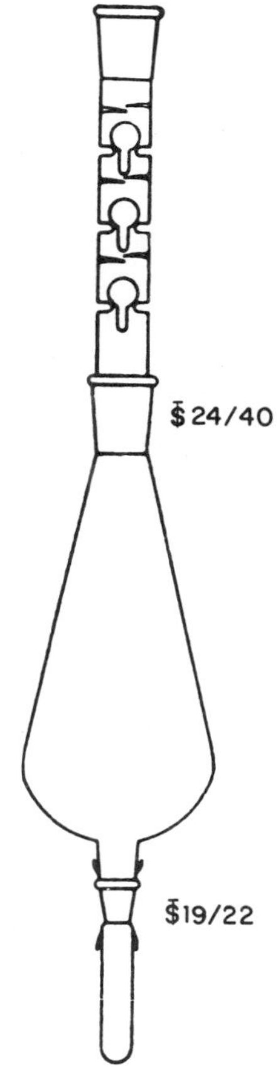

Fig. 8.2 Kuderna-Danish evaporation apparatus for concentration of solutions from several hundred ml down to 1 ml or less. Snyder column is attached to the top.

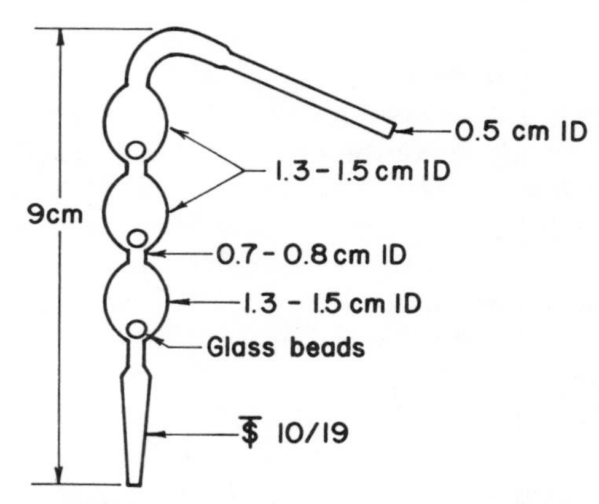

Fig. 8.3 Micro Snyder column for concentration of volatile organics in solutions of only a few milliliters.

small volume before LC analysis. In most cases evaporation under a stream of nitrogen or by rotary evaporation under vacuum at 30°–50°C will suffice. However, for compounds that may be lost through evaporation, special commercially available glassware such as a Kuderna-Danish apparatus (*2*) (Fig. 8.2) or a Snyder column (*2,3*) may be required to reduce losses. Figure 8.3 shows a micro Snyder column for concentration of volatiles. The column acts as a small fractionation tower in which any vapors of the analyte condense and are washed back down into the heated concentration flask.

II. SAMPLE EXTRACTION

A. Solvent Extraction of Moderately or Nonpolar Organics

Solvent extraction may be applied to the three physical types of samples—vapors, liquids, and solids. Organics in air or other gaseous mixtures can be extracted by passing the vapors through an appropriate solvent. The organics are "trapped" by the solvent and are thus ready for continued sample preparation or analysis. Liquid samples are normally aqueous, such as natural waters, or biological fluids, such as urine or plasma. Organics may be removed by shaking an aliquot of the aqueous sample with an immiscible organic solvent.

Solid samples such as soils are usually mechanically shaken with an appropriate solvent for a set period of time, and the solution is then decanted from the solid material. Plant material or biological tissues are homogenized or blended with a water-miscible organic solvent such as methanol, acetonitrile, or acetone. In these cases, the insoluble material is removed from the mixture by filtration. The filtrate can then be further processed before analysis by LC. For some solids, Soxhlet extraction may be employed. This apparatus uses refluxing solvents to extract the trace substances. The process can be repeated for many hours using a small volume of solvent and can be operated unattended. The organics, however, must be stable to the heat employed, which depends on the boiling point of the extracting solvent. In all of the above, the organic substance ends up in an organic solvent.

There are several points to be made concerning solvent extraction. The trace compound must be freely soluble in the extracting solvent. If some degree of selectivity is desired, then solvent selection can be of valuable assistance. For example, if polychlorinated biphenyls (PCBs) are to be extracted from river water, a very nonpolar liquid such as hexane would be preferred to a more polar one such as ethyl acetate or diethyl ether. The PCBs may be freely soluble in all three solvents, but the hexane would remove far less of the more polar substances which may interfere with PCB analysis at a later stage. Since the aim of sample cleanup is to remove as many coextractives as possible, the analyst should be aware of how best to accomplish this with each step of the procedure, including the initial extraction. Table 8.1 lists some common solvents with their relative polarities and boiling points.

The evaluation of a number of solvents for extracting organics from

TABLE 8.1

Solvent	Polarity	Boiling point (°C)
n-Hexane	0	69.0
Carbon tetrachloride	1.7	76.8
Toluene	2.3	110.8
Ethyl ether	2.9	34.6
Benzene	3.0	80.2
Methylene chloride	3.4	40.7
Ethyl acetate	4.3	77.2
Chloroform	4.3	61.3
Dioxane	4.8	101.3
Ethanol	5.2	78.4
Acetone	5.4	56.2
Acetonitrile	6.2	81.6
Dimethylformamide	6.4	155
Methanol	6.6	64.6
Water	9.0	100.0

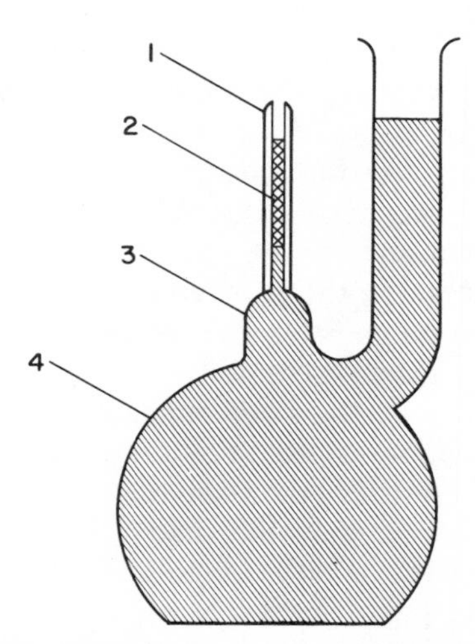

Fig. 8.4 Microextraction apparatus. (1) Capillary tube; (2) solvent layer; (3) modified volumetric flask; (4) water sample.

water has been reported (*4*). A microextraction apparatus was evaluated by Murray (*5*) and applied to the removal of phthalates and other organics from water samples. Figure 8.4 illustrates the apparatus. It consists of a modified 1-liter volumetric flask with a capillary tube. The 1-liter water sample is shaken for 2 min with only 200 μl of hexane. The hexane (about 50 μl recovered from the 200 μl added) rises to the top and enters the capillary tube, where it may be directly analyzed by GC or LC, or it may be removed to a clean test tube and the extraction repeated if necessary,

TABLE 8.2

Recoveries from Tap Water Using Microextraction Apparatus

Compound	Concentration	Recovery[a] (%)
aldrin	10 ng/liter	89.4
heptachlor epoxide	10 ng/liter	91.3
α-*cis*-Chlordane	10 ng/liter	92.0
Dieldrin	10 ng/liter	92.6
n-Decane	50 μg/liter	98.6
n-Dodecane	50 μg/liter	97.3
n-Tetradecane	50 μg/liter	96.7
n-Hexadecane	50 μg/liter	96.4
Dibutylphthalate	2.5 μg/liter	96.6
Butyl glycolyl butylphthalate	25 μg/liter	89.7

[a] Average of four extractions.

with second and third volumes of hexane. The partition ratios of phthalates and some nonpolar pesticides are such that three extractions yield about 90% or greater recoveries (see Table 8.2). This method is particularly valuable for volatile compounds which may be prone to evaporative losses during concentration of large volumes of organic solvents.

The equipment used for the solvent extraction of samples usually depends upon the sample type. For example, separatory funnels or the microextraction apparatus described above may be all that is needed for liquid samples. Volumes ranging from a few milliliters to several liters can be accommodated. Solid materials of plant or animal origin are usually blended (Waring blender or equivalent) or homogenized (Sorval Omnimixer or equivalent). This treatment is necessary to break the sample into minute particles to allow intimate contact between the sample and solvent. Tissue grinders are suitable for small samples of animal tissue (5 gm or less, depending on the quantity of connective tissue present). Ball grinding (*6*) and ultrasonic treatment (*7,8*) may be employed for soils and animal or plant tissue. These aid the contact between sample and solvent, allowing for better extraction of the trace organics into the extraction solvent. Soil samples and other powdery samples such as whole grains and flour may be extracted by mechanically shaking with solvent for 2 hr or more. The constant agitation of the samples is necessary for good recoveries.

B. Trace Enrichment

Trace enrichment is a useful extraction technique applied to liquid samples usually of an aqueous nature. It involves passing the sample through a column containing particles such as chemically bonded reversed-phase silica gel or macroreticular resins that have a strong affinity for organics. The trace substances are adsorbed onto the column and therefore removed from the aqueous sample. The organics may then be removed from the column with a small volume of organic solvent such as methanol or acetonitrile. The advantage of this type of extraction is that large volumes of water may be easily handled and large volumes of organic solvents are not required. This makes the approach particularly suited to field work. Many water samples may be collected in this manner then stored and returned to a central analytical laboratory, where the trace substances are eluted from the column for further cleanup and analysis. It has been shown, for example, that the organophosphate insecticide fenitrothion may be removed from water samples by employing an XAD-2 macroreticular resin; it may then be stored on the column for up to 5 weeks with greater than 90% recovery (*9*).

The use of trace enrichment for the removal of drugs of abuse from urine samples has been incorporated into a routine method (*10*). Trace enrichment may also be applied to air samples where high volumes of vapor are passed through a column containing, for example, a porous polymer such as Tenax GC (*11*). Afterward, the column may be eluted with an organic solvent to remove the adsorbed organics for analysis. Further discussion of trace enrichment for extraction and cleanup may be found in Sections IV,C and D on column cleanup techniques.

C. Extraction of Polar Organics and Conjugates

Polar organic compounds are much more difficult to extract from samples than are nonpolar substances. They have a high degree of water solubility and can chemically bind to proteins or carbohydrates. Most metabolic products of drugs or contaminants in both plant and animal systems are metabolized to more polar water-soluble forms, including conjugates. The analysis of these compounds by LC becomes much more difficult since the very nature of living organisms is based on water-soluble polar molecules including amino acids, nucleic acids, carbohydrates, and their numerous derivatives. The task becomes one of separating the analyte from a host of compounds similar in polarity and present in perhaps one-millionfold excess. As a result, the methodology normally employed is to use aqueous solutions for removal of these compounds from solid sample components. The resulting cleanup must be carried out in aqueous media.

If polar compounds are either strongly adsorbed for example to soil, or chemically bound as in the formation of a glucoside conjugate, acid treatment may be required in order to release them. This involves refluxing the sample for a sufficient period of time with dilute hydrochloric or other acid before continuing with the cleanup. Such rigorous treatment usually brings about the hydrolysis of many other sample components, compounding the cleanup problem. Certain types of conjugates may be enzymatically hydrolyzed under very mild conditions. These enzymes are very selective and can result in an extract much cleaner than produced by the acid hydrolysis procedure. Although enzymatic hydrolysis is highly recommended, the method is limited to the availability of enzymes (e.g., glucosidase, glucuronidase, and other hydrolyses are commercially available) and also by the uncertainty of the type of conjugation.

Cleanup techniques such as ion-exchange or reversed-phase chromatography are normally employed to further purify the extracts. However, in general, the techniques available for polar organic analysis are not nearly as satisfactory as those employed for nonpolar organic compounds. Research directed in this area would undoubtedly aid the analyst.

D. Distillation of Volatiles from Water Samples

Volatile polar organics such as low-molecular-weight alcohols, nitriles, aldehydes, and ketones are difficult to concentrate since they are particularly susceptible to evaporative losses. Distillation of these compounds from aqueous solution can be extremely efficient and provide a clean extract that may not need to be further concentrated before analysis. Figure 8.5 shows a diagram of an all-glass distillation–concentration apparatus suitable for the removal of volatile organics from aqueous solution (*12*). It consists of a distillation pot, condenser, a distillate collection chamber, a steam/water contact column, and an overflow return tube. The organics azeotropically distill into the distillate chamber and are preferentially retained there. This solution can then be sampled for chromatographic anal-

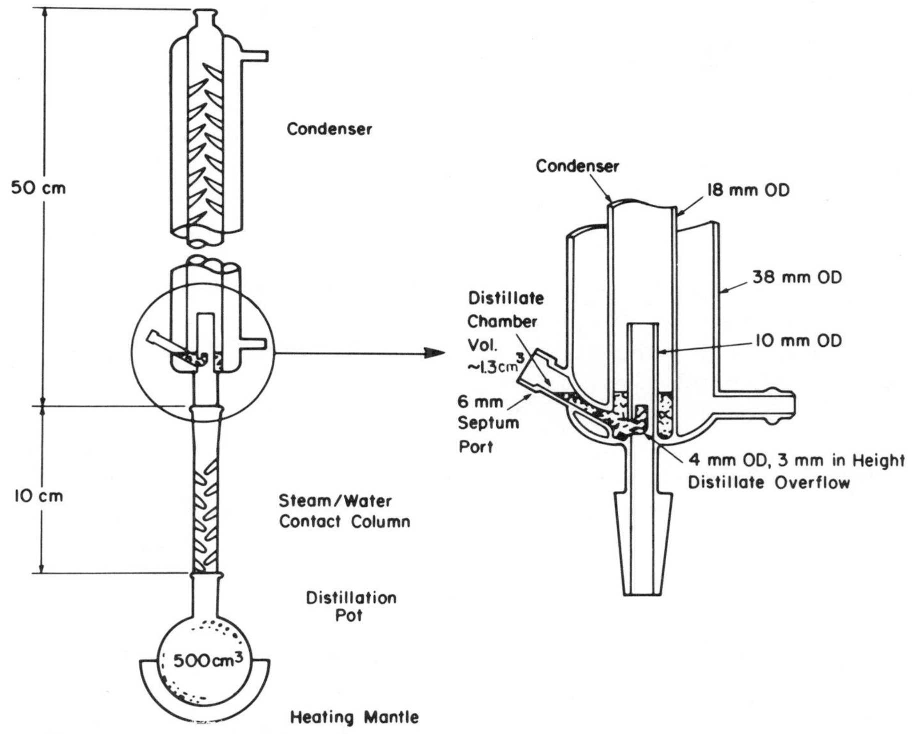

Fig. 8.5 All-glass distillation-concentration apparatus. From Peters (*12*), with permission from the American Chemical Society.

ysis. It should be possible to apply this approach to solid samples by first extracting the polar organics into aqueous solution, then distilling them out.

III. SAMPLE CLEANUP

A. Liquid–Liquid Partition

This approach to sample cleanup involves the distribution of the trace organic substance between two immiscible phases. The organic components of interest should partition favorably into one of the phases while much of the sample material remains in the other phase. As an example, consider the partition between water and chloroform of a sample that has been extracted with methanol. Methanol removes many organics over a wide polarity range, including very polar compounds. If a small aliquot of the methanol extract is added to the partitioning system, the analyte and all other substances in the extract will partition themselves between the two immiscible phases to establish an equilibrium depending on their relative polarities. Polar organics will favor the aqueous phase whereas nonpolar compounds will be predominantly in the organic layer. The distribution of these organics between the two liquids is constant for any given system. If the analyte is represented by A then the total quantity of A in the system would be

$$A = A_{aq} + A_{org} \tag{1}$$

where A_{aq} represents the quantity of A in the aqueous phase and A_{org} is the quantity in the organic layer. The ratio of the concentrations of these two is referred to as the distribution ratio and is constant even though the volumes of the two phases may change. Thus,

$$D = [A_{org}]/[A_{aq}] \tag{2}$$

where D = the distribution ratio. The actual quantity of A extracted into the organic phase may be given in terms of percent of the total A in the system:

$$\%\text{A extracted} = \frac{A_{org}}{A_{org} - A_{aq}} \times 100\% \tag{3}$$

Thus, if 5 μg of A were present in the system and 4 μg were found in the organic phase, then the extraction efficiency would be 80%. This recovery may be improved by increasing the volume of the organic phase, but the preferred method is to repeat the extraction with additional volumes of organic phase. Table 8.3 illustrates the difference in total recovery for a

TABLE 8.3

Extraction of Analyte *A* from Aqueous Solution

Extraction	Quantity of A in the System (μg)	Percent Recovered in Organic	Quantity in organic (μg)
First	10.0	50	5.0
Second	5.0	50	2.5
Third	2.5	50	1.25
		Total Recovered	8.75

compound that is extracted 50% with one extraction. It can be seen that three separate extractions remove 87.5% of the trace organic. One extraction performed with three times the volume of solvent would result in only 75% total recovery, according to calculations made using Eqs. 1, 2, and 3.

Bowman and Beroza (*13,14*) devised a method for identifying organics such as pesticides, esters, alcohols, aldehydes, ketones, acids, and hydrocarbons based on their partitioning behavior between two immiscible phases. They studied six different phase systems and calculated "p" values for a large number of compounds. These values are characteristic of each compound, and while particularly useful for confirming the identity of an analyte, they are also valuable in aiding the selection of a partition system for sample cleanup. The p values of a large number of compounds have been tabulated (*13,14*) employing phase systems consisting of hexane–acetonitrile, isooctane–dimethylformamide, hexane–90% ethanol, hexane–90% dimethylsulfoxide, etc. Table 8.4 lists some p values for a number of compounds with five different phase systems.

B. Acid–Base Effects

Acidic or basic organics can often be effectively purified by carrying out an organic–aqueous partition at a selected pH. For example, consider that parachlorophenol is to be determined in a particular environmental sample. If the sample extract is partitioned at neutral pH, the compound will be extracted into the organic phase with a host of other organic species which may require further sample treatment to remove. However, when the pH of the aqueous phase is adjusted so that it is basic, the phenol, being a weakly acidic compound, will ionize as follows:

$$\text{P—OH} + {}^{-}\text{OH} \longrightarrow \text{P—O}^{-} + \text{H—OH} \quad (4)$$

where P—OH represents parachlorophenol. In the ionized form, the com-

TABLE 8.4

***p* Values for Selected Compounds in Five Binary Solvent Systems**[a]

Compound	Solvent system				
	Hex/MeCN	Isooct/90% DMF	Isooct/90% DMSO	Hept/90% EtOH	Isooct/80% Acetone
Esters					
Methyl hexanoate	0.33	0.54	0.80	0.40	0.78
Methyl decanoate	0.58	0.89	0.97	0.60	0.94
Methyl benzoate	0.14	0.15	0.28	0.32	0.44
Dimethyl phthalate	0.03	0.02	0.03	0.18	0.27
Diethyl oxalate	0.03	—	0.08	0.16	0.41
Diethyl maleate	0.02	0.02	0.12	0.25	0.39
Dibutyl maleate	0.13	0.21	0.50	0.33	0.75
Diethyl adipate	0.09	0.19	0.35	0.21	0.61
Alcohols					
1-Hexanol	0.05	—	0.05	0.10	0.36
1-Octanol	0.13	0.14	0.16	0.14	0.56
1-Tetradecanol	0.47	0.58	0.82	0.32	0.91
Cyclohexanol	0.12	—	0.04	0.10	0.22
Benzyl alcohol	<0.01	<0.01	<0.01	0.05	0.12

Aldehydes and ketones					
Decanal	0.51	0.74	0.90	0.40	0.89
Cyclohexanone	0.13	0.17	0.27	0.24	0.46
3-Heptanone	0.33	—	0.74	0.37	0.75
Acids					
Hexanoic	0.04	—	<0.01	0.11	0.25
Octanoic	0.19	—	0.05	0.18	0.46
Stearic	0.65	0.64	0.92	0.59	0.98
Benzoic	<0.01	<0.01	<0.01	0.08	0.06
Phenylacetic	0.02	0.03	0.03	0.10	0.11
Hydrocarbons					
Decane	0.98	1.0	1.0	0.89	0.96
Octadecane	1.0	1.0	1.0	0.98	1.0
p-Xylene	0.58	0.67	0.81	0.67	0.87
Halogenated hydrocarbons					
1-Chlorohexane	0.62	0.75	0.92	0.71	0.87
1-Bromohexane	0.67	0.81	0.95	0.71	0.93
1,4-Dichlorobutane	0.22	0.26	0.37	0.47	0.75
Chlorobenzene	0.42	0.38	0.56	0.56	0.76
Bromobenzene	0.44	0.39	0.55	0.58	0.77
p-Chlorotoluene	0.51	0.52	0.71	0.62	0.83

[a] Determinations carried out by gas chromatography with flame ionization detection. Data are from Bowman and Beroza (*14*).

pound remains in the aqueous phase whereas all neutral and basic compounds are extracted into the organic phase. The organic phase could then be removed and the pH of the aqueous adjusted to acidic conditions so that:

$$P{-}O^- + H^+ \longrightarrow P{-}OH \quad (5)$$

Here the phenol is returned to its neutral form, enabling it to be extracted into fresh organic solvent free of neutral or basic organics that may have interfered. Substances such as phenols, carboxylic acids, and others may be treated in this manner. The same technique may be used for basic compounds as well. They may be kept in the aqueous phase by adjusting to a suitably acidic pH where the base, B, becomes ionized.

$$B + H^+ \longrightarrow BH^+ \quad (6)$$

After removal of neutral and acidic coextractives, the pH is readjusted so that B returns to its neutral form:

$$BH^+ + {}^-OH \longrightarrow B + H_2O \quad (7)$$

which is then extractable into organic solvent. Compounds usually con-

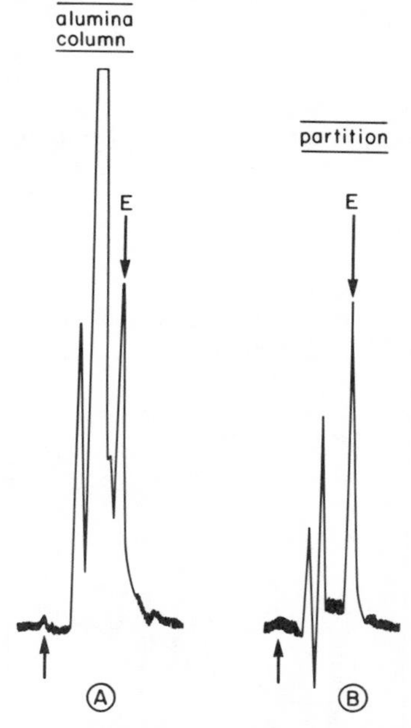

Fig. 8.6 Comparison of alumina column cleanup (A) and acid–base partition (B) for ethoxyquin in apples at 0.1 ppm. LC analysis employing a silica gel column with hexane–dioxane (85:15) as mobile phase; detection at 230 nm.

taining nitrogen, such as amines and heterocyclics, may be cleaned up in this manner. An example of how useful this can be is shown in Fig. 8.6. The antioxidant ethoxyquin (weakly basic) is cleaned up by an alumina column and by partitioning between 1 *M* hydrochloric acid and hexane, as described above. It can be seen that the partitioning resulted in a much cleaner extract as indicated by LC analysis at 230 nm UV detection. The partitioning is also less time-consuming than the column cleanup.

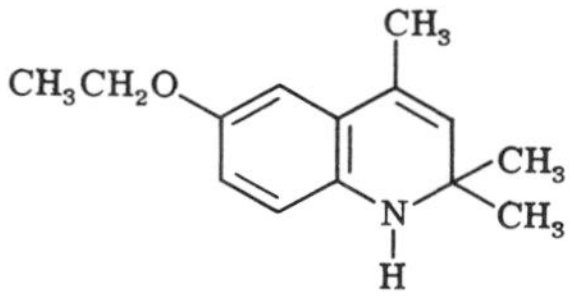

Ethoxyquin

C. Ion-Pair Extraction

Many organic compounds that are in an ionized state in aqueous solution can be extracted into certain organic solvents as ion pairs with suitable counter-ions. The degree or efficiency of the extraction is controlled by the nature of the organic solvent and the counter-ion, as well as the counter-ion concentration. Ion-pair extraction is most useful for the extraction of hydrophilic acidic or basic organics. The technique can be used selectively to aid cleanup, particularly in complex samples. For example, if a trace substance—say an organic base—is present in an aqueous sample extract in an ionized form, the solution may be partitioned with chloroform, resulting in the removal of many neutral or acidic sample components. After this a suitable counter-ion such as anthracene-2-sulfonic acid or picric acid may be added to the extract; then the aqueous phase is extracted again with chloroform. This time the organic base is extracted into the organic phase as a neutral ion pair with the counter-ion. The organic solution may then be treated further as required for analysis.

If we consider an organic base, B, in an ionized state in an aqueous phase, it would assume the form of HB^+. In order to extract this as an ion pair with X^-, the following reaction must occur.

$$HB^+ + X^- \longrightarrow HBX \qquad (8)$$

The two ionic components form a neutral ion pair. The equilibrium for this reaction may be defined by

$$k_{HBX} = \frac{[HBX]}{[HB^+][X^-]} \qquad (9)$$

where [HBX] is the concentration of the ion pair in the organic phase and $[HB^+]$ and $[X^-]$ are the concentrations of the ions in the aqueous phase. k_{HBX} is called the extraction constant.

The partition ratio, D_B, is a measure of the distribution of B between the two phases. If the only reaction involved in the ion-pair formation is that represented in Eq. 7, then:

$$D_B = \frac{[HBX]}{[HB^+]} = k_{HBX}[X^-] \tag{10}$$

The percent of *B* extracted into the organic can be calculated from Eq. (1) and the volume ratio q of the two phases where

$$q = \frac{V_{org}}{V_{aq}} \tag{11}$$

Thus, the percentage extracted, *P*, may be defined as:

$$P = 100\left(1 + \frac{1}{qD_B}\right)^{-1} \tag{12}$$

Equations (10) and (12) show that the extraction process is governed by the nature and concentration of the counter-ion and the nature of the phase system used, including pH and type of organic phase. Table 8.5 lists the extraction constants of picrate ion pairs of three organic bases in seven different solvents. It can be seen that the organic solvent can have a tremendous influence on the extraction of ion pairs. The great difference in chloroform and carbon tetrachloride in the ability to extract the ion pairs is attributed in part to the hydrogen-bonding capability of the former enabling a far greater extraction of the ion pairs. Schill (*15,16*) has com-

TABLE 8.5

Effect of Organic Solvent on Extraction of Picrate Ion Pairs

Solvent	Dielectric constant	log k_{HBX}[a] TBA	TMEA	Choline
1-Pentanol	13.9	—	1.24	1.19
Methylisobutyl ketone	13.1	—	0.64	0.98
Ethyl acetate	6.0	—	0.45	0.69
Methylene chloride	8.9	6.68	0.77	−0.32
Chloroform	4.8	5.91	0.04	−1.45
Benzene	2.3	3.59	—	—
Carbon tetrachloride	2.2	1.94	—	—

[a] TBA, tetrabutylammonium; TMEA, trimethylethylammonium; Choline, 2-hydroxyethyltrimethylammonium.

prehensively reviewed ion-pair extraction with applications to trace analysis of drug residues in biological fluids.

IV. COLUMN CHROMATOGRAPHY

This section discusses the applications of classical column chromatographic cleanup for trace analysis. The theory of chromatography as described in Chapter 6 applies equally well to this approach. The difference between this and high-performance LC is that the latter is much more efficient because of better-packed columns, superior packing materials, and precise control over flow rate. However, classical column chromatography requires little equipment other than a glass column and some suitable packing material. It can be effectively used for cleanup by performing a step-gradient type elution. For example, when employing adsorption-type materials, the nonpolar organics are eluted with a relatively nonpolar solvent combination. The polar strength of the eluting solvent is then increased in steps to elute the more polar organics. In this way an analyst can isolate the compound of interest from other sample coextractives which have different polarities.

Several factors must be kept in mind when employing column chromatographic cleanup. These are related to maintaining stable chromatographic systems that are reproducible from one day to the next and between one column and another prepared at the same time. The reproducibility of the elution profile will directly affect recoveries from the column. If an analyte is not eluted consistently in one fraction, then losses will occur because it has been partially eluted either before or after the collected fraction. There are a number of reasons for this occurrence. If the column is not properly packed, channels may result and cause an irregular flow profile, leading to severe band broadening and changes in elution volume. Usually, columns are dry-packed with gentle tapping of the side of the column, or slurry-packed by preparing a slurry, pouring it into the column, and permitting it to settle.

The flow rate of the eluting solutions should be as uniform as possible, usually about 1–2 ml/min. This is important since the chromatographic pattern may alter significantly if the column is eluted too quickly. When carrying out step-gradient elutions, it should be pointed out that the column should never be permitted to run dry before the next solvent mixture is added. Large quantities of air in the column will greatly affect the elution pattern of analyte molecules in subsequent fractions.

One very important consideration is the preparation and standardization of the column packing material, especially in adsorption chromatographic cleanup. This is discussed in detail in Section IV,A,1.

A. Adsorption Materials

Adsorption chromatography is one of the most common techniques employed for sample cleanup in organic trace analysis. Usually 5–10 gm of the adsorbent is packed into a glass column (about 1 cm i.d.) with an adjustable stopcock. The sample is dissolved in a very nonpolar solvent and allowed to percolate into the column. Then various combinations of organic solvents are passed through the column to selectively elute the compound of interest from sample coextractives. However, as mentioned above, the efficiency of the column is such that complete isolation of the analyte from other sample components is not possible. In spite of this, adsorption chromatography has found a useful place in sample cleanup, especially in the analysis of environmental contaminants. Details on the preparation of adsorbents for column chromatography as well as numerous applications directed toward trace analysis may be found in various publications (*17–19*). The following sections are not intended to detail the preparation and use of column materials, but they show that several types of materials are available to the analytical chemist. These are useful since the selectivities are different, enabling the analyst to select which would be most appropriate for his problem. Table 8.6 lists some sources of adsorbents for column chromatography.

1. Standardization of Adsorbent Activity

Adsorbent materials as received from suppliers are often in an unknown state of activation, i.e., the percent of adsorbed water is unknown. Even if the materials are fully activated by the manufacturer, they can adsorb moisture from the atmosphere during storage or shipping. The differences in adsorbent activity can adversely affect chromatographic reproducibility. Thus, it is now accepted procedure to prepare the adsorbent at a known activation before using it. This involves heating the material, usually at 130°C or higher, for several hours to remove all adsorbed water, a treatment that renders the adsorbent to its most highly active state. Often, the fully activated adsorbent retains the analytes too strongly, so that some degree of deactivation is employed. Deactivation refers to the addition of known quantities of water to the adsorbent, usually in the range of 2–10% water by weight. The extent of deactivation depends upon the analyte and type of sample matrix. For example, in order to retain very nonpolar compounds such as chlorinated dioxins, silica gel should be in a highly active state. Increasing the water content of the adsorbent decreases its adsorptive power. It has been found that deactivation greater than 2% with water increases the reproducibility of columns from different batches, the reason being that it is more difficult to

TABLE 8.6

Sources for Commonly Used Adsorbents

Substance	Source
Activated alumina (chromatographic grade)	Merck and Co., Darmstadt, Federal Republic of Germany
Woelm alumina	Research Specialties Co., Richmond, California
Darco carbon	Darco Corp., New York, New York
Nuchar carbon	West Virginia Pulp and Paper Co., New York, New York
Florisil (magnesium silicate)	Floridin Co., Pittsburg, Pennsylvania
Columbia activated carbon	Union Carbide and Carbon Co., New York, New York
Attapulgus clays (Attaclays, Attagels)	Attapulgus Clay Co., Philadelphia, Pennsylvania
Celites (diatomaceous earths)	Johns-Manville Corp., New York, New York
Silica gels	Merck and Co., Machery, Nagel and Co., Duren, Federal Republic of Germany
Magnesium oxide	Westvaco Chlorine Prod. Corp., New York, New York
Acid-washed Nuchar	Kensington Scientific Corp., Emeryville, California
Polyethylene–alumina	Kensington Scientific Corp., Emeryville, California

store highly activated adsorbents. When a batch of standardized deactivated adsorbent is prepared, it is necessary to determine the elution pattern of the analyte to ensure that it elutes in the designated fraction before beginning actual sample analyses.

2. Influence of Water, Lipids, and Oils in the Sample

The presence of water in the sample extract can influence the elution pattern of the organics of interest. The water can lead to increased deactivation of the adsorbent, causing an increase in speed at which the organics pass through the column. To prevent this, the water should be removed from the extract by adding solid anhydrous sodium sulfate, then filtering the solution. Also, it is recommended in many methods that a 0.5 to 1-cm layer of anhydrous sodium sulfate be added to the top of the column to further protect it from traces of water in the sample extract or in the eluting solvents. This can be especially critical with highly activated adsorbents.

The quantity of sample coextractives, e.g., lipid or other fatty material, that can be tolerated by column cleanup depends on the quantity of adsorbent being employed. About 1 gm of fat can be used with 10–15 gm of adsorbent and not interfere in the elution of nonpolar substances such as many halogenated hydrocarbons and pesticides. However, the fat will interfere with moderately polar substances, which require a more polar eluting solvent. With the latter, a solvent–solvent partition (see Section III,A) would be required to remove the fats or oils before column cleanup.

3. Silica Gel

Silica gel is probably the most commonly used adsorbent for chromatography. This includes high-performance LC, TLC, and classical column chromatography. The major differences in the three are the particle sizes and the particle-size distribution (for LC). Usually a mesh size of about 100 is used for column cleanup techniques. Smaller particles increase resistance to solvent flow, resulting in the need for pressurized flow systems.

Silica gel must be activated as mentioned in Section IV,A,1 and used in the fully activated form or deactivated with 5% water. The degree of deactivation required can be determined by how effectively it removes the analyte from coextractives. It should be pointed out that silica gel column cleanup may not be adequate to remove interferences from extracts that are to be analyzed by LC with a silica gel column. The reason is that any compound that will interfere in the determination of the analyte on an LC column consisting of silica gel likely will not be removed by preliminary cleanup with a classic column of the same material, even if different solvent combinations are used. Thus, it is recommended that another type of adsorbent, e.g., alumina, charcoal, or Florisil, be employed since the differences in selectivity may be beneficial for removing interferences from the final LC analysis on a silica gel column. Silica gel and most other adsorbents are useful for removing polar substances from the sample extract that can contaminate (often irreversibly) the analytical column, reducing its lifetime. Silica gel column cleanup is also suited to the removal of coextractives for reversed-phase chromatography since additional selectivity is gained by employing the two different chromatographic principles.

4. Alumina

Alumina (Al_2O_3, aluminum oxide) for column chromatography is available in either acidic, neutral, or basic forms, depending upon how it is prepared and activated by the manufacturers. This adsorbent is particularly useful in removing polar coextractives from a sample extract in much the same manner as does silica gel. Often, passing a sample extract through a few grams of activated alumina in a small column will suffice. It has been

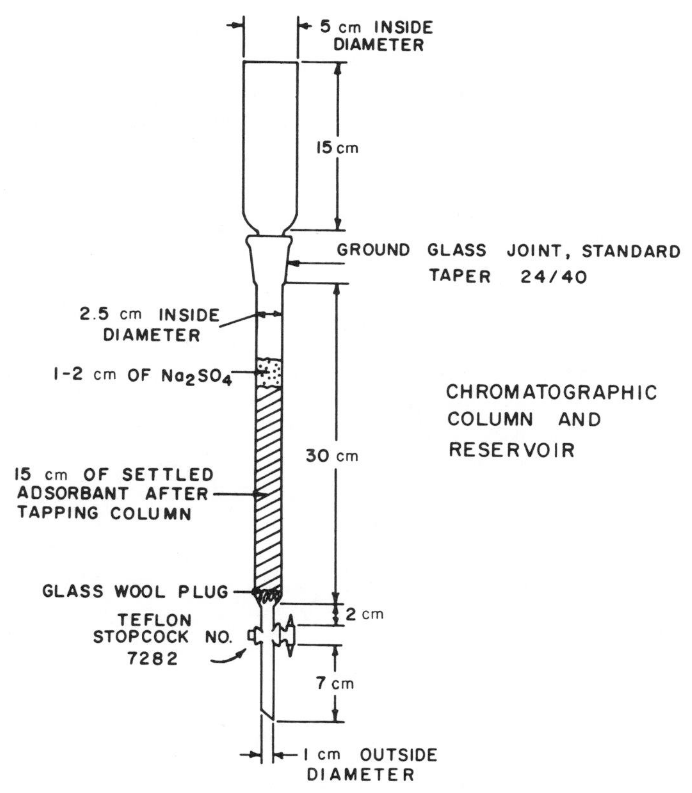

Fig. 8.7 Typical dimensions of a glass column employed for sample cleanup with Florisil.

found effective for removal of interferences from phenoxy acid herbicides in green vegetables where Florisil was not completely adequate (*20*).

5. Florisil

Florisil is an adsorbent composed of magnesium silicate (60/100 mesh) used for column chromatography, especially for the cleanup of pesticides. It is activated by the manufacturer by heating at 650°C. However, by the time it reaches the user, it is usually in an unknown state of deactivation; therefore, reactivation and standardization is necessary by the analyst. Figure 8.7 shows a typical column for sample cleanup employing Florisil as the adsorbent. Standardization of Florisil columns has been described by Mills (*21*).

6. Carbon–Cellulose

A type of column consisting of a mixture of carbon (Darco G60) and cellulose (Solka Floc BW40) has been described for the cleanup of nonpolar halogenated hydrocarbon compounds and compared to other carbon adsorbents (*22*). The mixture has a high capacity for retaining plant pigments and some waxes and oils. It may be prepared by mixing 20 gm of activated carbon and 64.2 gm of cellulose, using 3 gm of the mixture for each column.

B. Gel Permeation Chromatography

The advantages of gel permeation (or exclusion) chromatography for the cleanup of extracts for trace organic determinations was demonstrated by Stalling and co-workers (*23*) in 1972 in their work with pesticide residues. Other applications to polychlorinated and polybrominated biphenyls (*24*), styrenes (*25*), and industrial chemicals including dioxins and polychlorinated terphenyls (*26*) have since been carried out. The principle of the cleanup involves separation of molecules based on molecular size as described in Chapter 6. When gel permeation chromatography is used as a cleanup technique, it removes high-molecular-weight interferences

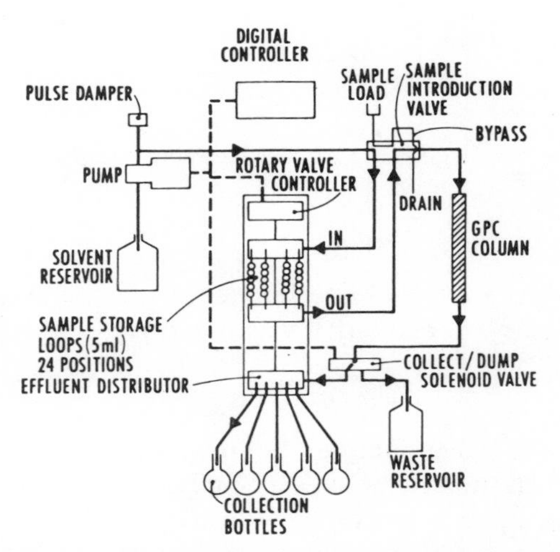

Fig. 8.8 Schematic of an automated gel permeation sample cleanup apparatus. Solid lines indicate solvent flow. Dashed lines indicate electrical connections. From Johnson *et al.* (*28*), with permission from the Association of Official Analytical Chemists.

from the compounds of interest, which are almost always of relatively low molecular weight (e.g., less than MW 400). Thus, during the elution of the column, the unwanted high-molecular-weight coextractives such as lipids and fats are removed first. Gel materials such as Sephadex LH-20 and Bio-Beads S-X2 (or S-X3) enable organic solvents to be used as mobile phases, which is advantageous for most trace organics—even nonpolar ones, which are poorly soluble in aqueous eluants.

One particular advantage of exclusion chromatography for sample cleanup is that the columns may be used repeatedly without noticeable changes in elution behavior or recoveries. Because of this, automation of the cleanup is possible and, in fact, a commercial automated gel permeation apparatus is available (*27*). Figure 8.8 shows a schematic (*28*) of the instrument, which is designed to process up to 23 samples completely unattended. Figure 8.9 shows the elution pattern of lipid substances of animal origin and Fig. 8.10 shows the elution of some vegetable oils. The analytes would be eluted after the majority of the high-molecular-weight substances were washed from the column. In Figs. 8.9 and 8.10, collection of the fraction containing the analyte would commence after the first 100 ml of eluant and would continue until all of the analyte is eluted (usually about an additional 60 ml).

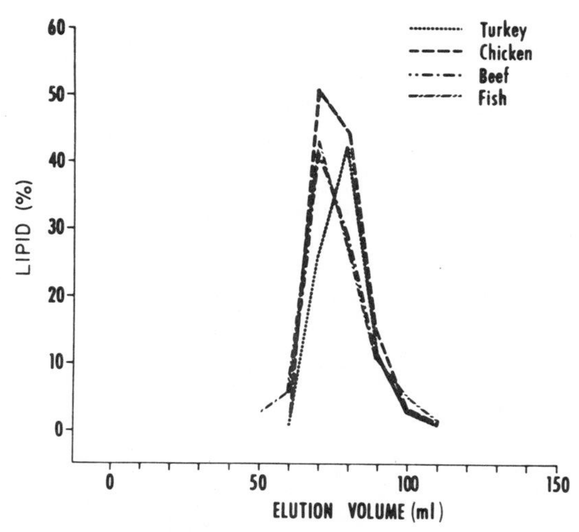

Fig. 8.9 Gel permeation elution pattern for a number of lipids of animal origin. Eluted through a column of Bio-Beads S-X3 with toluene/ethyl acetate (1 :3). From Johnson *et al.* (*28*), with permission from the Association of Official Analytical Chemists.

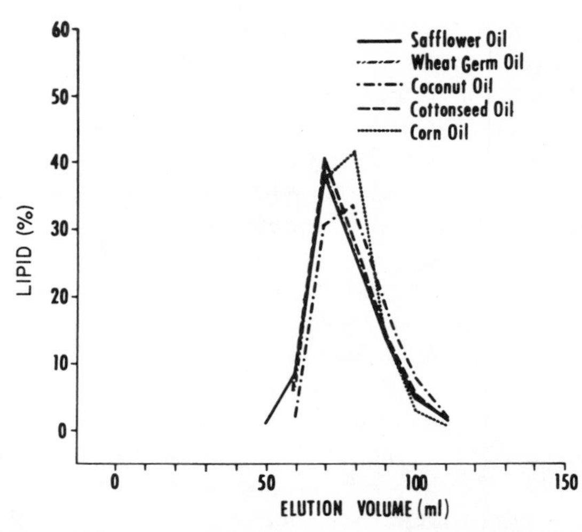

Fig. 8.10 Gel permeation elution pattern for a number of vegetable oils. Conditions as in Fig. 8.9. From Johnson *et al.* (*28*), with permission from the Association of Official Analytical Chemists.

C. Macroreticular Resins

Macroreticular beads (porous polymer beads) such as Amberlite XAD-2 or XAD-4, which are composed of polystyrene, have the ability to remove organics from aqueous solutions by means of adsorption (see Chapter 6 for discussion). The use of these for removal of organics from water samples has been mentioned in Section II,B. These resins have also been employed to advantage for sample extract purification. Ryan and Dupont (*29*) used XAD-2 to retain some tetracycline antibiotics from extracts of animal tissue. Figure 8.11 shows a flow diagram of the technique. The tissue homogenate, blended with 1.0 *N* HCl, was centrifuged and the supernatant was passed through the XAD-2 column. The column was washed with water to remove the water-soluble components and any remaining homogenate supernatant. The antibiotics were then eluted from the column with methanol before analysis by TLC.

XAD-2 resin has been used for the extraction and cleanup of drugs of abuse in urine samples (*10,30,31*). Centrifuged urine is passed through the column, which retains the drugs. The column is then washed with distilled water, which is discarded, while the drugs are eluted with methanol for analysis. The water wash removes excess urine and washes any remaining water-soluble coextractives from the column. Table 8.7 lists recoveries for some drugs of abuse through the XAD-2 column procedure (*10,31*). Junk *et al.* (*32,33*) have reviewed resin sorption methods for ap-

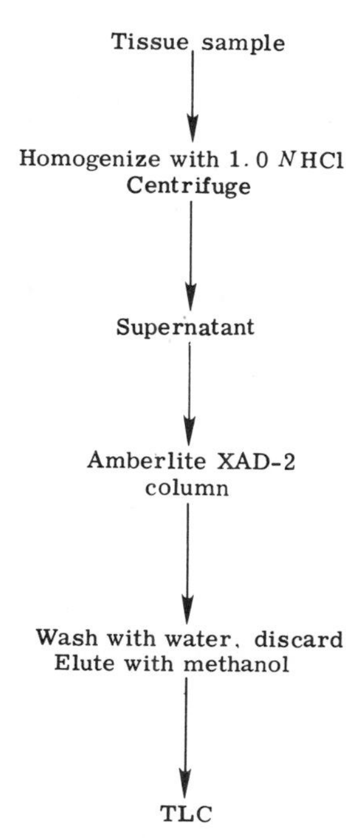

Fig. 8.11 Flow diagram of the Amberlite XAD-2 cleanup for tetracycline residues in animal tissues.

TABLE 8.7

Recoveries of Some Drugs of Abuse from Urine Using XAD-2 Resin[a]

Drug	Concentration in Urine (μg/ml)	% Recovery
Amphetamine	1.0	84
Morphine	1.25	84
Methadone	10.4	86
Phenobarbital	2.0	96
Caffeine	11.0	91
Cocaine	1.0	92
Meperidine	1.0	89
Mescaline	1.0	73
Meprobamate	1.0	74
Quinine	10.0	78
Nicotine	11.0	29

[a] From Mulé *et al.* (*10*) and Bastos *et al.* (*31*).

TABLE 8.8

Summary of Recoveries from Water Employing XAD-2 Resin[a]

Compound type	Average recovery (%)
Alcohols	94
Aldehydes	95
Ketones	95
Esters	93
Acids[b]	101
Phenols[b]	89
Ethers	90
Halogen compounds	87
Polynuclear aromatics	89
Alkylbenzenes	90
Nitrogen, sulfur compounds	89
Pesticides	90

[a] From Junk *et al.* (*32*).
[b] Water sample acidified with HCl.

plications to water pollutants. Table 8.8 summarizes the recoveries from water of various classes of compounds using XAD-2 resin (*32,33*).

D. Reversed-Phase Column Cleanup

Reversed-phase packing materials of the C_{18} or C_8 bonded-phase type have been employed for the extraction and cleanup of trace organics. Unlike the macroreticular resins, the reversed-phase materials are contained in very small columns such as LC guard columns or commercially available Sep-Pak disposable cartridges. They have the ability to remove dissolved organics from aqueous solutions passed through them in the same manner as the macroreticular resins.

Guard columns (or precolumns) are particularly suited to automation since they can withstand the high pressures involved in LC separations, thus enabling on-line analyses. Figure 8.12 shows a typical schematic of an on-line extraction and LC analysis as applied to aqueous samples (e.g., natural waters, industrial effluents). In the top diagram, pump A delivers a predetermined volume of sample through the injection valve to the reversed-phase precolumn, where the desired organics are retained. The remainder of the water passes through the column to waste. The volume of sample passed depends on the affinity of the organic compound for the

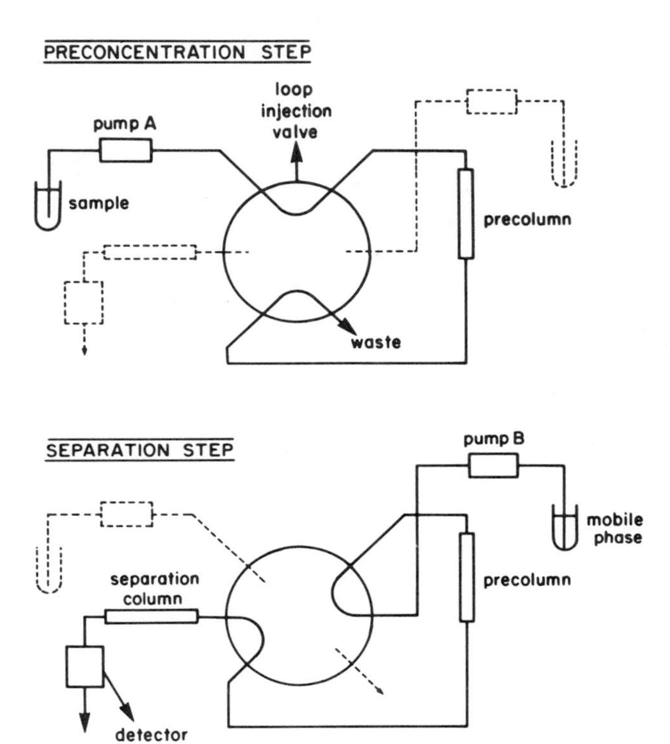

Fig. 8.12 Schematic for precolumn trace enrichment as applied to aqueous samples. See text for discussion. From Frei (*44*), with permission from Gordon and Breach Science Publ.

reversed-phase material. Too much sample may wash some of the analyte off the column. Thus, k' values for the sample in the aqueous solution should be very large to ensure good retention of the analyte by the precolumn, even with large sample volumes (e.g., several hundred milliliters or more). In the separation step of Fig. 8.12, pump A is stopped and the injection valve is turned so that mobile phase from pump B elutes the sample from the precolumn to the analytical column where separation of the organics is achieved either by gradient elution or with an isocratic system. Usually, gradient elution is required since the reversed-phase precolumns will indiscriminantly remove most lipophilic compounds from the aqueous samples.

The precolumns may also be used in an off-line mode for field work. A water sample from a lake or river may be pumped through a precolumn, which is then inserted into an LC system for analysis of the organics. The precolumns containing the adsorbed organics are more compact and, therefore, much easier to store than the water sample itself.

A similar on-line apparatus to that shown in Fig. 8.12 has been constructed for application to plasma analysis (*34*). However, the design is such that a precise volume of urine sample, 2 ml, is injected into a sample loop and is pumped into a precolumn as shown in Fig. 8.13. The organics of interest are retained by the column while the weakly retained interferences are flushed through with the appropriate mobile phase. The valve (V2) is then turned, and the mobile phase from pump 2 (P2), containing a suitable organic modifier, elutes the organics from the precolumn onto the analytical column for separation and detection. This system may also be automated.

Disposable cartridges such as the Sep-Pak type, which contain C_{18} reversed phase bonded to silica gel, are also suitable for trace enrichment. Although they cannot be used for on-line LC analyses, they are particu-

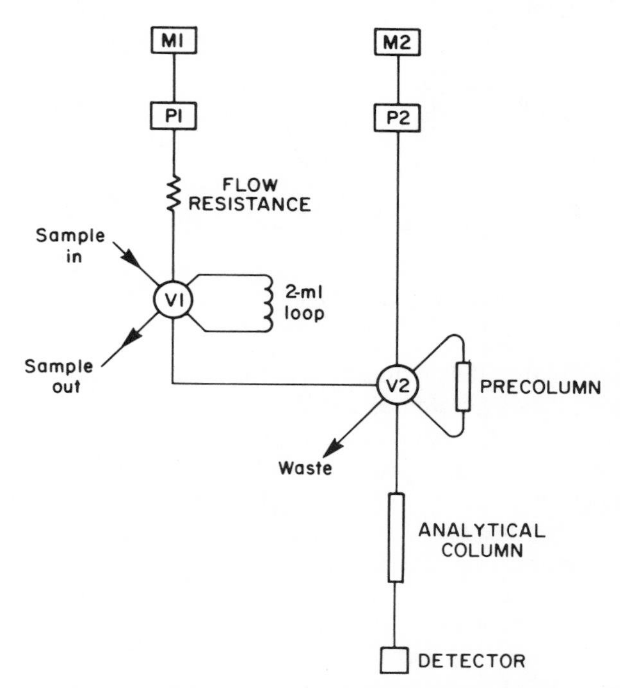

Fig. 8.13 Schematic for the trace enrichment of plasma samples. M1, mobile phase for trace enrichment; M2, mobile phase for analysis; P1 and P2 represent the pumps; V1 and V2 are injection valves. From Koch and Kissinger (*34*), with permission from the American Chemical Society.

larly useful for the collection of field samples. Since they have a high permeability, samples can be forced through them with a hand-held syringe. This is advantageous in the field since no special mechanical pump or power source is needed. The adsorbed organics must be eluted from the cartridges in a classical manner, then an aliquot of the eluant injected into an LC for separation.

Applications of trace enrichment with reversed-phase columns have been reported for the analysis of trace organics in plasma and serum (*34–36*), distilled water (*37*), drinking water (*38–40*), natural water (*40,41*), wastewater (*40,42*), chlorinated water (*43*), urine (*44–46*), and seawater (*47,48*).

In the above applications, the reversed-phase columns were used to isolate organics from the aqueous samples while the polar components passed through unretained and were discarded. On the other hand, if the polar constituents are the compounds of interest, the reversed-phase column also acts as a cleanup because it removes much of the organics from the aqueous sample. This can be especially helpful when the analysis is carried out by reversed-phase chromatography, since nonpolar organics are strongly retained and would be eluted as late peaks, which might not interfere in the determination of the analyte but which could interfere in subsequent chromatograms.

E. Ion-Exchange Column Chromatography

Polar organics of an ionic nature are often purified by passing the aqueous sample through an ion-exchange column. This is the classical approach to purifying extracts for both inorganic and organic ions and has been comprehensively treated (*49–52*). Discussion of the principles of ion-exchange chromatography appears in Chapter 6, Section VI. As a cleanup, the sample extract is passed through an appropriate ion-exchange column where the analyte is retained by the exchange material. All unretained coextractives pass through the column. The analyte may then be eluted with a suitable mobile phase. Whether performing anion- or cation-exchange chromatography, it is important to standardize the method well. The choice of the correct exchanger is very important, since losses will occur if the analyte is not sufficiently retained. Also, if the analyte is too strongly retained, then difficulty in removing it from the column may also lead to losses. An example of this problem is found in the development of ion-exchange cleanup methods for the quaternary ammonium herbicides paraquat and diquat (*53,54*).

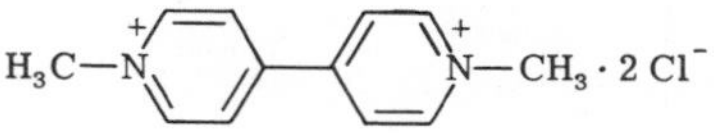

Paraquat dichloride

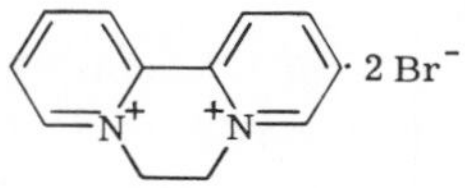

Diquat dibromide

The use of ion exchange for cleanup of samples where the final determinative step employs ion-exchange LC may not be entirely satisfactory unless the chromatographic selectivity of the cleanup step is significantly different from the analysis step. The reason for this is that any coextractives that would interfere in the final ion-exchange separation would not likely be removed with a cleanup procedure employing a similar ion-exchange system. If this poses a problem, then ion-pair chromatography or reversed-phase chromatography with ion suppression (see Chapter 6, Sections III–V) could be employed, or a different type of sample cleanup may be carried out (such as liquid–liquid partition with pH adjustment or ion-pair extraction; see Section III).

V. THIN-LAYER CHROMATOGRAPHY

Thin-layer chromatography (TLC) is very suitable for sample cleanup prior to LC analyses. The crude extracts are spotted near the bottom of the plate which is placed in a glass chamber containing the eluting solvent. The components of the sample spot migrate up the plate along with the eluting solvent. The relative distance each component moves depends upon its affinity for the adsorbent. Most TLC plates use adsorbents such as silica gel, alumina, or cellulose as chromatography layers, usually at thicknesses of 0.2–0.5 mm. The order of elution is related to polarity, the least polar substances migrating farthest up the plate. Figure 8.14 depicts a typical TLC apparatus showing some separated spots. For cleanup purposes, the appropriate spot is scraped off the plate and removed from the adsorbent particles by elution with an organic solvent. The resulting solution is usually very clean and can be analyzed directly by LC.

The major disadvantage of this technique is that most trace organics of current interest do not absorb light in the visible region; thus it is not possible to directly locate them on the plate after separation. The substances

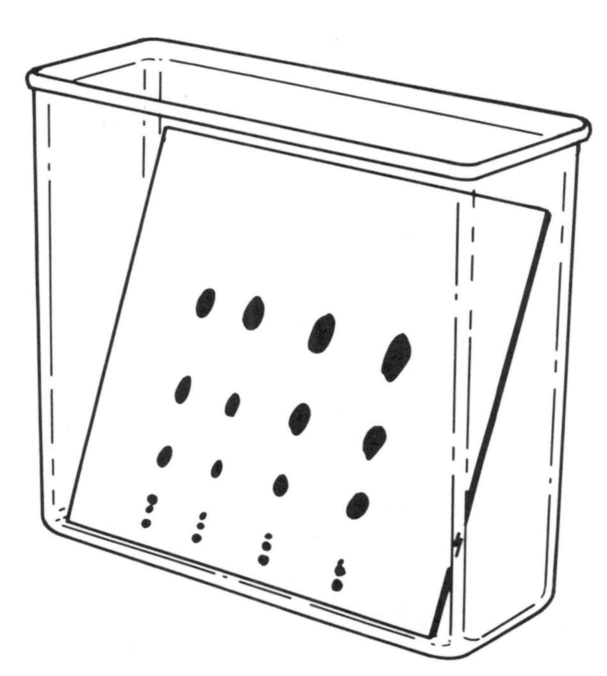

Fig. 8.14 Typical TLC set up with a 20- × 20-cm TLC plate in a chromatography chamber. Four sample extracts were spotted near the bottom. The black dots represent substances separated from the extract.

may be visualized by using spray reactions that create derivatives that are visible on the TLC layer. However, this changes the chemical structure of the analyte so it can only be removed from the layer and analyzed as the derivative and not the parent compound. However, a standard may be run beside the sample and it alone sprayed when the separation is complete to indicate the distance migrated. The corresponding area of the sample run can then be scraped off the plate and the analyte determined by LC as the intact compound. Several comprehensive volumes have appeared on the subject of TLC (*55–58*).

VI. DISTILLATION

Distillation not only affords a method whereby volatile organics may be extracted from aqueous samples (see Section II,D); it provides at the same time a very effective means of cleanup of sample extracts derived by other means. Sweep-codistillation (*59,60*) is particularly useful for volatiles in complex samples such as plant tissue. The solid samples are first extracted with a suitable organic solvent—for example, ethyl acetate—to remove the analyte. An aliquot is then concentrated and cleaned up by injection into a long heated glass tube packed with glass wool, followed at regular intervals (usually 3 min) by injections of the organic solvent. Nitrogen carrier gas sweeps the vaporized volatile components through the tube while the nonvolatile coextractives remain on the glass wool. The volatile components are then collected for analysis. Figure 8.15 shows a sweep-codistillation apparatus. Modifications to this, including automation, have been described (*60–62*). The approach has been shown to handle about 50 samples of soil or 10 samples of plant material before requiring cleaning (*61*).

This technique is basically a crude form of GC and is particularly suited as a cleanup for LC since the removal of interferences is based on volatility. Two nonpolar compounds (for example, an analyte and a sample coextractive) could have very different volatilities, enabling the purification of the more volatile analyte by sweep codistillation with little trouble. However, separation of the two by classical cleanup employing adsorption or reversed-phase chromatography may be difficult because of the similarity in the polarity of the two, especially if the less volatile sample coextractive is present in great excess compared to the analyte. No work has yet been reported on the application of sweep codistillation to LC analysis, although the potential is very good.

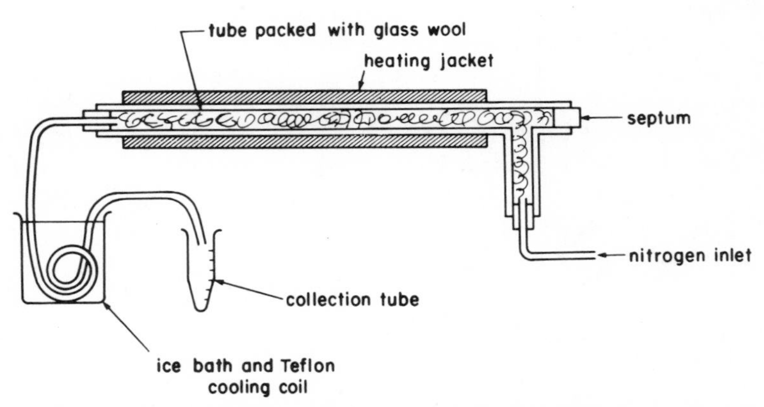

Fig. 8.15 Sweep-codistillation apparatus according to Storherr and Watts (*59*). Samples are injected via the septum. The organic solvent and volatile organics from the sample condense in the cooling bath and the solution is collected in the collection tube.

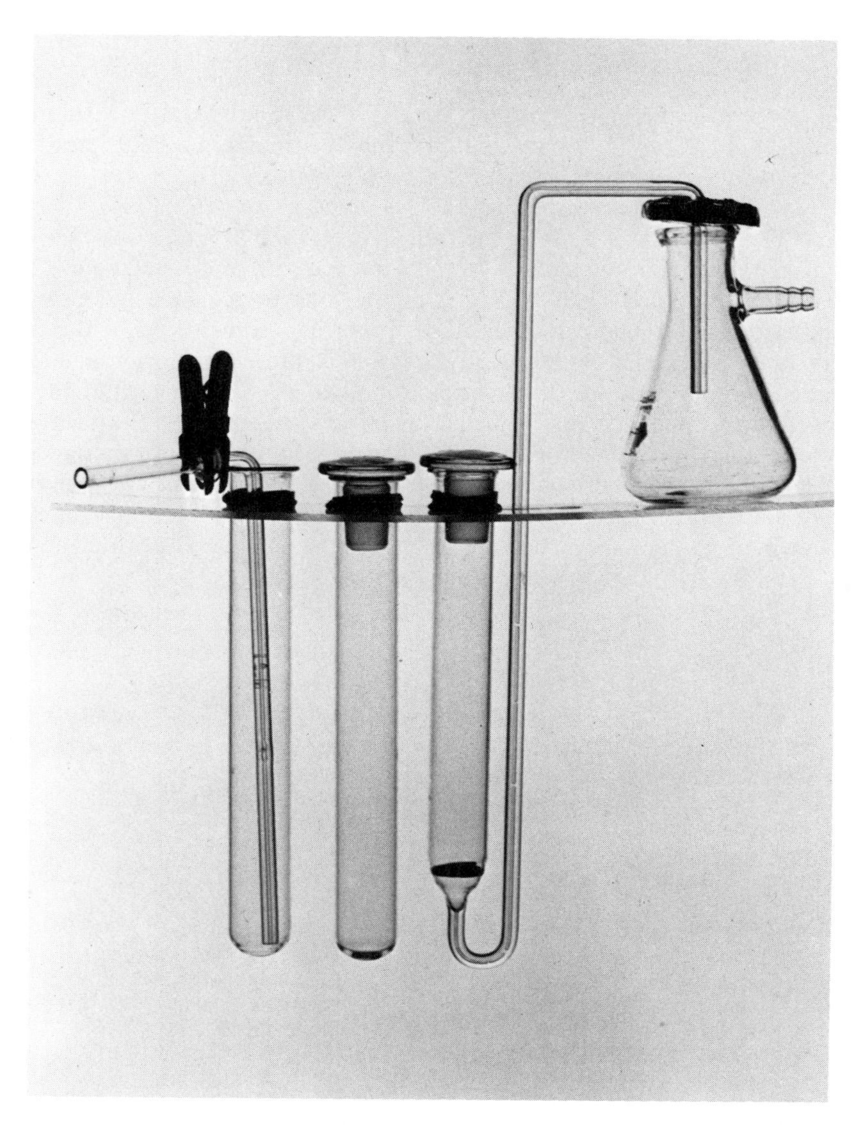

Fig. 8.16 Low-temperature precipitation set up. From left to right: (1) Capillary tube with ball and socket joint leading into a 50-ml test tube in which the precipitation occurs; (2) 50 ml test tube to hold wash solution; (3) filtration tube with capillary tube leading to a vacuum flask with tight-fitting seal.

VII. LOW-TEMPERATURE PRECIPITATION

Low-temperature precipitation involves the removal of coextractives such as lipids, waxes, and water from sample extracts in a single step. For example, a sample may be extracted with a mixture of acetone–benzene (19:1) (*63*) filtered, then placed in a test tube as shown in Fig. 8.16 (far left). The lower two-thirds of the tube is immersed in a methanol [or ethanol (*64*)] bath maintained at −78°C by the addition of dry ice (solid CO_2). Nitrogen is bubbled for 30 min through the solution to cause agitation and mixing during the precipitation. The mixture is transferred to the filter tube (also in the cold bath), where the liquid is filtered with suction and removed from precipitated water, lipid, and waxes. The precipitate is then rinsed with the cold wash solution (acetone–benzene, 19:1) and the filtrate is collected and concentrated to a small volume for chromatographic analysis. The method has been applied to the removal of coextractives from plant and animal tissues, including both fatty and nonfatty samples (*63–67*).

REFERENCES

1. B. Kolb, ed., "Applied Headspace Gas Chromatography." Heyden, London, 1980.
2. W. Thornburg, *in* "Analytical Methods for Pesticides and Plant Growth Regulators, (G. Zweig, ed.,), Vol. 1, p. 87. Academic Press, New York, 1963.
3. N. P. Sen and C. Dalpé, *Analyst* (*London*) **97,** 216 (1972).
4. R. G. Webb, *Int. J. Environ. Anal. Chem.* **5,** 239 (1978).
5. D. A. J. Murray, *J. Chromatogr.* **177,** 135 (1979).
6. O. W. Grussendorf, A. J. McGinnis, and J. Solomon, *J. Assoc. Off. Anal. Chem.* **53,** 1048 (1970).
7. R. E. Johnson and R. I. Starr, *J. Econ. Entomol.* **60,** 1679 (1967).
8. R. E. Johnson and R. I. Starr, *J. Econ. Entomol.* **63,** 165 (1970).
9. K. Berkane, G. E. Caissie, and V. N. Mallet, *J. Chromatogr.* **139,** 386 (1977).
10. S. J. Mulé, M. L. Bastos, D. Jukofsky, and E. Saffer, *J. Chromatogr.* **63,** 289 (1971).
11. R. H. Brown and C. J. Purnell, *J. Chromatogr.* **178,** 79 (1979).
12. T. L. Peters, *Anal. Chem.* **52,** 211 (1980).
13. M. C. Bowman and M. Beroza, *J. Assoc. Off. Agric. Chem.* **48,** 943 (1965).
14. M. C. Bowman and M. Beroza, *Anal. Chem.* **38,** 1544 (1966).
15. G. Schill, *in* "Ion Exchange and Solvent Extraction" (J. A. Marinsky and Y. Marcus, eds.), Vol. 6, p. 1. Dekker, New York, 1974.
16. G. Schill, *in* "Assay of Drugs and Other Trace Compounds in Biological Fluids" (E. Reid, ed.), p. 87. North-Holland Publ., Amsterdam, 1976.
17. "Official Methods of Analysis" (W. Horwitz, ed.). Assoc. Off. Anal. Chem., Washington, D.C., 1975.
18. "Pesticide Analytical Manual" (B. M. McMahon and L. D. Sawyer, eds.). U.S. Food Drug Adm., Washington, D.C., 1968.

19. "Analytical Methods for Pesticide Residues in Foods" (H. A. McLeod and W. Ritcey, eds.). Health Prot. Branch, Ottawa, 1973.
20. G. Yip, *J. Assoc. Off. Anal. Chem.* **45,** 367 (1962).
21. P. A. Mills, *J. Assoc. Off. Anal. Chem.* **51,** 29 (1968).
22. H. A. McLeod, C. Mendoza, P. J. Wales, and W. P. McKinley, *J. Assoc. Off. Anal. Chem.* **50,** 1216 (1967).
23. D. L. Stalling, R. C. Tindle, and J. L. Johnson, *J. Assoc. Off. Anal. Chem.* **55,** 32 (1972).
24. N. V. Fehringer, *J. Assoc. Off. Anal. Chem.* **58,** 978 (1975).
25. D. W. Kuehl, H. L. Kopperman, G. D. Veith, and G. Glass, *Bull. Environ. Contam. Toxicol.* **16,** 127 (1976).
26. L. W. Wright, R. G. Lewis, H. L. Crist, G. W. Sovocool, and J. M. Simpson, *J. Anal. Toxicol.* **2,** 76 (1978).
27. J. A. Ault, C. M. Schofield, L. D. Johnson, and R. H. Waltz, *J. Agric. Food Chem.* **27,** 825 (1979).
28. L. D. Johnson, R. H. Waltz, J. P. Ussary, and F. E. Kaiser, *J. Assoc. Off. Anal. Chem.* **59,** 174 (1976).
29. J. J. Ryan and J. A. Dupont, *J. Assoc. Off. Anal. Chem.* **57,** 828 (1974).
30. N. Weissman, M. L. Lowe, J. M. Beattie, and J. A. Demetriou, *Clin. Chem.* **17,** 875 (1971).
31. M. L. Bastos, D. Jukofsky, E. Saffer, M. Chedekel, and S. J. Mulé, *J. Chromatogr.* **71,** 549 (1972).
32. G. A. Junk, J. J. Richard, J. S. Fritz, and H. J. Svec, *J. Chromatogr.* **99,** 745 (1974).
33. G. A. Junk, J. J. Richard, J. S. Fritz, and H. J. Svec, *in* "Identification and Analysis of Organic Pollutants in Water" (L. H. Keith, ed.), p. 135. Ann Arbor Sci. Publ., Ann Arbor, Michigan, 1976.
34. D. D. Koch and P. T. Kissinger, *Anal. Chem.* **52,** 27 (1980).
35. J. Lankelma and H. Poppe, *J. Chromatogr.* **149,** 587 (1978).
36. D. Ishii, K. Hibi, K. Asai, M. Nagaya, K. Mochizuki, and Y. Mochida, *J. Chromatogr.* **156,** 173 (1978).
37. F. Eisenbeiss, H. Hein, R. Joester, and G. Naundorf, *Chromatogr. Newsl.* **6,** 8 (1978).
38. K. Ogan, E. Katz, and W. Slavin, *J. Chromatogr. Sci.* **16,** 517 (1978).
39. Waters Assoc., Tech. Bull. H63. Milford, Massachusetts, 1976.
40. C. G. Creed, *Res./Dev.* **27,** 40 (1976).
41. R. Kummert, E. Molnar-Kubica, and W. Giger, *Anal. Chem.* **50,** 1637 (1978).
42. Waters Assoc., Tech. Bull. H91. Milford, Massachusetts, 1977.
43. A. R. Oyler, D. L. Bodenner, K. J. Welch, R. J. Liukkonen, R. M. Carlson, H. L. Kopperman, and R. Caple, *Anal. Chem.* **50,** 837 (1978).
44. R. W. Frei, *Int. J. Environ. Anal. Chem.* **5,** 143 (1978).
45. P. Shauwecker, R. W. Frei, and F. Erni, *J. Chromatogr.* **136,** 63 (1977).
46. C. van Buuren, J. F. Lawrence, U. A. T. Brinkman, I. L. Honingberg, and R. W. Frei, *Anal. Chem.* **52,** 700 (1980).
47. W. E. May, S. Chesler, S. Cram, B. Gump, H. Hertz, D. Enagonio, and S. Dysezel, *J. Chromatogr. Sci.* **13,** 535 (1975).
48. W. A. Saner, J. R. Jadamer, R. W. Sager, and T. K. Killeen, *Anal. Chem.* **51,** 2180 (1979).
49. J. Inczedy, "Analytical Applications of Ion Exchangers." Permagon, New York, 1971.
50. O. Samuelson, "Ion Exchangers in Analytical Chemistry." Wiley, New York, 1953.
51. F. Helfferich, *Adv. Chromatogr.* **1,** 3.
52. A. Calderbank, *Residue Rev.* **12,** 14 (1966).
53. A. Calderbank and S. H. Yuen, *Analyst (London)* **90,** 99 (1965).

54. W. J. Kirsten, *Analyst (London)* **91,** 732 (1966).
55. I. Smith, ed., "Chromatographic and Electrophoretic Techniques," Vols. 1 and 2. Wiley, New York, 1969.
56. E. Heftmann, "Chromatography." Reinhold, New York, 1967.
57. J. G. Kirchner, "Thin-layer Chromatography." Wiley (Interscience), New York, 1967.
58. E. Stahl, "Thin-layer Chromatography." Springer-Verlag, Berlin and New York. 1967.
59. R. W. Storherr and R. R. Watts, *J. Assoc. Off. Agric. Chem.* **48,** 1154 (1965).
60. J. J. S. Kim and C. W. Wilson, *J. Agric. Food Chem.* **14,** 615 (1966).
61. B. A. Karlhüber and D. O. Eberle, *Anal. Chem.* **47,** 1094 (1975).
62. J. Pflugmacher and W. Ebing, *Fresenius' Z. Anal. Chem.* **263,** 120 (1973).
63. H. A. McLeod and P. J. Wales, *J. Agric. Food Chem.* **20,** 624 (1972).
64. J. F. Lawrence and H. A. McLeod, *Bull. Environ. Contam. Toxicol.* **12,** 752 (1974).
65. J. A. R. Bates, *Analyst (London)* **90,** 453 (1965).
66. K. A. McCully and W. P. McKinley, *J. Assoc. Off. Agric. Chem.* **47,** 652 (1964).
67. H. A. McLeod, *Anal. Chem.* **44,** 1328 (1972).

Chapter 9
Approach to Method Development and Routine Analysis

I. CHOOSING THE BEST CHROMATOGRAPHY SYSTEM

The selection of the most suitable chromatographic technique for resolving any given analytical problem is first governed by the equipment available to the analyst. If only gas chromatography instrumentation is on hand, then the analyst has no choice but to use a method employing that type of chromatography, even if it becomes necessary to derivatize the analyte in order for it to pass through a GC. For those who have only LC equipment, a reversed-phase column and perhaps a silica gel adsorption one will most likely be available. These two types of columns provide the analyst with great power for separating many compounds, and he needs only to select the most appropriate mobile phase. The major question in LC for trace analysis is whether the detection system is adequate. An LC that is to be employed for doing trace determinations must be equipped with at least a fixed-wavelength UV detector or other sensitive or selective detection device (see Chapter 5). The amount and type of equipment available to the analyst determine the flexibility with which he can approach an analytical problem.

Table 9.1 indicates the best first approach to take for LC analysis, depending on the nature of the analyte. In most instances, especially in routine monitoring, the chemical and physical nature of the analyte is known. Also, the type of sample matrix is known, although not necessarily its history.

Most trace analyses involve relatively low-molecular-weight compounds which, as seen in Table 9.1, may be divided into organic- or water-soluble species. Separations may be carried out either by re-

TABLE 9.1

Selection of Chromatography Mode

Nature of compound	Type of chromatography
High molecular weight	
Organic soluble	Exclusion chromatography with organic mobile phase
Water soluble	Exclusion chromatography with aqueous mobile phase
Low molecular weight	
Organic soluble	
Homologous series	Reversed-phase chromatography
Functional groups	Adsorption or reversed-phase chromatography
Water soluble	
Non-ionic	Reversed-phase chromatography
Ionic	
Acidic	1. Cation exchange chromatography 2. Ion-pair chromatography with a basic counter-in 3. Ion suppression with mobile phase at acidic pH
Basic	1. Anion exchange chromatography 2. Ion-pair chromatography with an acidic counter ion 3. Ion suppression with mobile phase at weakly basic pH

versed-phase chromatography, which is superior for the separation of a homologous series of compounds with the same functional group, or by adsorption chromatography, which is particularly suited to the separation of compounds with different functional groups. Reversed-phase chromatography can also be useful for the latter (in this case separations are based on differences in overall hydrophobicity).

Reversed-phase and adsorption chromatography should not be considered as being competitive but should be used when most appropriate. The former does have a wider application than adsorption chromatography since very nonpolar compounds such as dioxins, for example, can be analyzed using the same column that can be employed for polar, water-soluble nucleic acid bases. Adsorption chromatography is most useful for moderate- to nonpolar compounds, which includes most substances that are extractable from samples into organic solvents. Some types of polar compounds may also be chromatographed using silica columns and employing *in situ* partition systems where the silica gel becomes coated with the polar component of the mobile phase (*1–4*).

Water-soluble nonionic compounds (that is, relatively polar compounds) are best determined by reversed-phase chromatography. Ionic organics may also be determined by employing reversed-phase systems using ion-pair chromatography or ion suppression (see Table 6.6). Ion-exchange chromatography is also useful for ionic compounds.

It cannot be said that better results are obtained by one mode of chromatography than another without first trying. Figures 9.1 and 9.2 clearly point this out. Figure 9.1 compares chromatograms obtained using a C_{18} reversed-phase system with a silica gel adsorption system for the analysis of the insecticide carbaryl in the same apple extract at 1.0 ppm. Both chromatograms are relatively clean and serve equally well for the analysis. Figure 9.2 compares results obtained from a "dirty" extract of orange containing the fungicide thiophanate-methyl at 1.0 ppm. In this figure both systems are again much the same, i.e., the chromatograms contain many interferences and detection limits are similar. These two figures serve to illustrate that both modes can provide good results or can suffer from interferences, depending upon sample preparation and cleanup. One is not always better than the other.

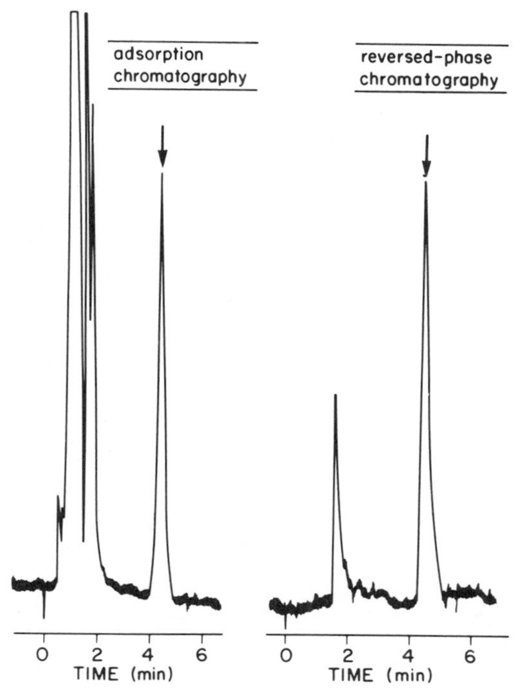

Fig. 9.1 A comparision of adsorption chromatography on silica gel (A) with reversed-phase (C_{18}) chromatography (B) for the LC analysis of carbaryl in an apple extract at 1.0 ppm. UV detection at 220 nm. Arrow indicates carbaryl peak.

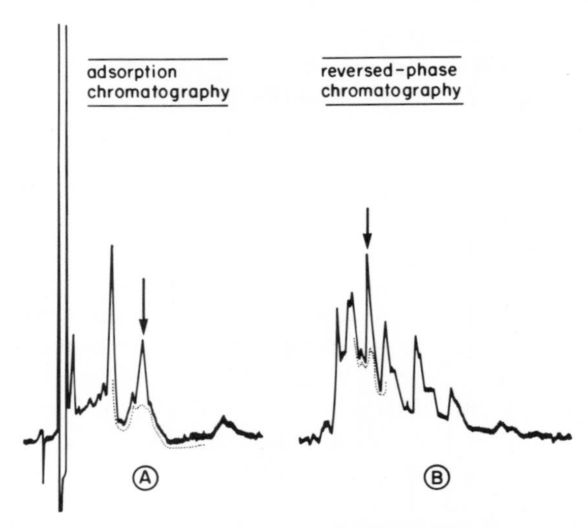

Fig. 9.2 A comparision of adsorption chromatography (A) and reversed-phase (C_{18}) chromatography (B) for the analysis of 1.0 ppm of thiophanate-methyl in orange. UV detection at 271 nm. Arrow indicates thiophanate-methyl peak. Dotted lines represent background from the sample blank.

An important factor in choosing a chromatographic system is the stability of the pure compound and solubility of the sample extract in the mobile phase. Some compounds tend to hydrolyze in the presence of water and thus cannot be stored in aqueous-based mobile phases for extended periods. If sample extracts contain lipid-soluble oils, for example, they may not be soluble in aqueous mobile phases and may cause cloudy solutions unsuitable for injection into a reversed-phase system. On the other hand, the extract would likely be very soluble in a moderately polar organic solvent and would thus be easily analyzed by adsorption chromatography.

II. ANALYSIS TIME AND INTERFERING PEAKS

Several factors are involved in determining the time of each chromatographic run. First, the k' of the analyte must be such that it is in a clean area of the chromatogram (i.e., an area suitable for quantitation), but not eluting so late that the time becomes impractical. This is particularly important if large numbers of samples are to be analyzed. Small k' values reduce analysis times and also increase detector response (see Chapter 6, Chromatography Theory). Large k' values also increase solvent consumption.

Late-eluting interferences can pose serious problems in a routine program. They can interfere in later chromatograms, nullifying the analyses. There are several approaches that may be taken to remove their effect. Usually additional cleanup (see Chapter 8, Sample Extraction and Cleanup) is required. However, it may be possible to change the detector parameters such that the interferences are not observed. This would mean for example, that one might change the wavelength of UV absorption. Figure 9.3 illustrates such an effect on the removal of interfering coextractives from an LC analysis of the wild oat herbicide asulam in a flour extract. The improvement from 254 nm to 280 nm is obvious, even though 280 nm is less sensitive.

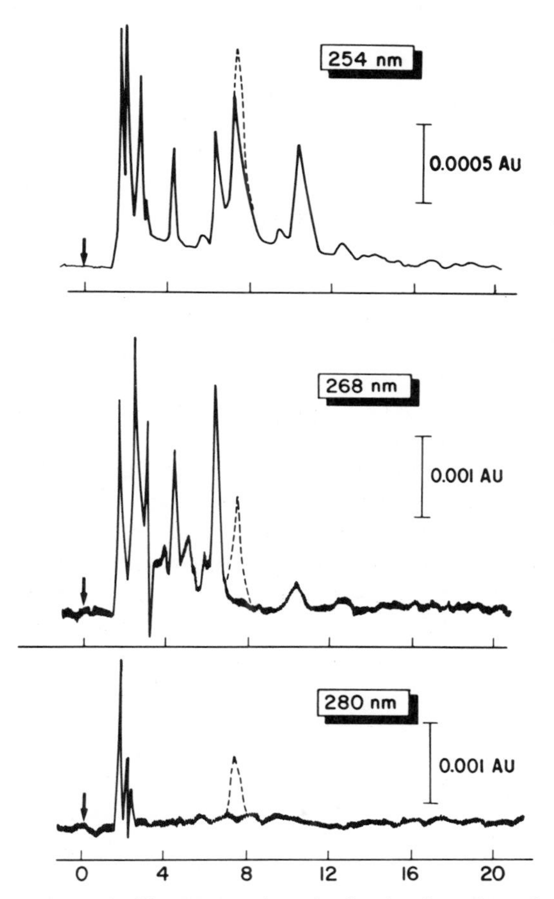

Fig. 9.3 Comparison of different wavelengths for the detection of asulam in whole wheat flour at 0.1 ppm. Dashed lines indicate the response to asulam. From Lawrence *et al.* (*16*), with permission from the American Chemical Society.

Late-eluting interferences may also be removed by changing the mode of chromatography. For example, later-eluting interferences in an adsorption chromatography system would elute faster than the analyte peak in a reversed-phase system, and vice versa.

If none of the above is satisfactory in removing late peaks, the analyst may have to use flow or gradient programming. Flow programming simply involves increasing the mobile phase velocity so that the late peaks will be flushed through the system as quickly as possible before the flow rate is returned to normal for the next analysis. This action reduces time but not mobile-phase consumption. Gradient elution is particularly useful for removal of late-eluting peaks by continuously changing the composition of the mobile phase. Although this approach is efficient at removal of such interferences, the LC system must be reequilibrated before the next analysis. Gradient elution is most useful for the analysis of a number of different analytes whose polarities vary over a wide range.

III. QUALITATIVE AND QUANTITATIVE ANALYSIS AND CONFIRMATION

By its nature LC is a qualitative technique before being a quantitative one. That is to say, the analyte must be qualitatively identified in the chromatogram before it can be quantitated. The only means to do this is by the measurement of the k' value. Thus it is essential that a chromatographic system produce reproducible k' values from one run to the next. Significant variation in retention can lead to misidentification of peaks and erroneous quantitation. Just how critical this may be depends upon the "cleanliness" of the chromatograms. In Fig. 9.1, for example, a slight change in the k' of carbaryl may not cause a serious problem. However, in chromatograms such as those shown in Fig. 9.2, a minor change in k' could lead to false results. Chromatographic systems should be equilibrated to the extent that the k' values do not significantly change over an extended period. Otherwise, standards would have to be repeatedly injected to account for the differences. The k' values are also directly related to the volume of mobile phase in which the peak is eluted. As the k' value increases, the volume of mobile phase in which the analyte is contained is diluted, causing a decrease in analyte concentration and a reduced peak height. This then would affect quantitation measured as peak height.

The quantitative reproducibility of an LC technique is related to several factors, including, as mentioned above, k' values. Detector response, of course, must be consistent. The effect of sample coextractives on the response of the analyte should be considered. These can influence quantitation by masking the analyte peak or by altering the chromatographic behavior. The extent of such effects can be examined by chromatographing a known mixture of analyte and sample, then observing the change in analyte peak compared to a standard injected afterward.

Confirmation of results obtained by LC is usually necessary in order to improve the certainty of the analysis. Since detection systems such as UV absorbance or fluorescence are nondestructive, the analyte may be collected after it passes through the detector and then analyzed by some other means, such as a different mode of LC (e.g., reversed-phase chromatography as opposed to an initial analysis by adsorption and vice versa), gas chromatography (GC), or mass spectrometry (MS). If necessary, the sample may be subjected to chemical reaction to form an appropriate derivative for either LC or GC (see Chapter 7, Chemical Derivatization).

IV. AUTOMATION

The ideal arrangement for routine analysis would be total automation, beginning with sample extraction and cleanup to chromatographic analysis, quantitation, and data printout. However, most analytical laboratories at present are not so fortunate, although the use of minicomputers—which not only process analytical results but also operate and control the chromatography, including sample injection—is increasing at a fast rate. The major advantages of this are that such systems can operate during the night, thus saving time; also, since they can operate unattended, they save on labor costs. The automation of sample handling prior to injection into the LC is somewhat more difficult because of the many complicated mechanical operations involved. Attempts at automating various aspects of sample preparation including extractions, partitions, filtrations, evaporations, column cleanup, and the necessary transfer steps have been reported in the literature (*5–12*). Figure 9.4 illustrates how an analysis may be automated from the initial extraction up to the final computer processing of the data (*13*). The only manual steps involve transferring the sample extracts to the automatic sampler of a gas chromatograph. The punched tape enables data storage or direct computer processing.

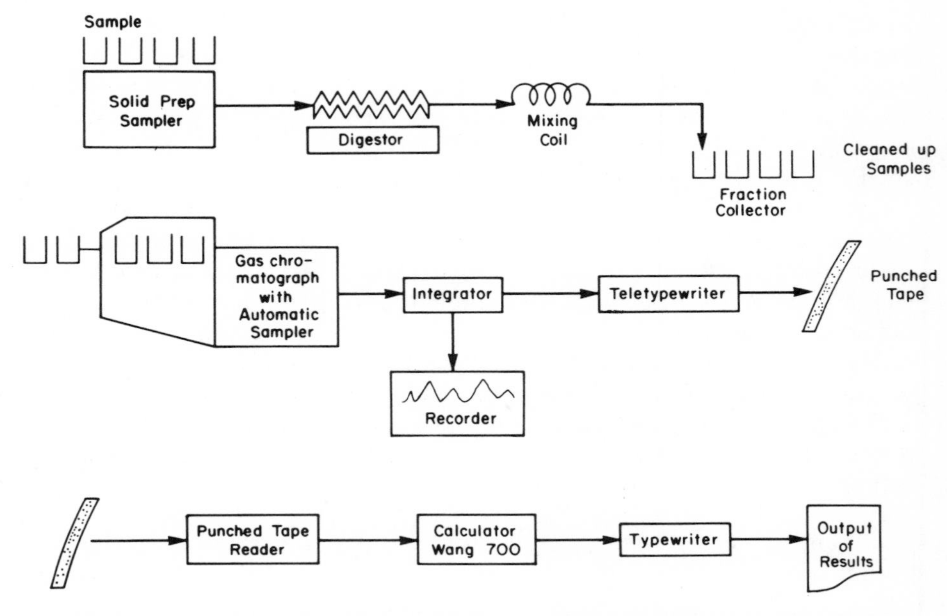

Fig. 9.4 Automated method for the analysis of triazine residues in soils. The top line represents sample extraction and cleanup. The second and third lines show the automated GC analysis and computer processing of the data, including final printout of results. From Hörmann *et al.* (*13*), with permission from the Association of Official Analytical Chemists.

V. INTEGRATION OF LC WITH OTHER ANALYTICAL TECHNIQUES

Routine trace analysis of a variety of compounds may require more than one analytical approach. In such cases it is preferable to incorporate the various determinative steps into one coordinated program. Figure 9.5 shows an example of the integration of LC with other techniques for trace analysis of organics. The approach depends on the type of analytical program desired. For example, in routine programs involving the determination of certain contaminants in foods, the vast majority of samples may prove to be negative and only a few percent may require quantitation. In this case a rapid semiquantitative (or even only qualitative) screening method would be useful to eliminate the negative samples. Only the few percent showing positive would be carried further for quantitation. This saves considerable time and effort by the analyst, who would otherwise attempt to quantitate all the samples knowing full well that most will be negative. However, in procedures where the analyte is likely to be in all

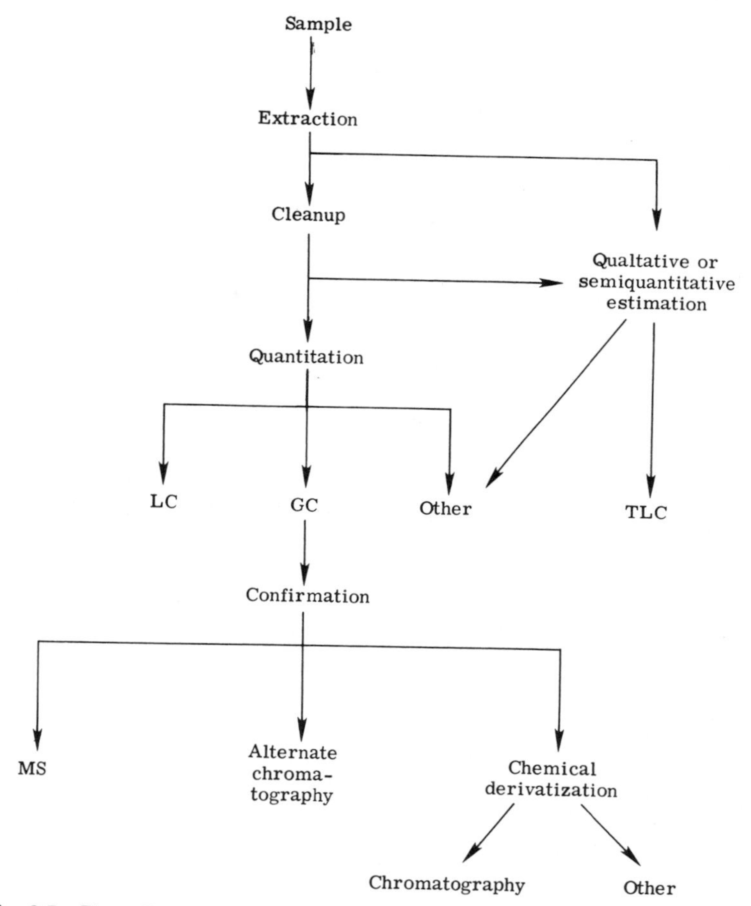

Fig. 9.5 Flow diagram of an integrated program for organic trace analysis.

samples, a semiquantitative screening program is of no use and all samples would have to be carried through the appropriate sample pretreatment for quantitation.

If a semiquantitative screening technique is desired, then thin-layer chromatography (TLC) should be attempted first. The advantage of TLC is that it requires a minimum of equipment and usually much less sample cleanup than for GC or LC. TLC may not be sensitive enough in many cases. However, selective spray reactions and derivatization techniques can enable an analyst to identify positively low nanogram quantities of substances on TLC plates. In fact, for some photosynthesis inhibiting herbicides, a TLC screening method has been developed that is capable of detecting 100 pg or less more selectively than either GC or LC, and with less cleanup (*14*).

Multiresidue trace analysis makes routine determinations much more difficult than analyzing for only perhaps one or two similar compounds. If a whole series of compounds must be routinely determined, problems may arise as to the efficiency of the sample cleanup. This depends on the nature of the compounds of interest. If they differ greatly in physical and chemical properties, then one extraction and cleanup procedure may not suffice. Even if they all pass through one method, it may not be possible to analyze for them in a single run or even with a single analytical system. It is in this area where the analyst can integrate LC with GC and other (for example, classic spectrophotometry) determinative steps. Those compounds that do not pass directly through a GC may be determined by LC. The results then could be confirmed by MS, an alternate chromatographic technique (LC or GC), or by chemical derivatization. The extraction and cleanup technique employed should be adequate for both LC and GC so that the analyst need only inject an aliquot of the final extract into each instrument for quantitation. It is impractical to develop a different cleanup for LC when great experience in the area of GC is available. Also, in general, cleanup techniques employed for GC, especially with EC detection, are directly applicable to many compounds that must be analyzed by LC. This already has been shown for a number of pesticides (*15–17*).

VI. LC AS A CLEANUP TECHNIQUE

LC has not only found use in organic trace analysis as a determinative step, but it is being used more and more for cleanup of samples for analysis by other techniques, especially GC and MS. With the current awareness of and interest in our environment, especially toward chemicals such as dioxins, nitrosamines, and other carcinogens which may occur at parts-per-trillion levels, the need for improved sample cleanup becomes of utmost importance. Those methods adequate for parts-per-million levels of organics become unsuitable at ultralow levels (1000- to 100,000-fold lower). The high resolution power of LC enables the analyst to collect very precise fractions containing the analyte in a relatively pure form. This would not be possible by classical open-column chromatography because of poor efficiency and lack of precise control over flow rate. Some examples of LC being employed as a cleanup have appeared in the literature for dioxins in chicken liver (*18*), and fish (*19*), polyaromatics in alcoholic beverages (*20*), and hexachlorocyclohexane isomers in wool fat (*21*), as well as for the mass spectrometric determination of some drugs in plasma (*22*).

REFERENCES

1. J. F. Lawrence and R. Leduc, *Anal. Chem.* **50,** 1161 (1978).
2. J. H. Knox and J. Jurand, *J. Chromatogr.* **103,** 311 (1975).
3. K. A. Ramsteiner and W. D. Hormann, *J. Agric. Food Chem.* **27,** 934 (1979).
4. J. H. Knox and J. Jurand, *J. Chromatogr.* **125,** 89 (1976).
5. M. E. Getz, G. W. Hanes, and K. R. Hill, *Natl. Bur. Stand. (U.S.), Spec. Publ.* No. 519.
6. J. H. A. Ruzicka and D. C. Abbott, *Talanta* **20,** 1261 (1973).
7. F. A. Gunther and D. E. Ott, *Residue Rev.* **14,** 12 (1966).
8. J. A. Ault, C. M. Schofield, L. D. Johnson, and R. H. Waltz, *J. Agric. Food Chem.* **27,** 829 (1979).
9. J. Solomon and W. L. Lockhart, *J. Assoc. Off. Anal. Chem.* **60,** 690 (1977).
10. J. Solomon, *Anal. Chem.* **51,** 1861 (1979).
11. L. P. J. Hoogeveen, F. W. Wilmott, and R. J. Dolphin, *Fresenius' Z. Anal. Chem.* **282,** 401 (1976).
12. B. A. Karlhuber and D. O. Eberle, *Anal. Chem.* **47,** 1094 (1975).
13. W. D. Hörmann, G. Formica, K. Ramsteiner, and D. O. Erberle, *J. Assoc. Off. Anal. Chem.* **55,** 1031 (1972).
14. J. F. Lawrence, *J. Assoc. Off. Anal. Chem.* **63,** 758 (1980).
15. J. F. Lawrence, *J. Chromatogr. Sci.* **14,** 557 (1976).
16. J. F. Lawrence, L. G. Panopio, and H. A. McLeod, *J. Agric. Food Chem.* **28,** 1323 (1980).
17. J. F. Lawrence and L. G. Panopio, *J. Assoc. Off. Anal. Chem.* **63,** 1300 (1980).
18. J. J. Ryan and J. C. Pilon, *J. Chromatogr.* **197,** 171 (1980).
19. L. L. Lamparski, T. J. Nestrick, and R. H. Stehl, *Anal. Chem.* **51,** 1453 (1979).
20. G. Toussant and E. A. Walker, *J. Chromatogr.* **171,** 448 (1979).
21. S. L. Ali, *J. Chromatogr.* **156,** 63 (1978).
22. J. J. DeRidder and H. J. M. Van Hal, *J. Chromatogr.* **146,** 425 (1978).

Chapter 10
Applications

I. INTRODUCTION

This chapter serves to illustrate, with examples from the recent literature, how different approaches taken toward organic trace analysis including sample preparation techniques, chromatography, and detection systems are integrated to yield a complete analytical method. The chapter is not intended to be a comprehensive source of trace analytical applications of LC. Numerous literature references may be found in the biennial reviews of *Analytical Chemistry* (*1*), *Journal of Chromatography Reviews*, in special issues of the *Journal of Chromatographic Science* (*2*) and in bibliographical sections of other analytical and chromatographic journals. Also, a recent book by Pryde and Gilbert (*3*) contains a comprehensive account of the applications of LC up to the end of 1977.

II. CLINICAL (BIOLOGICAL FLUIDS AND TISSUES)

A. Biogenic Amines

Catecholamines have been determined in tissue, plasma, and urine, employing electrochemical detectors (*4–6*). The methods involve adsorption of the catechols from the sample extract onto acidified alumina at high pH (*7*). This is carried out by adding a known quantity of alumina to a flask containing the sample, then shaking for about 20 min. The supernatant liquid is removed by aspiration; then the alumina is rinsed with water or 0.002 *M* sodium acetate to further remove impurities. Finally, the catecholamines are recovered from the alumina by adding 1 *M* acetic acid or 0.2 *M* $HClO_4$. An aliquot of this extract is then analyzed by LC. The method worked well for serum (*4*) and brain tissue (*6*); however, an additional cation-exchange column cleanup was required for urine samples (*5*). Chromatographic separations were carried out by reversed-phase

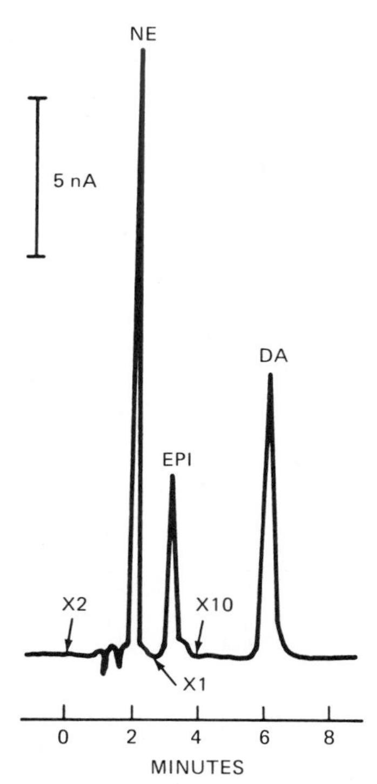

Fig. 10.1 Chromatogram of norepinephrine (NE), epinephrine (EPI), and dopamine (DA) in urine after ion exchange cleanup and adsorption on alumina. Concentrations: NE = 60 μg/liter; EPI = 22 μg/liter; DA = 510 μg/liter. From Riggin and Kissinger (*5*), with permission from the American Chemical Society.

ion-pair chromatography using sodium octylsulfate as the counter-ion (*5,6*) or by ion-exchange chromatography on a Vydac SCE column (*4*). Electrochemical detection was carried out at either +0.5 or +0.72V versus an Ag/AgCl reference electrode.

Figure 10.1 shows a chromatogram of a urine sample analyzed by the above technique. Figure 10.2 depicts results of a basal ganglia extract from mouse brain tissue. Fenn *et al.* (*4*) used a four-electrode configuration which provided additional information on the detection of the catecholamines based on their electrochemical reversibility. The compounds are detected first at the anode, where they are oxidized, then at the cathode, where they are reduced. The utility of this system is shown in Fig. 10.3 where both anodic and cathodic results were compared. The large peak "X" is not present in the cathodic chromatogram, indicating that it is not reversibly oxidized and thus does not contain a catechol

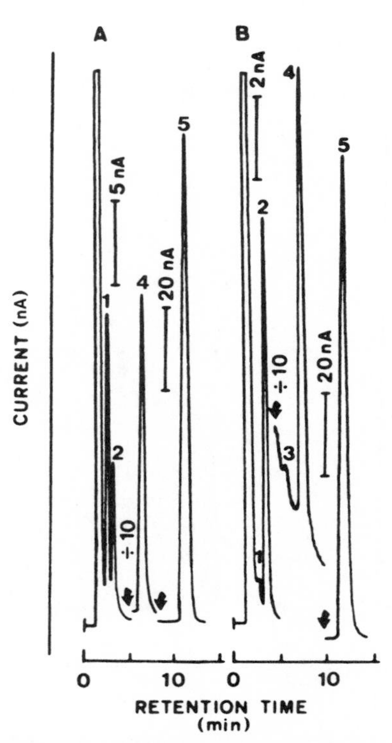

Fig. 10.2 Chromatogram from a basal ganglia extract. A, control; B, phargylin-treated. (1) 3,4-dihydroxyphenylacetic acid; (2) norepinephrine; (3) epinephrine; (4) dihydroxybenzylamine; (5) dopamine. From Mefford *et al.* (*6*), with permission from Academic Press, Inc.

moiety. Recoveries of some catecholamines through the alumina cleanup are presented in Table 10.1.

Electrochemical detection has also been employed for the sensitive and selective detection of serotonin in plasma after an ion-exchange column cleanup on Amberlite CG-50 followed by trace enrichment and reversed-phase LC (*8*). The trace enrichment process is carried out on-line with the LC system as shown in Fig. 8.12. The trace enrichment permits the analysis of large sample volumes, thus increasing sensitivity of the method. Figure 10.4 shows the results of analyzing 1 ml of plasma versus 100 μl. The 1-ml analysis is superior, even though 5-fold less detector sensitivity is required.

Derivatization with *o*-phthalaldehyde (OPA) as described in Chapter 7 (Section III,G) has been employed for the determination of biogenic amines in biological tissues and fluids (*9,10*). The OPA derivative of the amine is formed prior to chromatography, with the product being detected

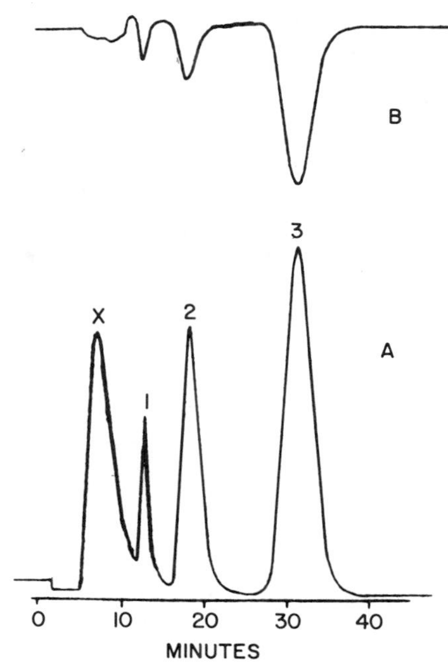

Fig. 10.3 Chromatographic results of mouse brain extract. (A) anodic detection; (B) cathodic detection. Four electrode system. Peaks: (x) unknown, (1) L-dopa; (2) norepinephrine; (3) dopamine. Conditions: Vydac SCX column; mobile phase, 0.04 *M* KH_2PO_4 buffer, pH 2.6; temperature, 24°C; potentials $E_1 = +0.8V$ and $E_2 = -0.2V$ versus SCE; sample, 100 μl of extract. From Fenn *et al.* (*4*), with permission from the American Chemical Society.

TABLE 10.1

Recovery of Catechol Standards from Alumina[a]

Compound	Recovery (n = 12) (%)
3,4-Dihydroxyphenylacetic acid	48.3 ± 1.2
Norepinephrine	60.9 ± 1.6
Epinephrine	57.7 ± 1.5
Dihydroxybenzylamine	58.3 ± 2.3
Dopamine	57.5 ± 1.3

[a] From Mefford *et al.* (*6*), with permission from Academic Press, Inc.

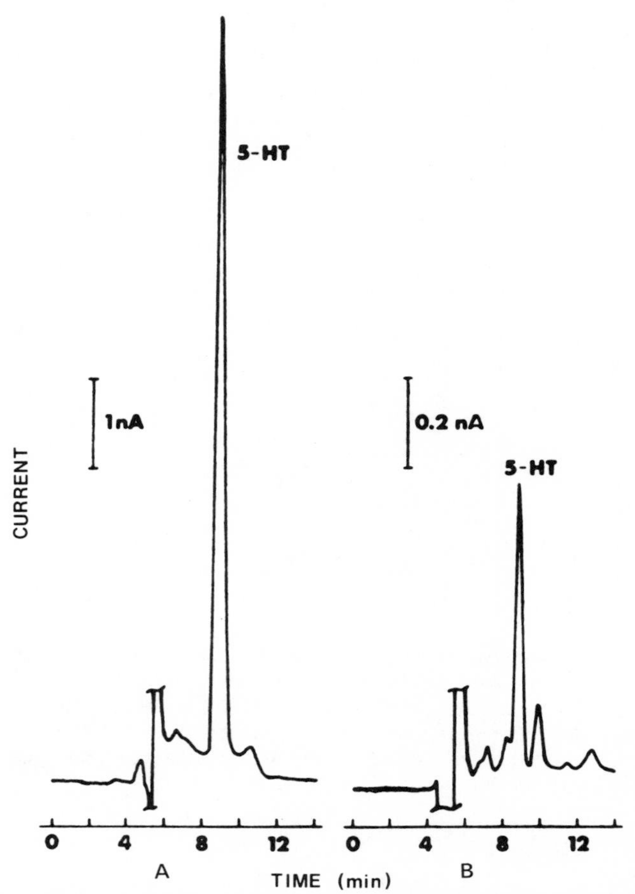

Fig. 10.4 Trace enrichment of human plasma. (A) 1 ml plasma; (B) 100 μl plasma. 5-HT = serotonin; E = +0.50V versus Ag/AgCl. From Koch and Kissinger (*8*), with permission from the American Chemical Society.

fluorometrically (excitation wavelength, 340 nm; emission, 480 nm). The methods involve direct reaction of deproteinized tissue or fluid extracts with OPA as a pH ~10.4. The mixtures are shaken for about 2 min, followed by the addition of solid NaCl, and extracted with ethyl acetate. The ethyl acetate is then removed after centrifugation, washed with phosphate buffer (pH ~10) and evaporated to about 100 μl under nitrogen for LC analysis.

Figure 10.5 outlines the procedure for plasma analysis. Because of the selectivity of the reaction (OPA reacts only with primary amines) and of fluorescence detection, no involved cleanup procedure was required

Take 2 ml deproteinated plasma. Add 100 ng IS (OCT). Adjust with 0.5 *M* KOH to pH 7.0 ± 0.2, centrifuge. Transfer supernatant from perchlorate salts.

↓

Derivatize with 400 μl o-phthalaldehyde (350 μg) at pH 10.4 and shake for 1 min.

↓

Add 2 gm NaCl, mix, extract derivatives twice with 2 ml ETOAc, and centrifuge.

↓

Add 2.0 ml of 0.05 *M* Na_2HPO_4 buffer, pH 10.0, to ETOAc extract, shake for 1 min., and transfer extract. Repeat.

↓

Reduce solvent volume to 100 μl under a sweep of dry nitrogen. Inject 10-50 μl on μbondapak phenyl HPLC column.

Fig. 10.5. Schematic representation of biogenic amine analysis in plasma via OPA derivatization and LC. From Davis *et al.* (9), with permission from the American Association of Clinical Chemists.

other than the normal deproteinization steps. Figure 10.6 illustrates the utility of the approach by showing a chromatogram of bovine plasma containing six added biogenic amines. Table 10.2 shows typical recovery values from pooled bovine plasma. The advantages of precolumn derivatization with OPA are that it stabilizes the labile amine molecules, facilitates extraction into organic solvent, and enables sensitive and selective detection by fluorescence.

B. Anticonvulsants in Serum and Plasma

Three different extraction techniques have been employed in the analysis of some anticonvulsant drugs from plasma or serum samples. Christofides and Fry (*11*) extracted 0.5 ml of plasma with 10 ml of ethyl acetate after the addition of 0.5 ml of 0.4 *M* phosphate buffer (pH = 7). The organic

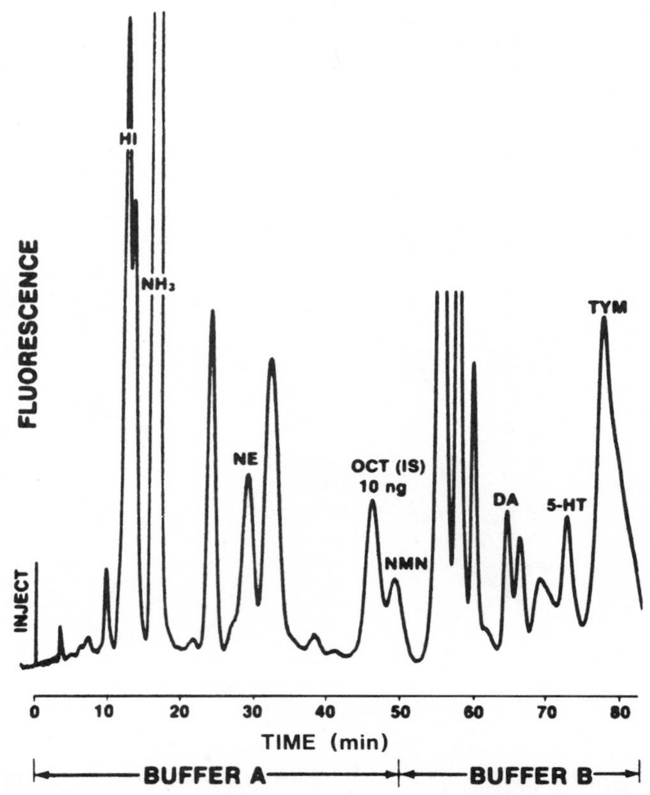

Fig. 10.6 Reversed-phase LC separation of biogenic amines in bovine plasma with the addition of six of the biogenic amines listed in Table 10.2. Buffer A, 0.05 *M* phosphate buffer, pH 5.1 containing 32% methanol. B, same as A but with 45% methanol. 5 ng of each amine. Octopamine (OCT) is the internal standard (IS). From Davis *et al.* (*9*), with permission from the American Association of Clinical Chemists.

extract was evaporated to dryness and then redissolved in 20 μl of methanol, of which 3 μl were used for LC analysis employing reversed-phase ion-pair chromatography with tetrabutylammonium ion as counter-ion. Table 10.3 lists recoveries of several anticonvulsants carried through the procedure. Detection of the peaks was carried out by UV absorbance at 200 nm.

Chu *et al.* (*12*) described an improved acetonitrile extraction which deproteinizes and extracts the drugs from serum simultaneously. The original method called for a serum–acetonitrile volume ratio of 1 : 1 (*13*,*14*). However, when reversed-phase chromatography was employed with a mobile phase consisting of acetonitrile–phosphate buffer combinations,

TABLE 10.2

Recovery of Biogenic Amines Added to Bovine Plasma at 25 μg/liter[a]

Amine	Control (μg/liter)	Control + 25 μ/liter	Found[b]	Average Recovery (%)
Histamine (HI)	30.0	55.0	13.6	54.4[c]
Norepinephrine (NE)	8.0	33.0	16.8	67.2
Normetanephrine (NME)	<0.5	25.0	19.4	77.6
Octopamine (OCT)	<0.5	25.0	15.8	63.2
Dopamine (DA)	1.0	26.0	16.3	65.2
Serotonin (5-HT)	15.0	40.0	16.5	66.0
Tyramine (TYM)	30.0	55.0	14.6	58.4[d]

[a] From Davis *et al.* (9), with permission of the American Association of Clinical Chemists.
[b] Corrected for control value.
[c] Average of four different runs.
[d] Unknown peak eluted with TYM; they were integrated together.

upon LC analysis, microprecipitates often formed, causing blocked frits and injection ports. By simply changing the serum–acetonitrile ratio to 2:3 (i.e., using more acetonitrile in the extraction), recoveries were not altered, but the supernatant was clear of any microprecipitate.

Perchalski and Wilder (*15*) removed benzodiazepine anticonvulsants from plasma by extracting alkaline plasma (pH 10.5) with benzene–dichloromethane (9:1). The organic layer is removed, evaporated to dryness, and then dissolved in mobile phase for analysis by adsorption chromatography employing a Partisil 5, silica gel column with cyclopentane–chloroform–acetonitrile–methanol (29:55.5:15:0.5) as the mobile phase. UV detection was made at 254 nm, enabling the use of

TABLE 10.3

Recovery of Drugs Added to Plasma[a]

Compound	Concentration (μmole/liter)	Absolute Recovery (n = 5) (%)
Ethosuximide	400	83
Pirimidone	40	85
Phenobarbital	80	81
Ethylphenacemide	40	81
Carbamazepine	20	82
Phenytoin	80	83

[a] From Christofides and Fry (*11*), with permission of the American Association of Clinical Chemists.

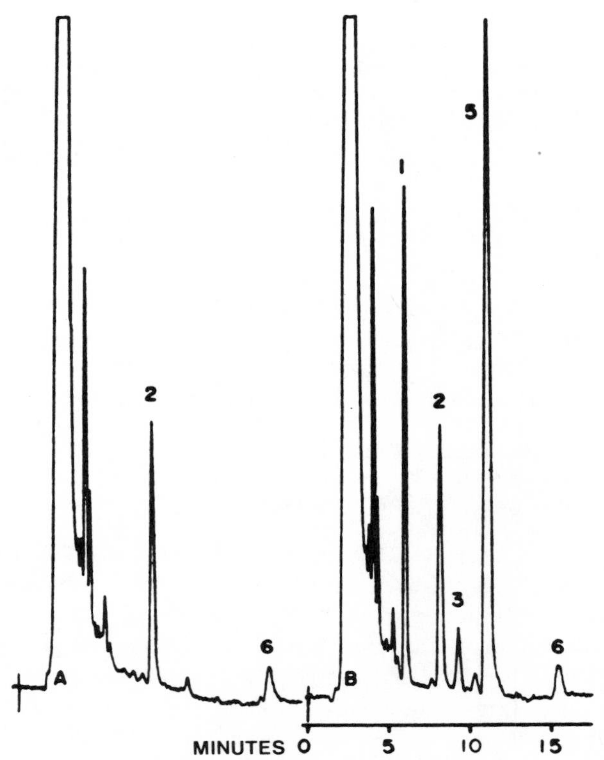

Fig. 10.7 Chromatograms of drug-free plasma (A) and spiked plasma (B). (1) Diazepam; (2) 4,5-dihydrodiazepam (internal standard); (3) clorazepam; (5) nordiazepam; (6) caffeine. Concentrations of 1, 2, and 5 were 172 ng/ml, 53 ng/ml, and 461 ng/ml, respectively. From Perchalski and Wilder (*15*), with permission from the American Chemical Society.

less expensive filter photometer type LC detectors. Figure 10.7 compares a control urine (containing internal standard, 4,5-dihydrodiazepam, and a small amount of caffeine) to the same sample spiked with diazepam, chlorazepam, and nordiazepam.

C. Restricted or Illicit Drugs

Morphine and its major metabolite, morphine-3-glucuronide, have been determined in blood by LC with electrochemical detection using glassy carbon (*16*) as the working electrode (+0.6V versus SCE). Morphine is extracted from whole blood (adjusted to pH 8.9) with ethyl acetate–isopropanol (9:1) after the addition of solid sodium chloride. The organic

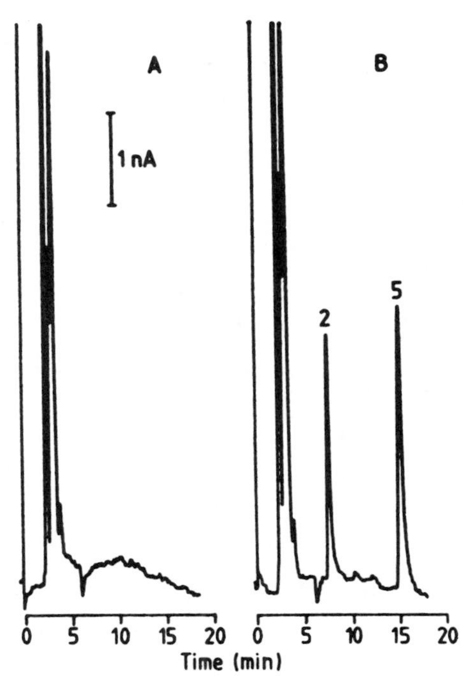

Fig. 10.8 Chromatograms of unhydrolyzed blood. A, Control blood; B, control spiked with 50 ng morphine/ml. Peak 2, morphine peak; Peak 5, dextrorphan (internal standard). Chromatography was carried out on silica (Syloid 74) with a mobile phase of methanol–ammonium nitrate buffer (pH 10.2) (9:1) at 1.0 ml/min. From White (*16*), with permission from Elsevier Scientific Publ. Co.

extract is evaporated to dryness and then redissolved in 50 μl 0.1 *M* HCl, saturated with NaCl. This is washed with diethyl ether, which is discarded. The remaining aqueous solution is basified to pH 8.9 and extracted twice with ethyl acetate–isopropanol (9:1). The organic extracts were combined and evaporated to dryness. The residue was dissolved in 0.02 *M* HCl for LC analysis. Approximately 81% of morphine is recovered through the method. Morphine-3-glucuronide is determined as morphine by carrying out a preliminary enzymatic hydrolysis of the blood (overnight incubation with β-glucuronidase) before organic extraction. In this manner total morphine and metabolite are determined. Figure 10.8 illustrates the selectivity of the method for morphine added to unhydrolyzed blood at 50 ng/ml.

Moffat and co-workers (*17,18*) carried out some very interesting comparison studies employing LC and radioimmunoassay (RIA) for LSD and some cannabinoids in plasma. They compared results of RIA analyses to

TABLE 10.4

The Plasma Concentration of Δ⁹-THC in Volunteers Who Had Smoked Δ⁹-THC Determined by RIA, LC, and GC–MS[a]

Subject	Time after smoking (min)	Plasma concentration of Δ⁹-THC (ng/ml)		
		Direct RIA[b]	RIA after separation by LC	GC–MS
1	Control	23[c]	0	0
	2	67	47	55
	12	48	15	18
	24	47	7	9
	34	48	5	8
	64	47	3	5
	126	48	1	2
2	Control[d]	ND[c]	ND	ND
	3	63	45	58
	13	44	20	30
	24	53	9	16
	34	47	6	9
	64	58	4	5
	124	50	2	1
3	Control	0	0	0
	2	37	26	26
	22	7	4	3
	31	5	1.1	ND
	60	9	0.9	ND
	120	6	0.8	ND

[a] From Williams *et al.* (*18*), with permission from Elsevier Scientific Publ. Co.
[b] These values include a contribution from THC metabolites.
[c] Subject 1 and 2 were both cannabis users and cannabinoid material may have been present in the plasma before the experiment.
[d] A control sample was taken for subject 2 but this was lost during centrifugation.

the same results with prior LC fractionation. The LC step enabled the detection of drug metabolites that interfered in the direct RIA method. Table 10.4 illustrates the differences observed for Δ9-THC when a chromatographic step is not incorporated. As a further comparison, gas chromatograph–mass spectrometry (GC–MS) was carried out. It can be seen that when the cannabinoid is isolated from plasma by LC the values are much lower and decrease with time, indicating metabolism of the drug. This is confirmed by the MS results. In this instance the RIA measures total Δ9-THC plus contributions from metabolites, and thus does not show the same degradation pattern with time.

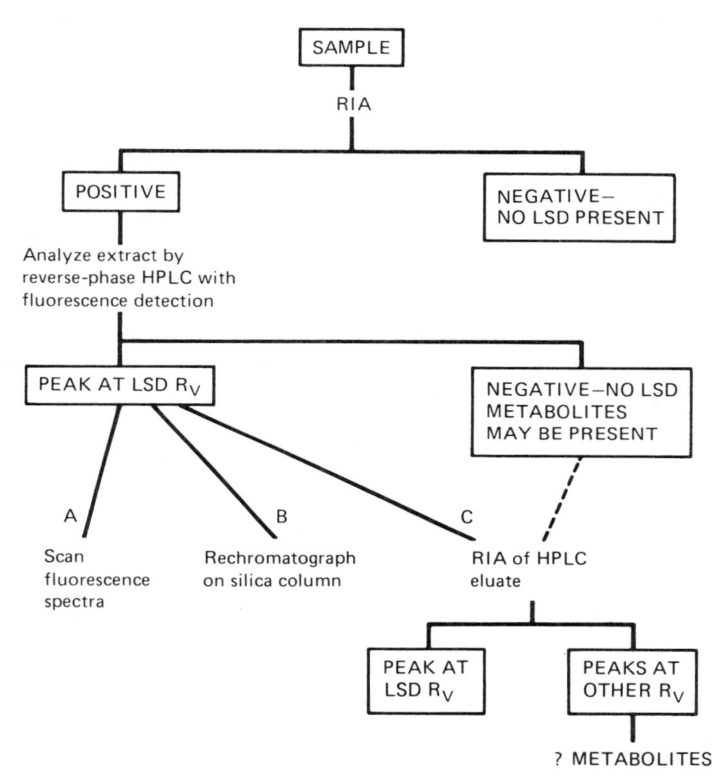

Fig. 10.9 Schematic for the RIA–LC analysis of LSD in biological fluids. A, B, and C are confirmatory procedures. From Twitchett *et al.* (*17*), with permission from Elsevier Scientific Publ. Co.

For LSD analyses the scheme shown in Fig. 10.9 was devised such that any RIA positive results would be confirmed by LC analysis (reversed-phase chromatography), either with fluorescence detection at 320 nm excitation (ex) and 400 nm emission (em), by a second RIA analysis of the appropriate LC fraction, or by normal-phase chromatography on silica.

D. Porphyrins

Free-acid porphyrins (*19*) have been determined in plasma samples by LC with either UV absorbance (403 nm) or fluorescence (~405 nm ex; ~600 nm em). The sample extraction and cleanup procedure is outlined in Fig. 10.10. It involves first of all the centrifugation of plasma with ethyl acetate–acetic acid (HOAC), upon which two layers of precipitate result. The two precipitates and two liquid phases are combined (respectively)

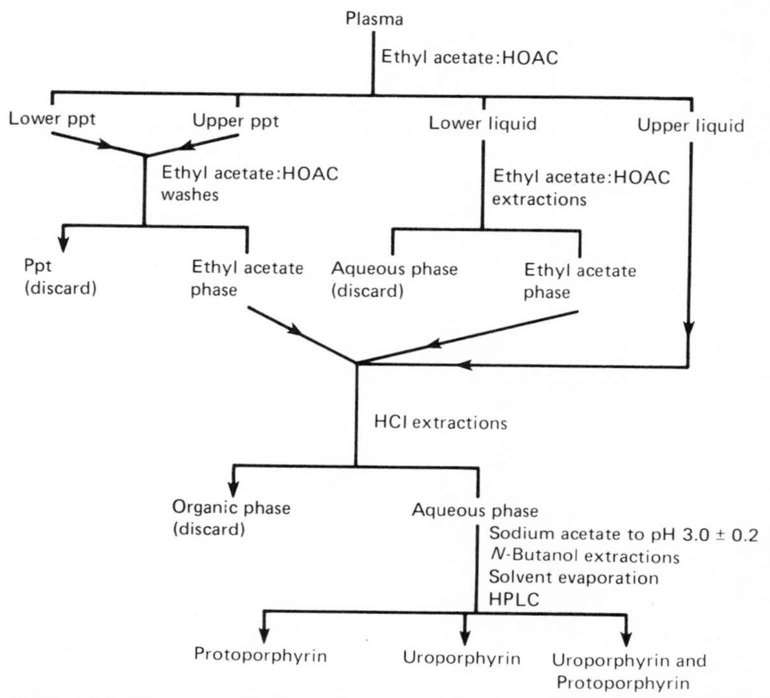

Fig. 10.10 Schematic of the extraction and cleanup of plasma samples for free-acid porphyrin determinations by LC. From Longas and Poh-Fitzpatrick (*19*), with permission from Academic Press, Inc.

and treated with ethyl acetate, and are then carried through the rest of the extraction procedure for LC analysis employing a silica column and gradient elution with a mobile phase of acetone and aqueous acetic acid.

Porphyrins have also been determined as their methyl esters (*20*) via chelation with copper (II). The chelates were analyzed by LC on a CN-bonded column with a mobile phase of ethyl acetate–*n*-heptane–isopropanol (40:60:0.5), with detection by UV absorbance at 400 nm.

E. Thiazide Diuretics

An LC method for thiazide diuretics in urine has been developed which is superior to the commonly used spectrophotometric technique in terms of sensitivity and selectivity (*21*). The approach involves extraction of the drugs from acidified urine (pH 5) with ethyl acetate. The organic layer is then washed with pH 8 phosphate buffer to remove interferences. The

TABLE 10.5

Recoveries of Thiazides from Urine[a]

Compound	Recovery (%)
Chlorothiazide[b]	55
Hydrochlorothiazide	53
Bendroflumethiazide	74
Methylclothiazide	93
Cyclothiazide	70
Polythiazide	85
Trichlormethiazide	63
Hydrofluthiazide	67
Benzthiazide	80

[a] From Tisdall *et al.* (*21*), with permission from the American Association of Clinical Chemists.

[b] Quantitated as hydrochlorothiazide after $NaBH_4$ treatment.

ethyl acetate is evaporated to dryness and the residue is dissolved in mobile phase for LC analysis employing reversed-phase chromatography (C_{18}) with acetonitrile–dilute acetic acid mixtures as mobile phases (two were required to analyze all the thiazide drugs). UV absorbance detection was performed at 271 nm. In order to analyze for chlorothiazide, the urine was first treated with $NaBH_4$ to convert it to hydrochlorothiazide. Table 10.5 lists recoveries for a number of the diuretics from urine.

F. Urinary Estriol

An LC method employing reversed-phase chromatography with fluorescence detection (220 nm ex; 608 nm em) has been developed for estriol in urine (*22*). A 2.0-ml sample of urine is treated with glucuronidase–arylsulfatase to hydrolyze estriol conjugates enzymatically. When hydrolysis is complete the urine sample is made basic and washed with diethyl ether, which is discarded. This treatment removes neutral and basic coextractives. The urine is then acidified with HCl and the estriol is extracted with diethyl ether. The ether layer is washed with carbonate buffer (pH 9.5) and water before being evaporated to dryness. The residue is dissolved in methanol for LC analysis with a mobile phase of acetonitrile–water containing a trace of KH_2PO_4. Figure 10.11 shows the selectivity of fluorescence compared to UV absorbance at 280 nm for the determination of estriol in urine. The estriol peak represents 34.0 mg/liter in the sample.

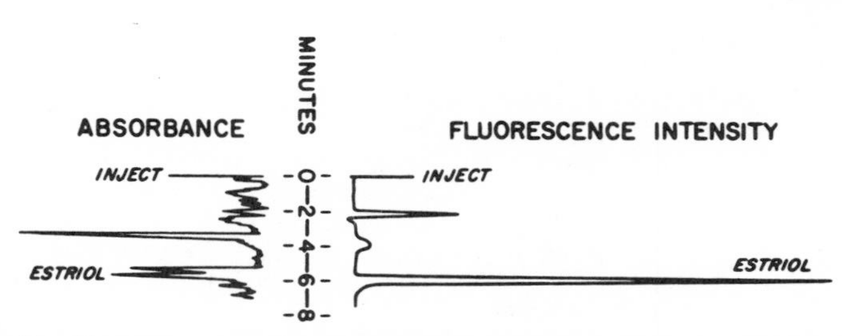

Fig. 10.11 Chromatogram of urine sample with fluorescence and UV absorption detection. From Taylor *et al.* (*22*), with permission from the American Association of Clinical Chemists.

G. Caffeine in Plasma and Tissues

Bonati *et al.* (*23*) compared GC with nitrogen–phosphorus selective detection to LC with UV detection at 273 nm for the determination of therapeutic levels of caffeine in various biological materials. Plasma or brain tissue homogenate was adjusted to pH 7 with borate buffer and was extracted with chloroform, which was removed and evaporated to dryness. The residue was dissolved in mobile phase for LC, or methanol for GC. Average recoveries of caffeine were 92% from plasma and 86% from brain tissue.

Both GC and LC were satisfactory for determining caffeine down to 0.25 μg/ml. However, the LC method was somewhat superior because of a larger linear range, slightly better sensitivity, and the ability to determine caffeine metabolites.

H. Theophylline in Serum and Plasma

A simple technique for the determination of theophylline in serum or plasma has been developed employing LC (*24*). The sample is treated with an equal volume of acetonitrile, which deproteinizes the serum or plasma. The mixture is shaken, then centrifuged for 5–10 min to remove solids. An aliquot of the supernatant liquid is analyzed directly by LC with a system consisting of a Partisil-10 ODS column and a mobile phase of acetonitrile–phosphate buffer (10 m*M*), 90:10 by volume. UV detection was carried out at 280 nm. No interferences were observed down to 5 μg/ml or less, even with sodium or lithium heparin or ethylenediaminetetraacetate present as anticoagulant. However, citrate proved to be unsatisfactory because of the presence of an interfering peak in the chromatograms. Table 10.6 shows recoveries obtained at different spiking levels.

TABLE 10.6

Recoveries (Day-to-Day) of Theophylline from Plasma and Serum[a]

n	μg/ml Added	Found	SD
8	5	4.91	0.21
11	10	10.16	0.23
9	15	15.12	0.22
6	16	16.01	0.09
4	20	19.52	0.20
10	30	30.36	0.52
4	40	39.55	0.43

[a] From Nelson *et al.* (*24*), with permission from the American Association of Clinical Chemists.

I. Gentamicin Antibiotic in Serum

Barends *et al.* (*25*) used precolumn derivatization of gentamicin with 2,4-dinitrofluorobenzene (DNFB) as the basis of an LC technique for serum analysis. The serum proteins are precipitated with acetonitrile; the antibiotic in the supernatant is reacted with DNFB (see Chapter 7, Section II,C) to yield the dinitrophenyl derivative, which is then analyzed by reversed-phase chromatography and detected at 365 nm. Figure 10.12 shows some chromatogram tracings of a control and spiked serum sample. Since gentamicin consists of three components, three DNFB derivatives are obtained (C_{1a}, C_1, C_2). The method is capable of detecting as low as 0.3 mg/liter of gentamicin in serum. Recoveries through the complete method were about 84% at the 4-mg/liter level.

J. Serum α-Tocopherol (Vitamin E)

A simple LC procedure for determining α-tocopherol in serum has been developed which can detect as low as 0.6 mg/liter (*26*). The α-tocopherol is removed from 200 μl of serum by extracting the deproteinized sample with two 0.5-ml volumes of hexane. The combined organic extracts are evaporated to dryness and the residue is dissolved in methanol for LC analysis. Chromatography is carried out with an RP-18 column and a mobile phase of 100% methanol. Detection was made by absorption at 292 nm. The α-tocopherol peak was collected and confirmed by GC–MS. Recoveries were reported to be greater than 89% on spiked samples of 5–20 mg/liter.

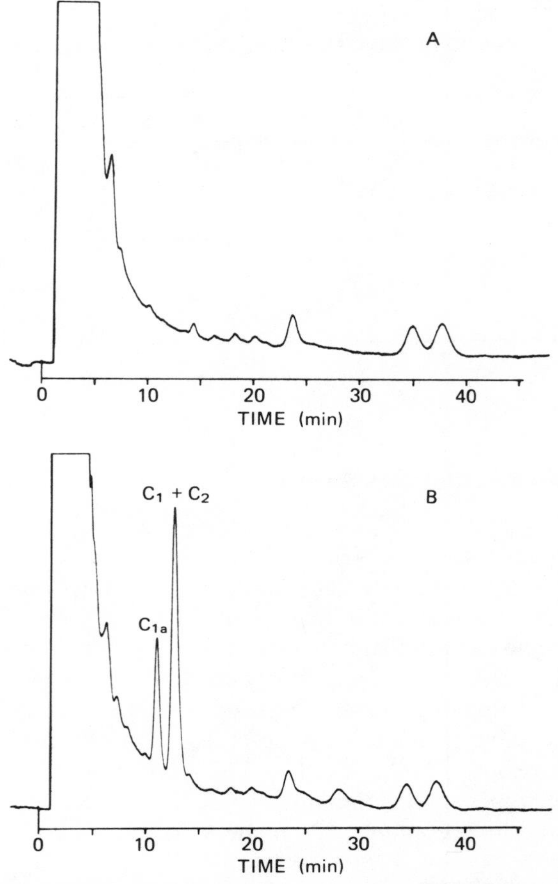

Fig. 10.12 LC analysis of serum samples. A, control; B, spike, 4 mg/liter gentamicin. 50 μl of equivalent serum injected. C_{1a}, C_1, and C_2 represent dinitrophenyl derivatives of the gentamicin components. From Barends *et al.* (*25*), with permission from Elsevier Scientific Publ. Co.

III. ENVIRONMENTAL

A. Atmospheres and Vapors

1. Polycyclic Aromatic Hydrocarbons

Many polycyclic aromatic hydrocarbons (PAH) are important because of their carcinogenicity, and many attempts have been made to analyze for them in atmospheric particulate matter. LC, particularly with fluores-

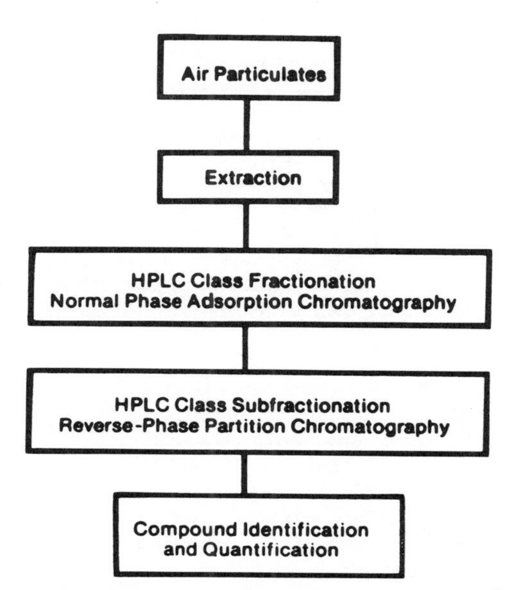

Fig. 10.13 Procedure for fractionating PAH extracted from atmospheric particulate matter. From Eisenberg (*27*), with permission from Preston Publications, Inc.

cence detection (*27–30*), has proved to be the technique of choice for many applications. Figure 10.13 shows a schematic employed by Eisenberg (*27*) for the analysis of PAH in particulate matter. The air particulate matter is collected by passing a large volume of air through glass fiber filters. The filters are Soxhlet-extracted (see Chapter 8, Section II,A) for 8 hr with cyclohexane, which is subsequently concentrated to a small volume using a Kuderna-Danish evaporator (see Chapter 8, Section I,C and Fig. 8.2). After this, the sample is fractionated by class, employing LC on silica gel with hexane–chloroform as mobile phase. Each fraction is then quantitatively analyzed by reversed-phase chromatography with fluorescence detection (excitation and emission depending upon the particular PAH).

Dong *et al*. (*29*) employed essentially the same sample preparation technique, except that they used TLC for prefractionation of the PAH. Figure 10.14 shows a schematic of their procedure. The PAH fraction was removed from the TLC plate and analyzed by reversed-phase LC with cross-checks made by GC–MS.

Das and Thomas (*28*) found that they could detect low to subpicogram quantities of nine major PAHs using fluorescence with a deuterium source. As a result they could dilute the Soxhlet extract sufficiently to eliminate interferences. Table 10.7 lists the fluorescence wavelengths

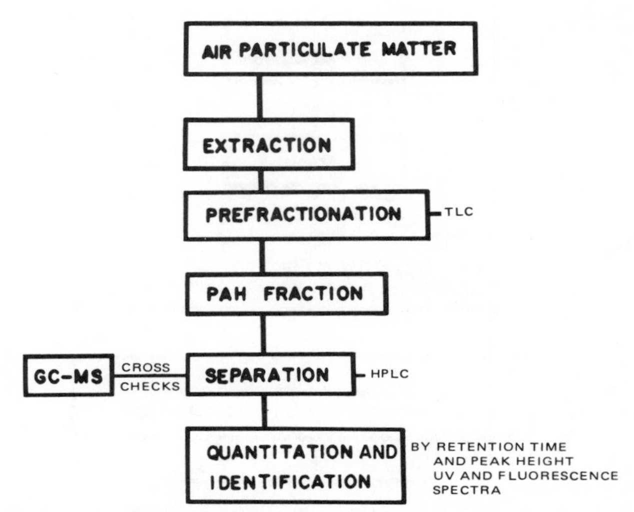

Fig. 10.14 Procedure including TLC fractionation and GC–MS cross-checks of PAH in air particulate matter. From Dong *et al.* (*29*), with permission from the American Chemical Society.

used and detection limits (2 : 1, signal/noise) for the nine PAHs examined. The technique was applied to the analysis of coke oven emissions as well as other types of substrates.

2. Isocyanates

Isocyanates may be analyzed in air samples by LC after derivatization with *N*-4-nitrobenzyl-*N*-*n*-propylamine hydrochloride (NR) or 1-

TABLE 10.7

Detection Limits of Standard PAH[a]

Compound	Fluorescence ex (nm)	Fluorescence em (nm)	Detection limit (pg)
Fluoranthene	280	>389	0.5
Benzo(*a*)anthracene	280	>389	0.6
Benzo(*k*)fluoranthene	280	>389	0.4
Benzo(*e*)pyrene	280	>389	5.1
Benzo(*a*)pyrene	280	>389	1.1
Dibenz(*ah*)anthracene	280	>389	2.3
Benz(*ghi*)perylene	280	>389	3.0
Chrysene	250	>370	2.3
Perylene	250	>370	0.6

[a] From Das and Thomas (*28*), with permission from the American Chemical Society.

naphthalenemethylamine (NMA), which yield respective ureas (*31–33*). However, the NMA reaction was found to be more sensitive by a factor of 40–50 compared to the NR derivative. This is mainly because of the fact that the NMA products are fluorescent whereas the NR ones are not (*31*). The method involves drawing air through a 25-ml Greenburg-Smith midget impinger containing 10 ml of 10^{-4} *M* NMA in toluene. When air sampling is complete, the contents of the impinger are flash-evaporated to dryness at about 60°C by means of a rotary evaporator. The residue is dissolved in dimethyl sulfoxide for LC analysis on a CN-bonded column (*31*), with a mobile phase consisting of 45% acetonitrile in water. Detection limits for hexamethylene diisocyanate-buiret trimmer, for example, was about 0.5 ng, which is equivalent to about 3 ppt in a 20-liter air sample.

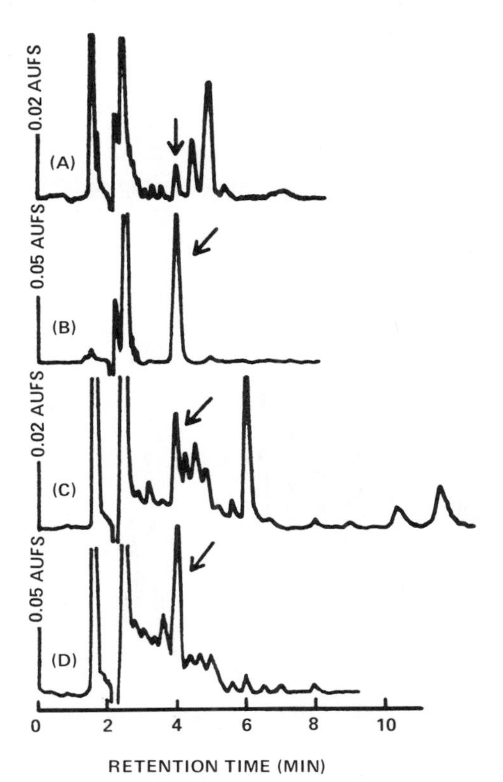

Fig. 10.15 Chromatograms of polluted air after formation of NBA derivative. A, urban air, 0.96 ppb; B, industrial emission (casting), 0.918 ppm; C, automobile exhaust, 0.295 ppm; D, tobacco smoke, 323 μg/cigarette. Arrow indicates NBA-phenol peak. From Kuwata *et al.* (*34*), with permission from the American Chemical Society.

3. Phenol

Phenol can be determined in polluted atmospheres such as urban air, industrial emissions, automobile exhaust, and tobacco smoke by LC via prechromatographic derivatization with *p*-nitrobenzenediazonium tetrafluoroborate to form the *p*-nitrobenzenazo (NBA) product (*34*). The procedure involves collection of phenol from air samples by bubbling through 10 ml of a 0.06% sodium hydroxide solution. The derivatizing reagent is then added to the solution and permitted to react for about 15 min at room temperature. An aliquot of this solution is then directly analyzed by reversed-phase LC (C_{18}) with a mobile phase of methanol–water (85 : 15). UV absorbance is monitored at 365 nm. Figure 10.15 shows resulting chromatograms from several sample types including urban air, industrial emissions, automobile exhaust, and tobacco smoke. The method is applicable to other phenolics such as those shown in Fig. 10.16.

4. Formaldehyde

Formaldehyde in air can be determined by LC after formation of the corresponding 2,4-dinitrophenylhydrazone (*35*,*36*) (see Chapter 7, Section II,F) either by bubbling the sample through impinger solution containing 2,4-dinitrophenylhydrazine (DNPH) or by passing it through a silica column

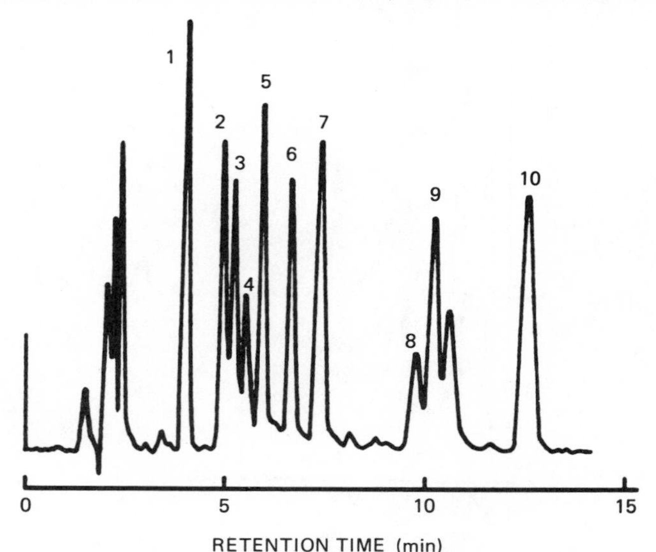

Fig. 10.16 LC chromatogram of the NBA derivatives of phenolic compounds. (1) phenol; (2) *m*-cresol; (3) *o*-cresol; (4) α-naphthol; (5) 3,5-xylenol; (6) 2,3-, 2,5-, 2,6-xylenol; (7) *p*-cresol; (8) β-naphthol; (9) 3,4-xylenol; (10) 2,4-xylenol. From Kuwata *et al.* (*34*), with permission from the American Chemical Society.

in which DNPH is coated onto the surface of the particles. In the latter case the formaldehyde reacts on the column to form the derivative, which is then easily removed from the silica gel by treatment with acetonitrile. The resulting solution can then be analyzed by LC (C_8) with absorption detection at 340 nm.

B. Water Samples

1. Polycyclic Aromatic Hydrocarbons

LC with fluorescence detection has been applied to the analysis of PAHs in aqueous samples (*37*,*38*). The sample treatments prior to determination are somewhat different than for air samples as described in Section III,A,1. The water samples can be directly extracted with an immiscible organic solvent such as dichloromethane (*37*) or cyclohexane (*38*). The organic phase may then be evaporated to a small volume in a Kuderna-Danish or rotary evaporator and directly analyzed by LC. If the samples require further cleanup, an alumina column can be employed (*38*). The PAH fraction is eluted with a mixture of cyclohexane–benzene (1:1 v/v), which is then concentrated and analyzed by LC. For analysis, Ogan *et al*. (*37*) used reversed-phase chromatography with gradient elution as shown in Fig. 10.17. Also, in order to detect as many PAHs as possible, they included a detector wavelength program for monitoring the fluorescence. This program is a compromise since all PAHs do not fluoresce equally under the three different detector conditions selected. Figure 10.18 shows a typical chromatogram of a river water sample with

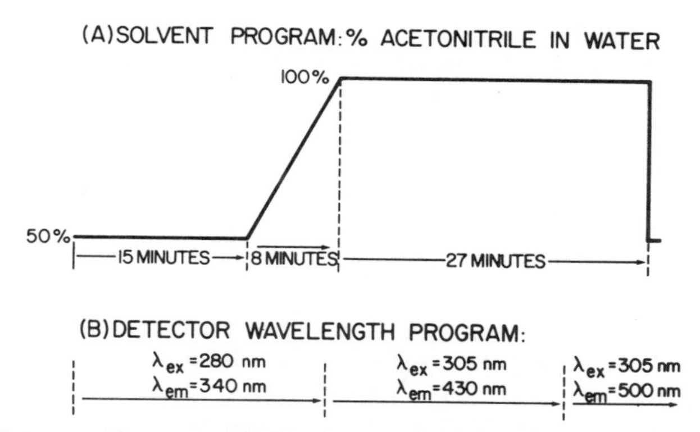

Fig. 10.17 Programs for analysis of PAHs. A, solvent gradient program; B, detector wavelength program. From Ogan *et al.* (*37*), with permission from the American Chemical Society.

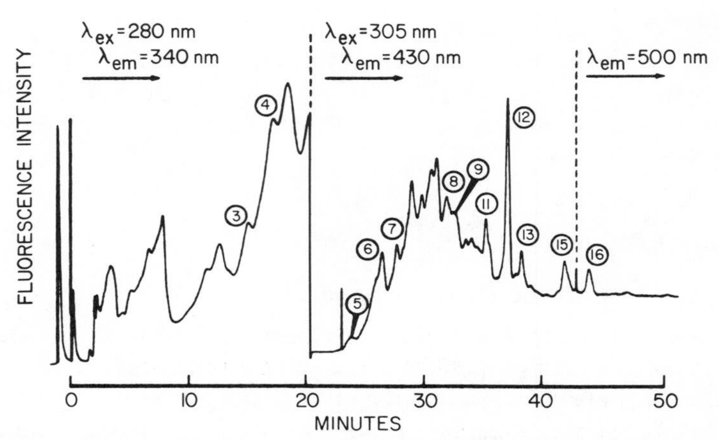

Fig. 10.18 Analysis of a river water extract. Numbers correspond to the PAHs in Table 10.8. From Ogan *et al.* (*37*), with permission from the American Chemical Society.

the detector program indicated. Table 10.8 lists the PAHs included in the study.

2. Phthalate Esters

Porous polymer beads of the cross-linked polystyrene type (see Chapter 6, Section IV and Chapter 8, Section IV,C) have been used for both adsorption and reversed-phase chromatography of phthalate esters in river water (*39*). UV absorption at 224 nm was employed for detection. Concentrations as low as 2 ppb in the samples could be detected (absolute detection limit was about 2 ng of di-2-ethylhexylphthalate). Sample preparation included only a filtration step to remove suspended particulate matter from the river water, followed by an organic extraction with *n*-hexane, which was then directly analyzed by LC.

3. Fatty Acids in River Water

Figure 10.19 shows a detailed schematic for isolating free and bound fatty acids from river water samples (*40*). Hydrolysis is employed to release conjugated or bound acids. The free acids are extracted from the combined acidified aqueous solutions into organic solvent. This extract is purified by employing a 9% deactivated Florisil column (see Chapter 8, Section IV,A,5). The fatty acid fractions are then derivatized with phenacyl–bromide (see Chapter 7, Section II,C). The resulting esters are analyzed by reversed-phase chromatography (C_{18}) with gradient elution

TABLE 10.8

Detector Wavelengths Used for PAH Analysis[a]

Excitation, 280 nm; emission, 340 nm
1. Naphthalene
2. Acenaphthene
3. Fluorene
4. Phenanthrene
Excitation, 305 nm; emission, 430 nm
5. Anthracene
6. Fluoranthene
7. Pyrene
8. Benzo(*a*)anthracene
9. Chrysene
10. Benzo(*e*)pyrene
11. Benzo(*b*)fluoranthene
12. Benzo(*k*)fluoranthene
13. Benzo(*a*)pyrene
14. Dibenz(*a,h*)anthracene
15. Benzo(*ghi*)perylene
Excitation, 305 nm; emission, 500 nm
16. Ideno(1,2,3-cd)pyrene

[a] From Ogan *et al.* (*37*), with permission from the American Chemical Society.

from 40% acetonitrile–water to 100% acetonitrile. Average recoveries of fatty acids in the range of 20–400 ppb are 68%.

4. Pesticides

The analysis of phenylurea herbicides in river water as well as in other samples such as soils and grains has been carried out by LC (*41*). The herbicides are extracted from the water with dichloromethane, which is then evaporated to dryness and the residue dissolved in methanol for LC employing a Spherisorb ODS column with a mobile phase of 60% methanol–water, containing 0.6% (by volume) concentrated ammonium hydroxide. The compounds were detected by UV absorbance at 240 nm. Figure 10.20 shows typical results of a river water sample spiked with various ureas at 0.1 mg/kg.

Rather than use organic solvent extraction of water samples, Paschal *et al.* (*42*) employed a column of XAD-2 macroreticular resin to remove the herbicide atrazine from agricultural runoff waters. The sample ($\simeq$100 ml) is passed through the column, after which the remaining water is removed by aspiration. Following this, the atrazine is eluted from the resin with diethyl ether. This was evaporated to dryness and then dissolved in ace-

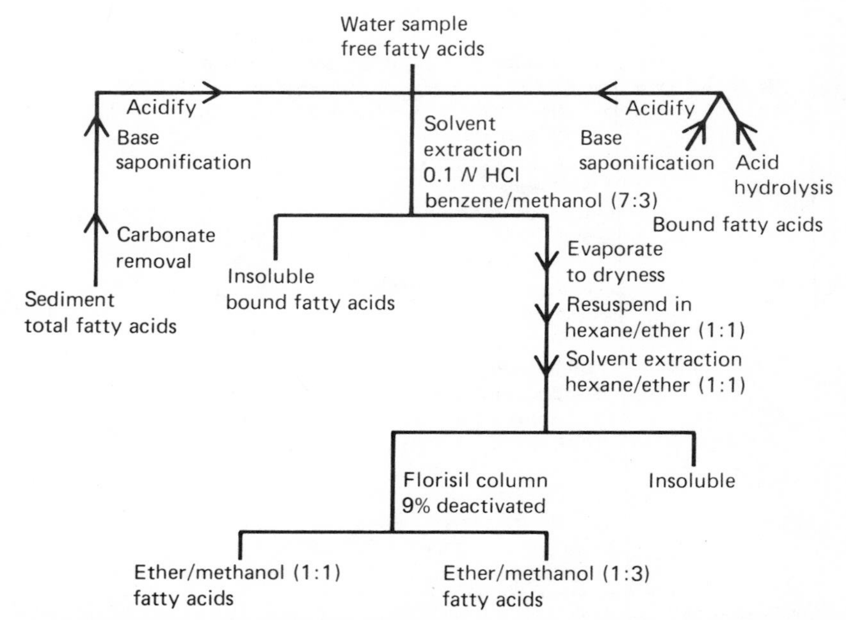

Fig. 10.19 Schematic of the isolation and concentration of fatty acids from river water samples. From Hullett and Eisenreich (*40*), with permission from the American Chemical Society.

tonitrile for reversed-phase LC with detection at 230 nm. The detection limit was about 2 ppm under the conditions used.

Trace enrichment for the organophosphate *abate* was used for its determination in water by LC (*43*). The aqueous sample was first passed through a Bondapak–phenyl column where *abate* was retained. A linear gradient from 100% water to 100% acetonitrile was used to elute the pesticide from the column. As low as 5 ppb of *abate* could be detected in pond water samples when monitored at 280 nm (absorbance).

C. Solids and Oils

1. Polycyclic Aromatic Hydrocarbons in Coal Tar Pitch

A special TLC cleanup technique has been incorporated into a method for the LC analysis of polycyclic aromatic hydrocarbons (PAHs) in coal tar pitch and other bituminous products (*44*). The TLC adsorbent is a mixture of silica gel and the organo-clay Bentone 34 (dimethyldioctadecyl ammonium bentonite). The clay gives a selective separation of PAHs different to that achieved with LC employing silica gel adsorption chromatog-

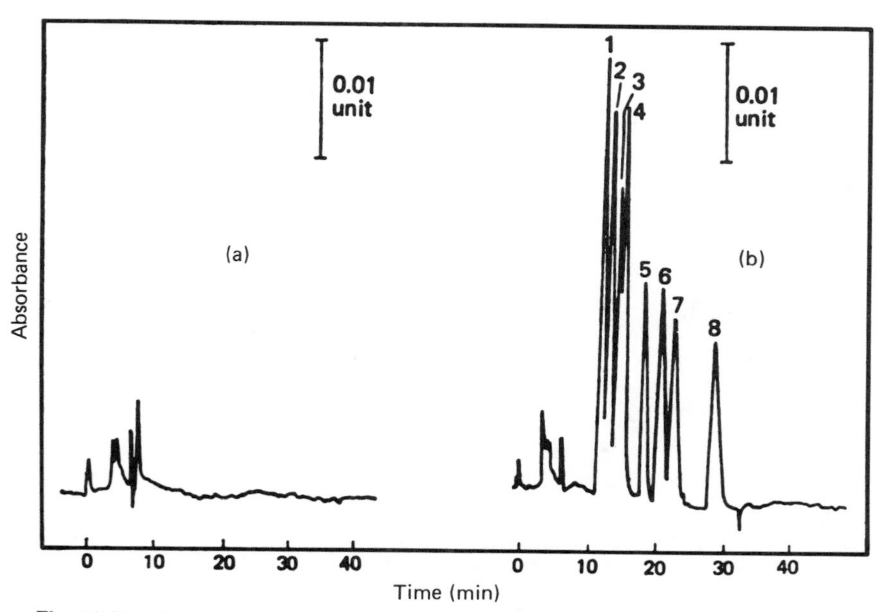

Fig. 10.20 Chromatograms of (a) blank river water and (b) river water spiked at 0.1 mg/kg with (1) monuron; (2) monolinuron; (3) metobromuron; (4) chlortoluron; (5) diuron; (6) linuron; (7) chlorbromuron; and (8) chloroxuron. From Farrington *et al.* (*41*), with permission from the Royal Society of Chemistry.

raphy. The tar sample is extracted with tetrahydrofuran, then spotted on the TLC plate and developed with toluene. The appropriate area of the layer is then scraped off the plate and the PAH fraction washed from the adsorbent particles for LC analysis using adsorption chromatography on silica gel with hexane as the mobile phase and UV detection at designated wavelengths, depending upon the individual PAH determined. Boden (*45*) compared both adsorption and reversed-phase LC for the analysis of coal tar pitch volatiles and found that both were successful in detecting and quantitating benzo(a)pyrene.

2. Pesticides in Soil

Oryzalin and prosulfalin, two sulfanilamide-type herbicides, have been determined in soil by a technique using reversed-phase chromatography with UV detection at 254 nm (*46*). The soil is extracted with methanol by mechanical shaking. An aliquot of this is partitioned between water and hexane, the latter being discarded. The herbicides are extracted from the aqueous phase with methylene chloride, which is then concentrated and cleaned up on a Florisil column (see Chapter 8, Section IV,A,5.). The

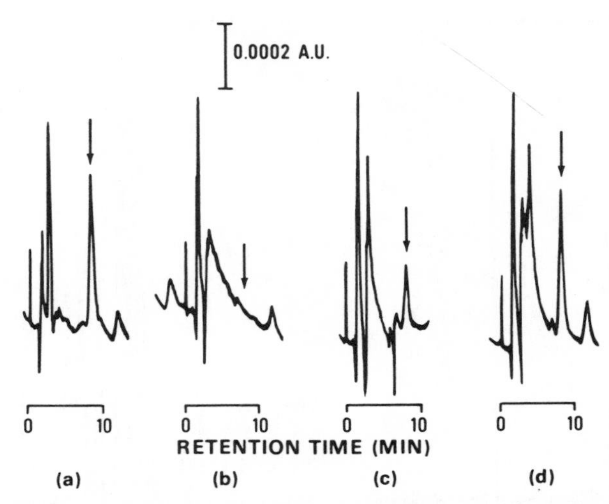

Fig. 10.21 Chromatograms of soil spiked with prosulfalin: (a) standard, 17.5 ng; (b) control soil; (c) spiked soil, 0.02 ppm; (d) spiked soil, 0.05 ppm. From Macy and Loh (*46*), with permission from the American Chemical Society.

fraction containing the sulfanilamides is evaporated to dryness and the residue is dissolved in mobile phase. Figure 10.21 shows a typical soil analysis of prosulfalin spiked in soil at 0.02 and 0.05 ppm.

The carbamate insecticide carbofuran was analyzed in soil and water after its conversion to 7-hydroxybenzofuran by base hydrolysis (*47*). The soil samples were extracted with a methanol–water mixture. The carbamate was removed from the solution by partitioning with methylene chloride. The latter, which contains the pesticide, was evaporated to near dryness and the residue treated with 0.5 *N* sodium hydroxide to convert the carbamate to its corresponding phenol. The phenol was extracted from the acidified reaction mixture and dissolved in acetonitrile for reversed-phase chromatography with UV detection at 280 nm. The average recovery for loam and silt-loam soils was 82.4%.

IV. FOODS

A. Agricultural Chemicals

1. 3,5-Dinitro-*o*-toluimide (Coccidiostat) in Feed

The coccidiostat 3,5-dinitro-*o*-toluimide has been determined in poultry feedstuffs by an LC method employing silica gel adsorption chromatography and UV detection at 270 nm (*48*). The feed sample is extracted by mechanical shaking with acetonitrile–water (17 : 3). The extract is then filtered and a 20-μl aliquot is injected directly into the LC system and determined with a mobile phase of acetonitrile–chloroform (1 : 1).

2. Pesticides

Carbamate insecticides may be analyzed directly in crops by LC either by adsorption (*49–51*) or reversed-phase chromatography (*52*). The adsorption chromatographic methods involve solvent extraction of the foods by homogenizing, followed by an aqueous–organic partition. The organic phase is concentrated and then passed through a Florisil column (see Chapter 8, Section IV,A,5); the elution fraction containing the carbamates is analyzed by LC on a silica gel column with UV detection at 254 nm. Figure 10.22 shows typical chromatograms of three carbamate insecticides spiked at 0.1 ppm in corn and cabbage samples. Recoveries of ten different carbamates from a variety of crops were generally greater than 70%.

Methomyl and oxamyl, two carbamate insecticides, have been analyzed in a number of vegetables by reversed-phase LC employing a C_{18} column with a mobile phase consisting of acetonitrile–phosphate buffer (11 : 89) and UV detection at 240 nm. The compounds are extracted from the crops with ethyl acetate, the organic solvent is evaporated, and the residue is partitioned between water and hexane. The hexane is discarded

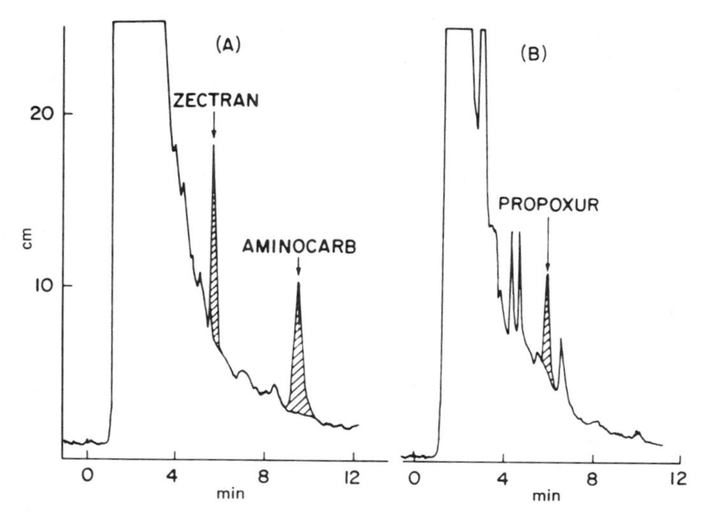

Fig. 10.22 Chromatograms of (A) Zectran and aminocarb in corn at 0.1 ppm; (B) propoxur in cabbage at 0.1 ppm. From Lawrence (*49*), with permission from the American Chemical Society.

TABLE 10.9

Percent Recovery of Oxamyl and Methomyl in Crop Extracts Spiked at 2.0 ppm[a]

Sample	Average Recovery ± SD		Number of Detn's
	oxamyl	methomyl	
Standards	93 ± 2	94 ± 2	6
Celery	65 ± 14	81 ± 2	4
Cabbage	76 ± 5	77 ± 3	4
Collard	63	75	1
Turnip greens	62 ± 2	77 ± 4	4
Mustard	61 ± 7	75 ± 9	4
Grapefruit (1 ppm)	69 ± 4	76 ± 4	3
Nectarines	77 ± 5	79 ± 3	4
Tomatoes (1 ppm)	74 ± 5	75 ± 7	4
Apples	73 ± 4	79 ± 8	4
Corn	61 ± 10	70 ± 12	3

[a] From Thean *et al.* (*52*), with permission from the Association of Official Analytical Chemists.

and the carbamate residues are removed from the aqueous phase with three partitions of chloroform. The combined organic extracts are concentrated and analyzed by LC. Table 10.9 lists recoveries through the method for both carbamates in ten different fruits and vegetables spiked at 1 or 2 ppm. Figure 10.23 shows typical results for a celery sample.

Urea herbicides are particularly amenable to LC with UV detection, and an LC method for their analysis in foodstuffs using adsorption chromatography has been developed (*53*). The extraction and cleanup technique is the same as that mentioned above for carbamate insecticides (*49–51*), except that some of the ureas are eluted from the Florisil column in a more polar fraction. The method is easily capable of detecting most common urea herbicides at the 0.1-ppm level in a variety of foods. Recoveries through the procedure have been reported to be generally greater than 80% between 0.1 and 1.0 ppm. Figure 10.24 shows typical chromatographic results for three common ureas (liuron, diuron, and monuron) in a corn sample spiked at 0.1 ppm.

The determination of some fungicide residues on citrus fruits by reversed-phase LC has been accomplished (*54*). The method involves refluxing the citrus sample with dilute hydrochloric, followed by steam distillation (see Chapter 8, Section VI) for the isolation of biphenyl and *o*-phenylphenol. Benomyl is converted to its hydrolysis product, carbendazim by the acid refluxing and is determined as such. Both carbendazim and thiabendazole are extracted from the acid solution and cleaned up by

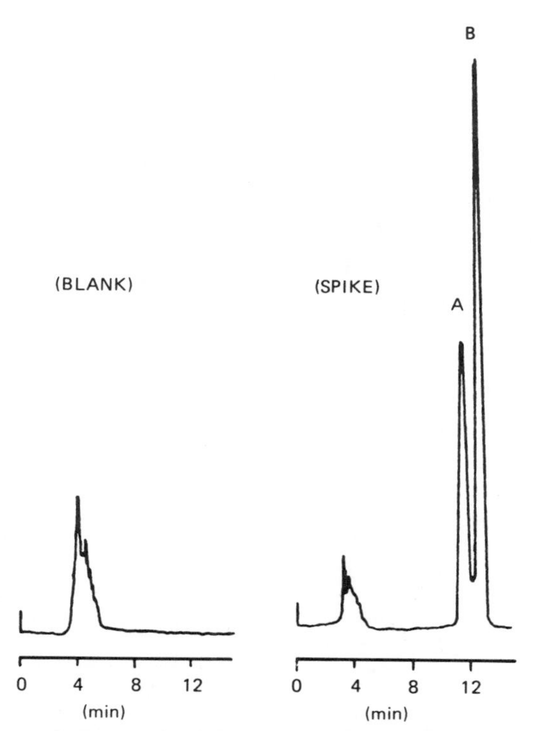

Fig. 10.23 Chromatograms of celery: A, oxamyl; B, methomyl spiked at 2.0 ppm. From Thean *et al.* (*52*), with permission from the Association of Official Analytical Chemists.

partitioning with chloroform under acidic and basic conditions (see Chapter 8, Section III,B). The purified extracts are then analyzed under several LC conditions, including both adsorption and reversed-phase chromatography. Table 10.10 lists the conditions found most useful for analysis of the fungicides in citrus crops.

B. Aflatoxins

Liquid chromatography with fluorescence detection has been found to be particularly useful for the analysis of aflatoxins (*55–58*). Panalaks and Scott (*55*) designed a silica gel-packed flow-cell (see Chapter 5, Section III,B,3 and Fig. 5.11). The fluorescence of the aflatoxins is significantly enhanced when they are in an adsorbed state (*59*). Figure 10.25 compares results of the packed flow cell versus UV detection (made in series) for a

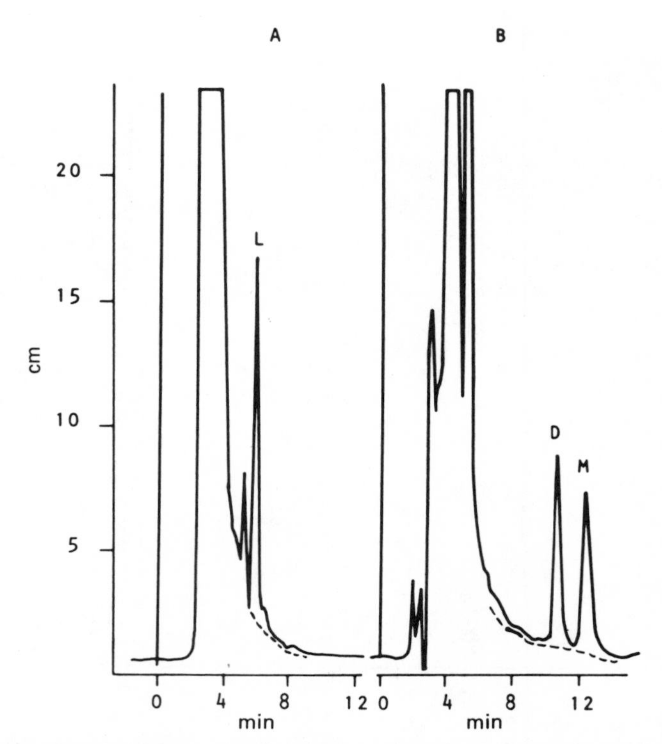

Fig. 10.24 Chromatograms of corn spiked at 0.1 ppm with linuron (L), diuron (D), and monuron (M). A and B represent different Florisil elutions from the same column, B being more the polar fraction From Lawrence (*53*), with permission from the Association of Official Analytical Chemists.

peanut butter extract spiked with 5 ppb of aflatoxins G_1, G_2, B_1, and B_2. The advantage of the fluorescence technique can be clearly seen.

The extraction of the above four aflatoxins from peanut samples may be carried out using acidified aqueous methanol (*58*). The compounds are partitioned into dichloromethane and cleaned up on a silica gel column before LC determination by adsorption chromatography on microparticulate silica gel with a mobile phase of water-saturated dichloromethane–cyclohexane–acetonitrile (25:7.5:1.0) with either 1.5% ethanol or 2% 2-propanol. As low as 0.3–1.0 ppb of aflatoxins may be determined in peanut products.

Aflatoxin M_1 is a hydroxylated metabolite of aflatoxin B_1, which may appear in the milk of dairy cattle. Two methods have appeared recently that include reversed-phase LC and fluorescence detection (*56*,*57*). Winterlin *et al.* (*56*) used a C_{18} Sep-Pak cartridge (see Chapter 8, Section

TABLE 10.10

Conditions of LC of Fungicides[a]

Fungicide	Sample extract solution	Column packing	Mobile phase
1. Biphenyl	Hexane	Silica gel	2,2,4-Trimethylpentane
2. 2-Phenylphenol	Hexane	Silica gel	1% Ethanol/2,2,4-trimethylpentane
3. Thiabendazole carbendazim	Chloroform	Silica gel	1% Ethanol + 0.2% morpholine/chloroform
4. Biphenyl[b] 2-Phenylphenol Thiabendazole Carbendazim	Reconstituted in methanol	C_{18} Reversed-phase	Methanol–water–ammonia (60:40:0.6)

[a] From Farrow *et al.* (*54*), with permission from the Royal Society of Chemistry.
[b] Used for confirmation purposes; excellent separation of the four compounds.

IV,D) to isolate the aflatoxin from the diluted milk samples. The cartridge is washed with 10% acetonitrile–water, which is discarded, followed by elution of the M_1 with 30% acetonitrile–water, which is then directly determined by LC. As little as 0.1 ppb of M_1 can be determined. Figure 10.26 illustrates a typical result for M_1-contaminated milk.

Beebe and Takahashi (*57*) developed an LC method for aflatoxin M_1 based on its trifluoroacetic acid (TFA) reaction product. The TFA reaction is known to cause the addition of water across the double bond of the terminal furan ring of aflatoxins B_1 and G_1, causing a greatly enhanced fluorescence. The same treatment with M_1 causes about a three- to fourfold increase in fluorescence intensity. The milk is blended with an acetone–water mixture, filtered, and treated with lead acetate followed by sodium sulfate. The solution is washed with hexane, which is discarded; the M_1 is removed by partition with chloroform. The organic extract is evaporated to dryness and the residue treated with TFA. The reaction product is diluted with acetonitrile–water (1:9) for LC analysis. Figure 10.27 compares results of a milk sample with and without the TFA reaction.

C. Vitamins

1. Vitamin A

Vitamin A (retinol) has been determined in dairy products (milk and cheese) by reversed-phase LC (C_{18}) with acetonitrile–water (95:5) and UV detection at 328 nm (*60*). The samples are saponified in a brown flask

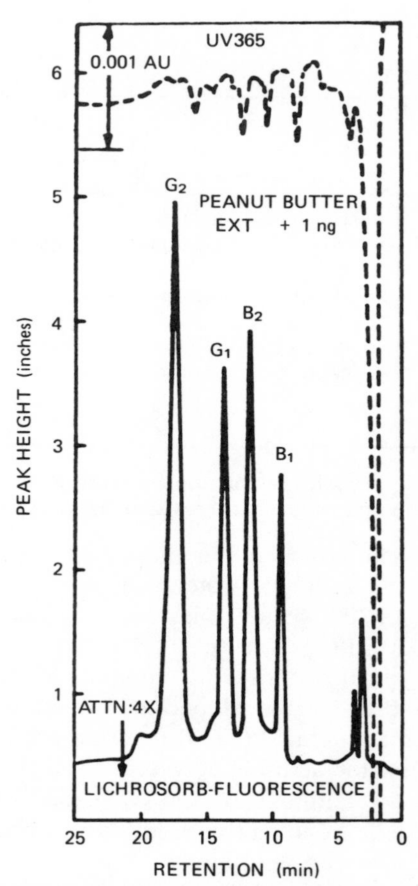

Fig. 10.25 Chromatograms of a peanut butter extract spiked at 5 ppb with each aflatoxin. Upper trace is UV detection; lower one is fluorescence with a silica gel-packed flow cell. From Panalaks and Scott (*55*), with permission from the Association of Official Analytical Chemists.

(vitamin A is light-sensitive). The mixture is cooled and the retinol extracted by partitioning with petroleum ether. The organic phase is evaporated to dryness and the residue is dissolved in acetonitrile for LC analysis. The method can detect 20 ng of vitamin A. In a comparison to the standard colorimetric procedure, the LC results were not significantly different. However, the latter proved to be much simpler.

2. Vitamin B_6

Gregory (*61*) modified an LC method (*62*) for vitamin B_6 and applied it to cereals. The vitamin was extracted from the cereal by sonicating with

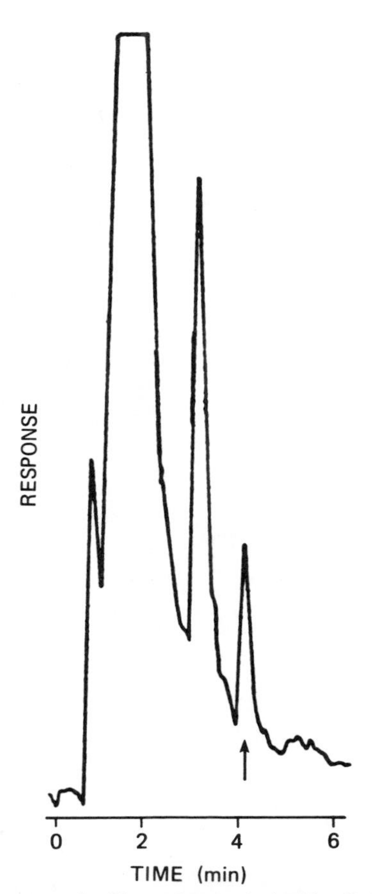

Fig. 10.26 Chromatogram of milk contaminated with aflatoxin M_1 (0.38 ppb). From Winterlin *et al.* (*56*), with permission from the American Chemical Society.

aqueous potassium acetate (pH 4.5). After centrifugation the supernatant was treated with trichloroacetic acid and centrifuged again. The supernatant was then filtered and analyzed by LC on a C_{18} column with a mobile phase of phosphate buffer (pH 2.2) with detection by fluorescence. In a comparison with a conventional microbiological method, the LC technique was more satisfactory in terms of simplicity, rapidity, accuracy, and precision.

3. Vitamin D_3

Vitamin D_3 was analyzed in animal feeds by LC (*63*) employing an NH_2 bonded column and a mobile phase of 30% $CHCl_3$ in hexane. UV detec-

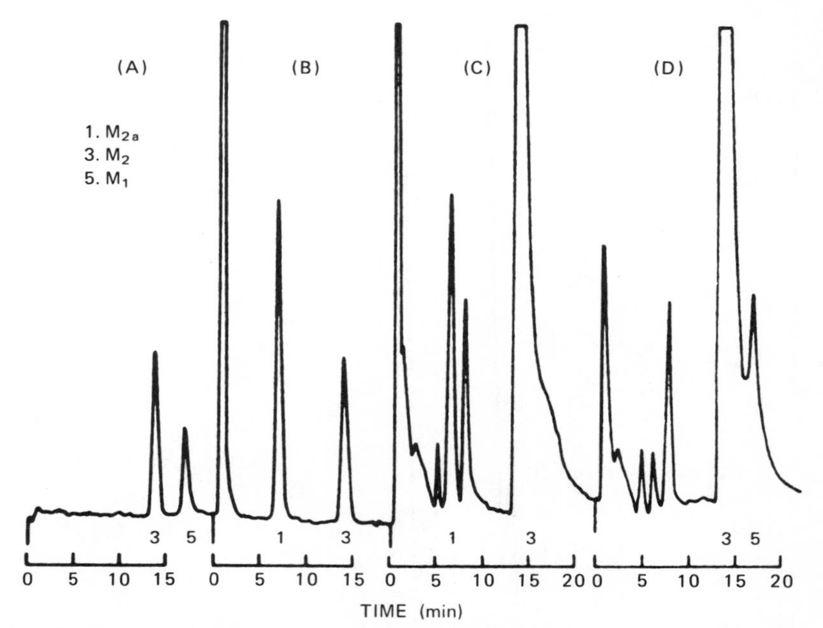

Fig. 10.27 LC determination of 1 ppb aflatoxin M_1 in milk. A, standards untreated with TFA; B, standards, TFA; C, TFA-treated milk sample; D, untreated milk. M_{2a} represents the TFA reaction product of M_1. From Beebe and Takahashi (*57*), with permission from the American Chemical Society.

tion was carried out at 264 nm. The feed samples were extracted with organic solvent, then concentrated and cleaned up on a silica gel column before analysis by LC. The column cleanup was essential for removal of interfering coextractives. Figure 10.28 shows a typical chromatogram of vitamin D_3 in 20 gm of beef supplement at the level of 10,000 IU/kg.

4. Vitamin E

Vitamin E (which consists of four tocopherols and four tocotrienols) has been analyzed in foods and tissues by LC (*64*). The samples were homogenized with isopropanol and acetone. After the addition of water, the vitamin E was extracted with hexane and analyzed by adsorption chromatography with fluorescence detection. As little as 4 μg/gm could be detected in the samples studied. Figure 10.29 compares the fluorescence detection to UV detection of the same extract. It can be seen that the fluorescence trace is superior for the analysis. Butylated hydroxyanisole, an antioxidant food additive, was also found in some samples.

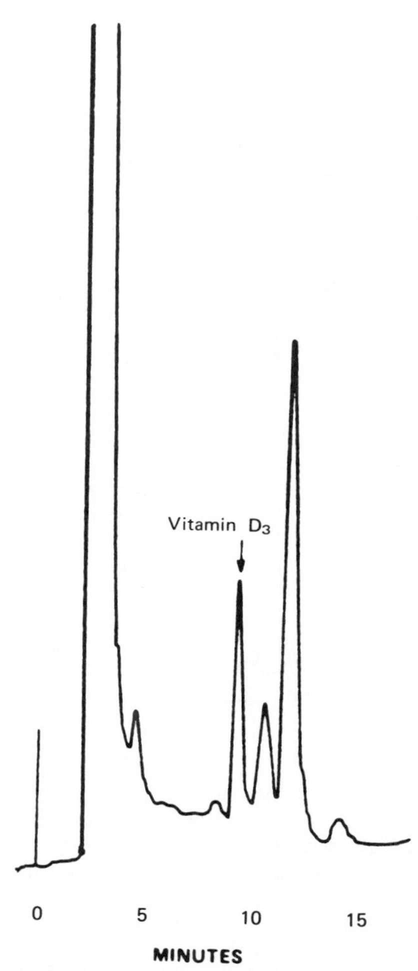

Fig. 10.28 Chromatogram of vitamin D_3 found in beef supplement at 10,000 IU/kg. From Cohen and Lapointe (*63*), with permission from Preston Publications, Inc.

5. Niacin

An LC method for determining niacin in cereal products has been developed which is simpler than the official microbiological or chemical methods (*65*). The cereal sample is extracted with an aqueous suspension of $Ca(OH)_2$, which is then heated for 30 min, cooled, and centrifuged. The supernatant is passed through an ion-exchange column (AGl-X8 anion-

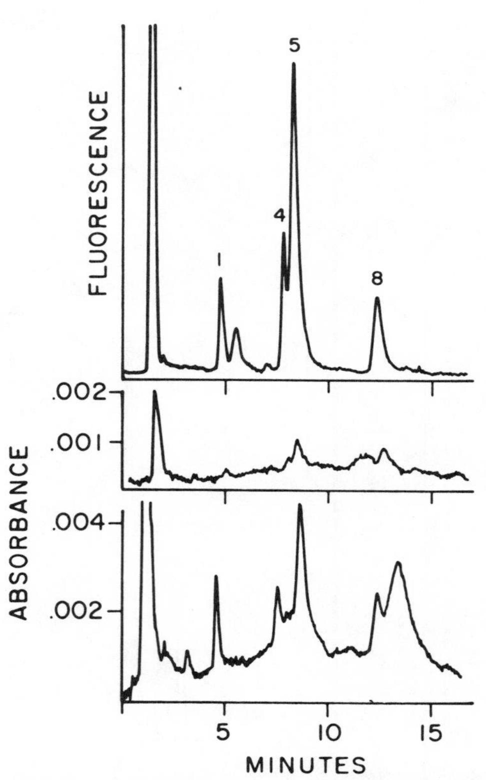

Fig. 10.29 A comparison of fluorescence (ex, 290 nm; em, 330 nm) with absorbance (295 nm) detection of vitamin E constituents in a TV dinner extract. Upper trace is the fluorescence result. Middle trace represents same injection passed through the absorbance detector. Lower chromatogram represents a fivefold concentration of the extract. Peaks: (1) α-tocopherol; (4) butylated hydroxyanisole; (5) γ-tocopherol; (8) δ-tocopherol. From Thompson and Hatina (*64*), with permission from M. Dekker, Inc.

exchange) for cleanup. Permanganate oxidation of the sample provides further cleanup without affecting the niacin. The sample extracts are analyzed on a C_{18} column with a mobile phase of 5% methanol in water containing a trace of tetrabutylammonium ion. Detection is made at 254 nm (UV absorption).

REFERENCES

1. *Anal. Chem.* **52,** No. 5 (1980); **51,** No. 5 (1979); **50,** No. 5 (1978); **49,** No. 5 (1977); **48,** No. 5 (1976); **47,** No. 5 (1975).
2. *J. Chromatogr. Sci.* **17,** No. 3 (1979); **15,** No. 9 (1977); **12,** No. 8 (1974).

3. A. Pryde and M. T. Gilbert, "Applications of High Performance Liquid Chromatography." Chapman & Hall, London, 1979.
4. R. J. Fenn, S. Siggia, and D. J. Curran, *Anal. Chem.* **50,** 1067 (1978).
5. R. M. Riggin and P. T. Kissinger, *Anal. Chem.* **49,** 2110 (1977).
6. I. N. Mefford, M. Gilberg, and J. D. Barchas, *Anal. Biochem.* **104,** 469 (1980).
7. A. H. Anton and D. F. Sayre, *J. Pharmacol. Exp. Ther.* **138,** 360 (1962).
8. D. D. Koch and P. T. Kissinger, *Anal. Chem.* **52,** 27 (1980).
9. T. P. Davis, C. W. Gehrke, C. W. Gehrke, Jr., T. D. Cunningham, K. C. Kuo, K. O. Gerhardt, H. D. Johnson, and C. H. Williams, *Clin. Chem.* **24,** 1317 (1978).
10. L. D. Mell, Jr., R. N. Hawkins, and R. S. Thompson, *J. Liq. Chromatogr.* **2,** 1393 (1979).
11. J. A. Christofides and D. E. Fry, *Clin. Chem.* **26,** 499 (1980).
12. J. Y. Chu, L. Oliveras, and S. Deyasi, *Clin. Chem.* **26,** 522 (1980).
13. J. J. Soldin and J. G. Hill, *Clin. Chem.* **22,** 856 (1976).
14. P. K. Kabra, B. E. Stafford, and L. J. Marton, *Clin. Chem.* **23,** 1248 (1977).
15. R. J. Perchalski and B. J. Wilder, *Anal. Chem.* **50,** 554 (1978).
16. M. W. White, *J. Chromatogr.* **178,** 229 (1979).
17. P. J. Twitchett, S. M. Fletcher, A. T. Sullivan, and A. C. Moffat, *J. Chromatogr.* **150,** 73 (1978).
18. P. L. Williams, A. C. Moffat, and L. J. King, *J. Chromatogr.* **155,** 273 (1978).
19. M. O. Longas and M. B. Poh-Fitzpatrick, *Anal. Biochem.* **104,** 268 (1980).
20. V. Miller and M. Malina, *J. Chromatogr.* **145,** 290 (1978).
21. P. A. Tisdall, T. P. Moyer, and J. P. Anhalt, *Clin. Chem.* **26,** 702 (1980).
22. J. T. Taylor, J. G. Knotts, and G. J. Schmidt, *Clin. Chem.* **26,** 130 (1980).
23. M. Bonati, D. Castelli, R. Latini, and S. Garattini, *J. Chromatogr.* **164,** 109 (1979).
24. J. W. Nelson, A. L. Cordry, C. G. Aron, and R. A. Bartell, *Clin. Chem.* **23,** 124 (1977).
25. D. M. Barends, J. S. F. van der Sandt, and A. Hulshoff, *J. Chromatogr.* **182,** 201 (1980).
26. A. P. De Leenheer, V. O. De Bevere, A. A. Cruyl, and A. E. Claeys, *Clin. Chem.* **24,** 585 (1978).
27. W. C. Eisenberg, *J. Chromatogr. Sci.* **16,** 145 (1978).
28. B. S. Das and G. H. Thomas, *Anal. Chem.* **50,** 967 (1978).
29. M. Dong, D. C. Locke, and E. Ferrand, *Anal. Chem.* **48,** 368 (1976).
30. M. A. Fox and S. W. Staley, *Anal. Chem.* **48,** 992 (1976).
31. S. P. Levine, J. H. Hoggatt, E. Chladek, G. Jungclaus, and J. L. Gerlock, *Anal. Chem.* **51,** 1106 (1979).
32. C. Sango, *J. Liq. Chromatogr.* **2,** 763 (1979).
33. K. L. Dunlop; R. L. Sandridge, and J. Keller, *Anal. Chem.* **48,** 497 (1976).
34. K. Kuwata, M. Uebori, and Y. Yanasaki, *Anal. Chem.* **52,** 857 (1980).
35. R. K. Beasley, C. E. Hoffman, M. L. Ruppel, and J. W. Worley, *Anal. Chem.* **52,** 1110 (1980).
36. K. Kuwata, M. Uebori, and Y. Yanasaki, *J. Chromatogr. Sci.* **17,** 264 (1979).
37. K. Ogan, E. Katz, and W. Slavin, *Anal. Chem.* **51,** 1315 (1979).
38. D. Kasiski, K. O. Klinkmuller, and M. Sonneborn, *J. Chromatogr.* **149,** 703 (1978).
39. S. Mori, *J. Chromatogr.* **129,** 53 (1976).
40. D. A. Hullett and S. J. Eisenreich, *Anal. Chem.* **51,** 1953 (1979).
41. D. S. Farrington, R. G. Hopkins, and J. H. A. Ruzicka, *Analyst* (*London*) **102,** 377 (1977).
42. D. Paschal, R. Bicknell, and K. Siebenmann, *J. Environ. Sci. Health,* **B13,** 105 (1978).
43. A. Otsuki and T. Takaku, *Anal. Chem.* **51,** 833 (1979).
44. D. W. Grant and R. B. Meiris, *J. Chromatogr.* **142,** 339 (1977).

45. H. Boden, *J. Chromatogr. Sci.* **14,** 392 (1976).
46. T. D. Macy and A. Loh, *Anal. Chem.* **52,** 1381 (1980).
47. T. R. Nelson and R. F. Cook, *J. Agric. Food Chem.* **27,** 1186 (1979).
48. I. W. Burns and A. D. Jones, *Analyst* (*London*) **105,** 509 (1980).
49. J. F. Lawrence, *J. Agric. Food Chem.* **25,** 211 (1977).
50. J. F. Lawrence and R. Leduc, *J. Agric. Food Chem.* **25,** 1362 (1977).
51. J. F. Lawrence and R. Leduc, *J. Assoc. Off. Anal. Chem.* **61,** 872 (1978).
52. J. E. Thean, W. G. Fong, D. R. Lorenz, and T. I. Stevens, *J. Assoc. Off. Anal. Chem.* **61,** 15 (1978).
53. J. F. Lawrence, *J. Assoc. Off. Anal. Chem.* **59,** 1066 (1976).
54. J. E. Farrow, R. A. Hoodless, M. Sargent, and J. A. Sidwell, *Analyst* (*London*) **102,** 752 (1977).
55. T. Panalaks and P. M. Scott, *J. Assoc. Off. Anal. Chem.* **60,** 583 (1977).
56. W. Winterlin, G. Hall, and D. P. H. Hsieh, *Anal. Chem.* **51,** 1873 (1979).
57. R. M. Beebe and D. M. Takahashi, *J. Agric. Food Chem.* **28,** 481 (1980).
58. W. A. Pons, Jr. and A. O. Franz, Jr., *J. Assoc. Off. Anal. Chem.* **61,** 793 (1978).
59. J. B. F. Lloyd, *Analyst* (*London*) **100,** 529 (1975).
60. M. H. Bui-Nguyen and B. Blanc, *Experientia* **36,** 374 (1980).
61. J. F. Gregory, *J. Agric. Food Chem.* **28,** 486 (1980).
62. J. F. Gregory and J. R. Kirk, *J. Food Sci.* **43,** 1801 (1978).
63. H. Cohen and M. Lapointe, *J. Chromatogr. Sci.* **17,** 510 (1979).
64. J. N. Thompson and G. Hatina, *J. Liq. Chromatogr.* **2,** 327 (1979).
65. T. A. Tyler and R. R. Shrago, *J. Liq. Chromatogr.* **3,** 269 (1980).

Index

A

Absorbance detection, *see* Detectors
Acetates, 160
Acid–base effects, 209
Adsorbent activity, 216
Adsorption chromatography, 117, 216
Aflatoxins, 275
Agricultural chemicals, 272
Aldehydes, *see* Carbonyl compounds
Alkylation, 165
Alumina, 218
Amines, 157, 160, 165, 167, 172, 174, 246
Antibiotics, 261
Anticonvulsants, 251
Atmosphere, 262
Automation, 16, 241

B

Benzoate esters, 156
Benzoylation, 156
Benzoyl chloride, 156
Biological fluids and tissues, 246
Bonded phases, 123
Buffers, 128

C

Caffeine, 260
Capacity factor, 9, 112
Carbamates, 170
Carbon–cellulose column, 220
Carbonyl compounds, 161, 224
Carboxylic acids, 159, 175, 224, 268
Chelates, 162
Charge-transfer chromatography, 140
Chromatographic system optimization, 235
Chromatography mode, 236
Chromatography theory, 110
Chromophoric groups, 144
Cleanup, 11, 198
Clinical analysis, 246
Column chromatography, 215
Column packing techniques, 50
Columns, 47, 49
Column switching, 225
Confirmatory tests, 145, 240
Counter ions, 135

D

Dansyl chloride, 167
Derivatization
  absorbance, 154
  fluorescence, 167
  general, 14, 143
  reactions, 149
Detectors, 3, 7, 55
  absorbance, 8, 59
  atomic absorption, 81, 183
  electrochemical, 84, 182, 247
  flame, 83
  fluorescence 72, 267, 278, 282
  ion selective electrodes, 92
  laser, 84
  miscellaneous, 102
  radiochemical, 76, 183
  refractive index, 98
  solute transport, 93
  suppliers, 58
Diazomethane, 165
2,4-Dinitrofluorobenzene, 157

2.4-Dinitrophenylhydrazones, 161
Distillation, 229
Drugs, 223, 248, 253
Dry packing, 50

E

Efficiency, 9, 10, 113
Electrochemical detection, *see* Detectors
Enantiomers, 164
Environmental analysis, 262
Enzyme reactions, 187
Esterification, 156, 159
Exclusion chromatography, *see* Gel permeation
Extraction, 11, 198, 202

F

Florisil, 219
Fluorescamine, 174, 186
Fluorescence, 8
Fluorescence detection, *see* Detectors
Fluorescent derivatives, 143, 145
Foods, 272
Formulations analysis, 6
Fungicides, *see* Pesticides

G

Gas chromatography versus liquid chromatography, 1, 4, 13, 143
Gel permeation,
  chromatography, 138
  cleanup, 220

H

Herbicides, *see* Pesticides
Hydrazones, 161, 178
Hydrophobic effect, 124

I

Injection ports, 7, 34
  auto injectors, 44
  loop, 42
  septa, 38
  stopped-flow, 40
  syringe-loop, 43
Insecticides, *see* Pesticides
Interfering peaks, 238
Ion chromatography, 136
Ion exchange, 135, 227
Ion-pair
  chromatography, 131
  extraction, 213
  formation, 166, 168
Isocyanates, 157, 264
Isotherms, 111
Isothiocyanates, 157, 178

K

k', 9, 112
Ketones, *see* Carbonyl compounds

L

Ligand-exchange chromatography, 139
Ligand structures, 163
Lipids, 221
Liquid chromatography
  as a cleanup, 244
  integration with other techniques, 242
  versus gas chromatography, 13, 143
Liquid–liquid partition, 208
Low temperature precipitation, 232

M

Macroreticular resins, 129, 222
Matrix effects, 14
McReynolds constants, 1
Methyl iodide, 165
Microextraction apparatus, 204
Microprocessors, 16
Minimum detectable levels, 12

Mobile phase
  role, 2, 120
  selectivity, 126

N

NBD-chloride, 172, 187
Non-segmented reactors, 185

O

Oils, 270
O-Phthalaldehyde, 176, 187, 248
Oxidation–reduction reactions, 188
Oximes, 161

P

Packing materials
  microparticulate, 47
  pellicular, 47
  suppliers, 52, 216
Partition chromatography, 122, 128
Partition values, 210
Pesticides, 169, 204, 224, 269, 271, 273
Phenols, 156, 157, 167, 176, 179, 224, 266
Photochemical reactions, 189
Phthalates, 268
Picrate ion pairs, 214
Plasma, 249, 260
Plate height, 115
Polar organics, 206
Polyaromatic hydrocarbons, 262, 267
Porous particle, 114
Porphyrins, 257
Postcolumn reactions, 146, 184
Precolumn reactions, 146
  applications, 186
Pumps, 7, 18
  comparison, 31
  constant pressure, 25
  constant volume, 19
p-Values, *see* Partition values

Q

Qualitative analysis, 240

R

Radioimmunoassay, 255
Residue analysis, 6
Resolution, 116
Retention, 112
Reversed-phase chromatography, 123, 224
Routine analysis, 235

S

Sample-concentration, 201
  preparation, 200
  solution, 34
Sampling, 34, 198
Segmented reactors, 185
Serum, 260
Silica gel, 218
Slurry packing, 51
Soil, 271
Solid samples, 270
Solvent strength, 119, 130
Soxhlet extraction, 198, 263
Stationary phases, 52; *see also* Packing materials
Sulfonamides, 161
Sulfonates, 161
Sweep-condistillation, 230

T

Temperature effects, 126
Thiazide diuretics, 258
Thin-layer chromatography, 5, 169, 228, 270
Thiohydantoins, 158
Tissues, 260
Tocopherols, 280
Trace enrichment, 205, 224
Tubular reactors, 185

U

Urine, 258

V

Vitamins, 277
Volatiles, 207

W

Water samples, 267
Wavelength selection, 239, 269